AF360722

Process–Structure–Properties in Polymer Additive Manufacturing

Process–Structure–Properties in Polymer Additive Manufacturing

Editors

Swee Leong Sing
Wai Yee Yeong

MDPI • Basel • Beijing • Wuhan • Barcelona • Belgrade • Manchester • Tokyo • Cluj • Tianjin

Editors
Swee Leong Sing
Nanyang Technological University
Singapore

Wai Yee Yeong
Nanyang Technological University
Singapore

Editorial Office
MDPI
St. Alban-Anlage 66
4052 Basel, Switzerland

This is a reprint of articles from the Special Issue published online in the open access journal *Polymers* (ISSN 2073-4360) (available at: https://www.mdpi.com/journal/polymers/special_issues/process_structure_polymer_additive_manufacturing).

For citation purposes, cite each article independently as indicated on the article page online and as indicated below:

LastName, A.A.; LastName, B.B.; LastName, C.C. Article Title. *Journal Name* **Year**, *Volume Number*, Page Range.

ISBN 978-3-0365-1371-3 (Hbk)
ISBN 978-3-0365-1372-0 (PDF)

Contents

About the Editors

Swee Leong Sing is a Presidential Postdoctoral Fellow at the Singapore Centre for 3D Printing (SC3DP) and School of Mechanical and Aerospace Engineering (MAE), Nanyang Technological University (NTU), Singapore. He has been active in the 3D printing field for more than 8 years. He obtained his BEng (Hons) in Aerospace Engineering and PhD in Mechanical Engineering with a topic in additive manufacturing (AM) in 2012 and 2016, respectively. Swee Leong's research focuses on using advanced manufacturing techniques as enablers for materials development and to create strategic values for the industries. He is also active in inter-disciplinary research and translational work. His research has been awarded the Best PhD Thesis Award by MAE, NTU, Singapore as well as the Springer Theses Award from Springer Nature, Germany in 2017. Swee Leong has worked on numerous 3D printing projects with government agencies, universities, research institutes and industrial collaborators. Swee Leong has filed five patents pertaining to 3D printing processes and materials. He has published 1 book, 4 book chapters and more than 50 peer reviewed journal and conference articles. Swee Leong currently has a h-index of 25 with more than 3000 citations (Web of Science).

Wai Yee Yeong Active in 3D printing research since 2004, Associate Professor Wai Yee Yeong has created multiple frontiers in 3D printing, taking the leads in 3D bioprinting, electronics printing and metal printing research. Prof Yeong is currently serving as Associate Chair (Students) at School of Mechanical and Aerospace Engineering (MAE), Nanyang Technological University (NTU), Singapore. She has been awarded with research grant in the order of >USD 10 million and her works have been featured on media such as CNA, The Straits Times and other channels. Her portfolio includes serving as Programme Director (Aerospace and Defence) at Singapore Centre for 3D Printing (SC3DP), Program Director (3D Printing) at NTU-HP Digital Manufacturing Lab and Co-director of NTU Institute for Health Technologies. Her research and innovative leadership are well-recognized internationally, serving as Technical Chair for international conferences and Associate Editor for the top tier journal, Virtual and Physical Prototyping. She is the inaugral winner for the TCT-Women in 3D Printing Innovator Award in 2019 for her overall achievement in 3D printing. She has also been named as a finalist for Lush Prize (Science Category) in 2018 for her work in bioprinting of tissue models to replace animal testing. She has filed more than eight patents and know-hows, creating new processes, 3D printable metal alloy and hydrogel. At scholarly front, she has co-authored 3 textbooks, 7 book chapters, and over 150 technical papers with more than 5000 citations and h-index of 38.

polymers

Editorial

Process–Structure–Properties in Polymer Additive Manufacturing

Swee Leong Sing * and Wai Yee Yeong

Singapore Centre for 3D Printing, School of Mechanical & Aerospace Engineering, Nanyang Technological University, 50 Nanyang Ave, Singapore 639798, Singapore; wyyeong@ntu.edu.sg
* Correspondence: sing0011@e.ntu.edu.sg or slsing@ntu.edu.sg

Citation: Sing, S.L.; Yeong, W.Y.
Process–Structure–Properties in
Polymer Additive Manufacturing.
Polymers **2021**, *13*, 1098. https://
doi.org/10.3390/polym13071098

Received: 1 March 2021
Accepted: 29 March 2021
Published: 30 March 2021

Publisher's Note: MDPI stays neutral
with regard to jurisdictional claims in
published maps and institutional affiliations.

Additive manufacturing (AM) methods have grown and evolved rapidly in recent years. AM for polymers is an exciting field and has great potential in transformative and translational research in many fields, such as biomedical, aerospace, and even electronics. Current methods for polymer AM include material extrusion, material jetting, vat polymerisation, and powder bed fusion [1,2].

With the promise of more applications, detailed understanding of AM, from the processability of the feedstock to the relationship between the process-structure-properties of AM parts, have become more critical. More research work is needed in material development to widen the choice of materials for polymer additive manufacturing [3]. Modelling and simulations of the process will allow the prediction of microstructures and mechanical properties of the fabricated parts while complementing the understanding of the physical phenomena that occurs during the AM processes.

In this Special Issue, state-of-the-art review and current research, which focus on the process–structure–properties relationships in polymer additive manufacturing, are collated. In their review article, Dickson et al. looked into using fused filament fabrication, a type of material extrusion AM, for the fabrication of parts with fibre-reinforced thermoplastic composites [4]. For original research, Bere et al. looked into the manufacturing of complex parts with fibre-reinforced polymers using moulds fabricated by material extrusion AM [5]. Chen et al. looked into the crystallisation and thermal behaviours of poly(ethylene terephthalate)/biphenols complexes through melt post-polycondensation [6]. Abdolmaleki and Agarwala looked into using polyvinylidene difluoride added with barium titanate (PVDF–BaTiO$_3$) for printed electronics [7]. Nagarajan et al. explored a methodology for the manufacture of magnetic composites using an in-house developed material jetting AM machine [8]. Sahay et al. demonstrated the use of a parallel plate for electrospinning-based AM in the fabrication of helicoidally arranged polyacrylonitrile fibre-reinforced polyvinyl alcohol polymer thin films [9]. Platek et al. analysed the deformation process of regular cell structures, manufactured using fused filament fabrication under quasi-static loading conditions [10]. Udroiu and Braga explored a new methodology for process capabilities analysis for material jetting AM [11]. Luis et al. studied the fabrication of silicone meniscus implants using a novel heat cured material extrusion technique [12]. Stoia et al. studied the mode I fracture toughness of polyamide and alumide samples fabricated using selective laser sintering [13]. Lastly, Andreaczyk et al. have developed a novel method for the experimental validation of numerically optimised turbomachinery components that are fabricated using AM [14].

Acknowledgments: This research is supported by the National Research Foundation, Prime Minister's Office, Singapore under its Medium-Sized Centre funding scheme.

Conflicts of Interest: The authors declare no conflict of interest.

References

1. Goh, G.D.; Sing, S.L.; Yeong, W.Y. A review on machine learning in 3D printing: Applications, potential, and challenges. *Artif. Intell. Rev.* **2021**, *54*, 63–94. [CrossRef]
2. Yap, Y.L.; Sing, S.L.; Yeong, W.Y. A review of 3D printing processes and materials for soft robotics. *Rapid Prototyp. J.* **2020**, *26*, 1345–1361. [CrossRef]
3. Goh, G.D.; Yap, Y.L.; Tan, H.K.J.; Sing, S.L.; Goh, G.L.; Yeong, W.Y. Process-structure-properties in polymer additive manufacturing via material extrusion: A review. *Crit. Rev. Solid State Mater. Sci.* **2019**, *45*, 113–133. [CrossRef]
4. Dickson, A.; Abourayana, H.; Dowling, D. 3D Printing of Fibre-Reinforced Thermoplastic Composites Using Fused Filament Fabrication—A Review. *Polymers* **2020**, *12*, 2188. [CrossRef] [PubMed]
5. Bere, P.; Neamtu, C.; Udroiu, R. Novel Method for the Manufacture of Complex CFRP Parts Using FDM-based Molds. *Polymers* **2020**, *12*, 2220. [CrossRef] [PubMed]
6. Chen, S.; Xie, S.; Guang, S.; Bao, J.; Zhang, X.; Chen, W. Crystallization and Thermal Behaviors of Poly(Ethylene Terephthalate)/Bisphenols Complexes Through Melt Post-Polycondensation. *Polymers* **2020**, *12*, 3053. [CrossRef] [PubMed]
7. Abdolmaleki, H.; Agarwala, S. PVDF-BaTiO3 Nanocomposite Inkjet Inks with Enhanced β-Phase Crystallinity for Printed Electronics. *Polymers* **2020**, *12*, 2430. [CrossRef] [PubMed]
8. Nagarajan, B.; Schoen, M.A.; Trudel, S.; Qureshi, A.J.; Mertiny, P. Rheology-Assisted Microstructure Control for Printing Magnetic Composites—Material and Process Development. *Polymers* **2020**, *12*, 2143. [CrossRef] [PubMed]
9. Sahay, R.; Agarwal, K.; Subramani, A.; Raghavan, N.; Budiman, A.S.; Baji, A. Helicoidally Arranged Polyacrylonitrile Fiber-Reinforced Strong and Impact-Resistant Thin Polyvinyl Alcohol Film Enabled by Electrospinning-Based Additive Manufacturing. *Polymers* **2020**, *12*, 2376. [CrossRef]
10. Płatek, P.; Rajkowski, K.; Cieplak, K.; Sarzyński, M.; Małachowski, J.; Woźniak, R.; Janiszewski, J. Deformation Process of 3D Printed Structures Made from Flexible Material with Different Values of Relative Density. *Polymers* **2020**, *12*, 2120. [CrossRef]
11. Udroju, R.; Braga, I.C. System Performance and Process Capability in Additive Manufacturing: Quality Control for Polymer Jetting. *Polymers* **2020**, *12*, 1292. [CrossRef] [PubMed]
12. Luis, E.; Pan, H.M.; Sing, S.L.; Bajpai, R.; Song, J.; Yeong, W.Y. 3D Direct Printing of Silicone Meniscus Implant Using a Novel Heat-Cured Extrusion-Based Printer. *Polymers* **2020**, *12*, 1031. [CrossRef]
13. Stoia, D.I.; Marsavina, L.; Linul, E. Mode I Fracture Toughness of Polyamide and Alumide Samples obtained by Selective Laser Sintering Additive Process. *Polymers* **2020**, *12*, 640. [CrossRef] [PubMed]
14. Andrearczyk, A.; Konieczny, B.; Sokołowski, J. Additively Manufactured Parts Made of a Polymer Material Used for the Experimental Verification of a Component of a High-Speed Machine with an Optimised Geometry—Preliminary Research. *Polymers* **2020**, *13*, 137. [CrossRef]

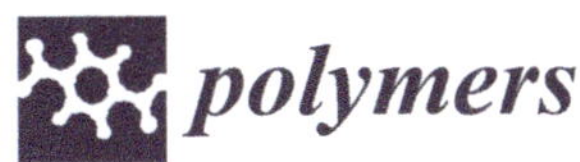 *polymers*

Review

3D Printing of Fibre-Reinforced Thermoplastic Composites Using Fused Filament Fabrication—A Review

Andrew N. Dickson *, Hisham M. Abourayana and Denis P. Dowling

School of Mechanical and Materials Engineering, University College Dublin, Belfield, D04 V1W8 Dublin, Ireland; hisham.abourayana@ucdconnect.ie (H.M.A.); denis.dowling@ucd.ie (D.P.D.)
* Correspondence: andrew.dickson@ucdconnect.ie

Received: 26 August 2020; Accepted: 20 September 2020; Published: 24 September 2020

Abstract: Three-dimensional (3D) printing has been successfully applied for the fabrication of polymer components ranging from prototypes to final products. An issue, however, is that the resulting 3D printed parts exhibit inferior mechanical performance to parts fabricated using conventional polymer processing technologies, such as compression moulding. The addition of fibres and other materials into the polymer matrix to form a composite can yield a significant enhancement in the structural strength of printed polymer parts. This review focuses on the enhanced mechanical performance obtained through the printing of fibre-reinforced polymer composites, using the fused filament fabrication (FFF) 3D printing technique. The uses of both short and continuous fibre-reinforced polymer composites are reviewed. Finally, examples of some applications of FFF printed polymer composites using robotic processes are highlighted.

Keywords: fused filament fabrication; polymers; fibre reinforcement; mechanical properties

1. Introduction

Three-dimensional (3D) printing, also known as additive manufacturing (AM), can be used to print a range of metallic, polymer and composite parts with complex geometries and great design flexibility, while minimising processing waste [1,2]. Applications of this processing technology have included parts fabricated for use in the biomedical, automotive and aerospace sectors [3]. Three-dimensional printing was first introduced during the early 1980s, using the process of stereolithography, in which UV lasers are used to cure layers of polymer into 3D shapes [4]. These methods can be used to process materials such as epoxies. For example, a hydrogel combined with a UV curable adhesive to form a composite exhibiting properties similar to organic tissues, such as cartilage, was demonstrated [5]. A range of other polymer 3D printing technologies are also available, including Selective Laser Sintering (SLS), Laminated Object Modelling (LOM), Multi Jet Fusion Printing and Fused Filament Fabrication (FFF) processes [6–8]. The latter technique, which is also known by the trade name Fused Deposition Modelling, is one of the most widely used amongst all the 3D printing techniques, showing great potential for fabricating 3D geometry parts with the capacity to compete with conventional processing techniques [9–11]. In this technique, the polymer filament is extruded through the nozzle that traces the part's cross sectional geometry layer by layer, as shown in Figure 1 [12]. The nozzle contains resistive heaters that keep the polymer at a temperature just above its melting point, so that it flows easily through the nozzle and forms the layer [13]. The extruding apparatus is typically mounted onto an X–Y computer numerical control (CNC) gantry, allowing the printing of complex geometric patterns. Once a pattern is deposited, the build platform is lowered, or the extruding orifice is raised-up, to deposit the next layer [14,15].

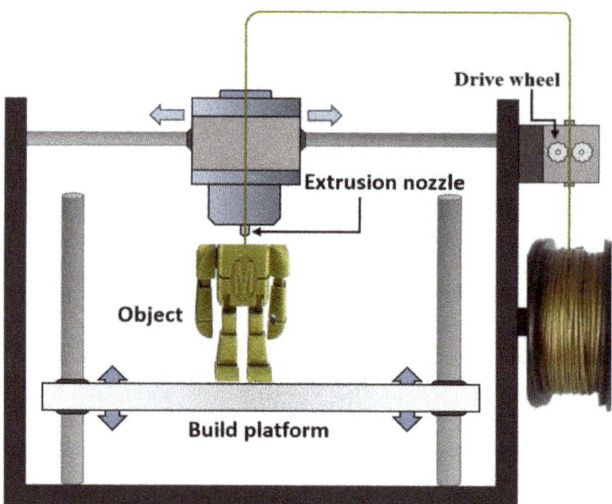

Figure 1. Schematic of FFF process for the printing of parts using the melted polymer filament.

At present, thermoplastic polymers are the most frequently utilised feed-stock materials for the FFF process, due to their relatively low cost as well as their low melting temperatures [16]. These polymers include acrylonitrile butadiene styrene (ABS), polylactic acid (PLA), polycarbonate (PC), polyether ether ketone (PEEK) and nylon [17]. The resulting pure polymer products, however, can often lack the strength to produce fully functioning engineering parts, which has restricted the wider adoption of this technology [18]. In order to address this issue, reinforcing materials, such as fibres, are added into the polymer matrix during printing, in order to produce a composite structure which typically exhibits improved mechanical properties [19].

Reinforcing fibres in composite materials can be in the form of continuous fibres or discontinuous (short) fibres [20]. Continuous fibres have long aspect ratios and have a preferred orientation, while short fibres have short aspect ratios and generally have a random orientation [21]. Due to the fibre orientation, continuous fibre composites offer higher strength and stiffness qualities than those of discontinuous fibre composite [22]. This paper will initially introduce and discuss short fibre-reinforced polymer composites, before reviewing the rapidly expanding field of 3D printed continuous fibre composites. Commercial developments of FFF, including the use of robotic printing techniques for larger scale printing, are also reviewed.

2. 3D Printing of Short Fibre-Reinforced Composite

Composites fabricated reinforced using short fibres are attractive because of their ease of fabrication, economy and superior mechanical properties [23]. They are typically produced by extrusion compounding, injection, or compression moulding processes [22]. For FFF processing, the filaments are fabricated in a two-step process; this firstly involves mixing the polymer pellets and fibre in a blender and secondly extruding the compound to create the filament [24]. Typical short fibres used as reinforcements include carbon and glass fibres. Recently, basalt fibre has also received attention [25].

As detailed in Table 1, several authors have investigated the addition of short fibres into a thermoplastic polymer to provide composite filaments used as a feedstock for FDM process. The reported studies included investigations of the effect of fibre content, as well as its orientation and length, on the processability and performance of the resultant fibre-reinforced composites. Some studies involved a comparison between the properties of 3D printed composites and those fabricated by traditional compression moulding techniques.

Table 1. A summary of materials used for 3D printing of short fibre-reinforced polymer composites.

Matrix	Reinforcement	wt %	Studying	Ref.
PLA	Carbon	15	Experimental characterisation and micrography of 3D printed PLA and PLA reinforced with short carbon fibres.	[26]
ABS	Carbon	5	Effects of process parameters on the tensile properties of FDM fabricated carbon fibre composite parts.	[27]
ABS	Carbon Graphite	5 5	Investigate the effects of reinforcements on porosities and tensile properties of FFF-fabricated parts built at two raster angles.	[28]
PEEK	Carbon Glass	5–15 5–15	Explore the potential engineering application of FDM-3D printing short fibre-reinforced PEEK composites	[29]
PLA	Basalt		Analytical study of the 3D printed structure and mechanical properties of basalt fibre-reinforced PLA composites using X-ray microscopy.	[30]
ABS	Basalt	0–60	Development and mechanical properties of fibre-reinforced ABS for in-space manufacturing applications.	[31]
ABS	Glass	10–18	Study of short fibre-reinforced ABS polymers for use as FFF feedstock material.	[32]
PP	Glass	30	Investigate the effects of fibre reinforcements on the physical and mechanical properties of FFF-fabricated parts.	[33]

Fibre content plays an important role in determining the properties of FFF composite filaments. Generally, tensile strength increases with increasing fibre content. Composite filaments, however, with high fibre content, can be very difficult to print, due to issues with nozzle clogging, in addition to the excessive viscosity of the melted composite filament [34,35]. Therefore, the determination of an appropriate fibre content in the composite used as a filament for FFF is often a compromise between processing difficulty and the performance characteristics of the resulting composites [32].

Ning et al. [36] investigated the effect of carbon fibre content and length on the mechanical properties and porosity of FFF printed ABS/carbon fibre composite parts. The composite filaments were fabricated with different fibre contents (3 to 15 wt %) and different fibre lengths (100 and 150 μm). This study demonstrated that the best performing FFF printed parts were obtained for samples reinforced with 5 wt % carbon fibre, which achieved 22.5% and 30.5% increases in tensile strength and Young's modulus, respectively, compared with the ABS only specimens. A further increase in the fibre content to 10% or higher resulted in a decrease in tensile strength due to the higher porosity. Moreover, the composite specimens reinforced with longer carbon fibres (150 μm) exhibited higher tensile strength and Young's modulus values, and lower toughness as well as ductility, compared with those reinforced with shorter carbon fibres (100 μm).

A study by Tekinalp et al. [34] investigated short fibre (10–40 wt %)-reinforced ABS composites as a feedstock for FFF printing, in order to report on their processability, microstructure and mechanical performance. The additive components were also compared with traditional compression moulded (CM) composites. The results showed that FFF 3D printing yielded samples with very high fibre orientation, lower average fibre length and high porosity levels (16–27%) compared with those obtained using the CM process. Tensile test results demonstrated that tensile strength and modulus were increased, with increasing fibre content for both the FFF and compression moulded samples (Figure 2). However, the improvement in the mechanical properties of FFF 3D printed samples is close to that of those fabricated with CM process, attributed to the high degree of fibre alignment in FFF 3D printed samples compared to the random orientation of the fibres in moulded samples. The authors compensated for some of the loss of strength due to high porosity and decreased fibre length.

For load bearing applications, the composite filaments used in the FFF process must exhibit adequate mechanical properties, such as strength, stiffness, ductility and flexibility [19,32]. The addition of short fibres into pure thermoplastics polymers, however, while improving the resulting printed parts' tensile and flexural strengths, can be at the cost of the reduced flexibility and handleability of the resulted composite filaments. Authors have addressed this issue through the addition of a small amount of plasticiser and compatibility agents. For example, the processability values of short glass fibre-reinforced ABS composites with three different glass fibre contents (10.2, 13.2 and 18 wt %) were investigated in relation to their use as a feedstock filament for FFF [32]. The ABS was mixed with glass fibre in a twin-screw extruder, and then granulated into small pellets. These pellets were then fed into a single screw extruder and extruded into a filament. The glass fibres were found to reduce the flexibility of the resulting filament and make it impossible to feed into the FFF printer, therefore plasticiser (linear low-density polyethylene) and compatibiliser (hydrogenated Buna-N) were added to improve the ability to process the filaments through the printer nozzle and the properties of the FFF parts. The properties of the resulting composite filaments showed they would work well as a feedstock for FFF processes. This study also demonstrated that adding short glass fibre to ABS polymer resulted in a reduction in adhesive strength between the layers in the resulting FFF 3D printed samples, while the tensile strength was increased with increasing fibre contents. The authors reported that this may be due to enhanced fibre bridging across layers during printing, as fibre content increased.

A study by Sodeifian et al. [33] reported on how the flexibility of glass fibre-reinforced polypropylene composite filaments was enhanced by adding maleic anhydride polyolefin (POE-g-MA) as a modifier. POE-g-MA was added with three different weight percentages, namely 10, 20 and 30 wt %. The filaments were used to produce test specimens using FFF printing. The test specimens were also provided using the CM method, to compare the results with those of the FFF method. This study demonstrated that the tensile strengths of the specimens with 10 wt % POE-g-MA were the same irrespective of the manufacturing method used, but the FFF 3D printed specimens exhibited higher flexibility. The FFF 3D printed specimen with 20 wt % POE-g-MA showed superior mechanical properties, compared with those prepared by using the CM method. Increasing POE-g-MA to 30 wt % yielded an increase in the strength and a decrease in flexibility. X-ray diffraction analysis indicated the higher crystallinity of the specimens prepared by compression moulding, compared with that obtained using 3D printing. Figure 3 helps to demonstrate the interlayer and intralayer adhesion of the 3D printed specimens, as compared to that prepared using the CM method.

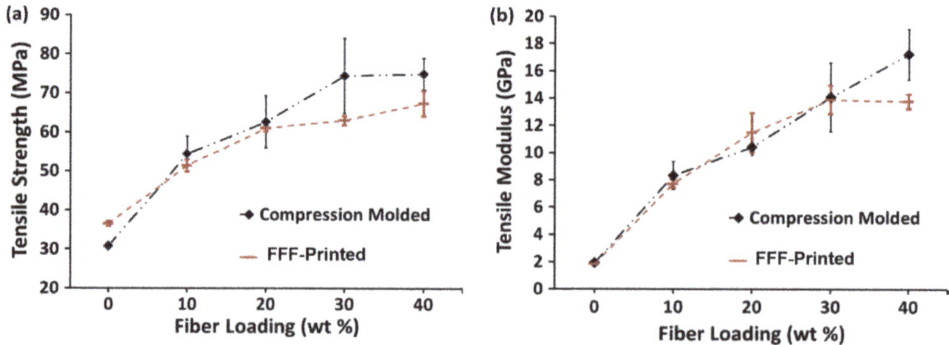

Figure 2. Effect of fibre content and preparation process on (**a**) tensile strength and (**b**) modulus of ABS/carbon fibre composites [34].

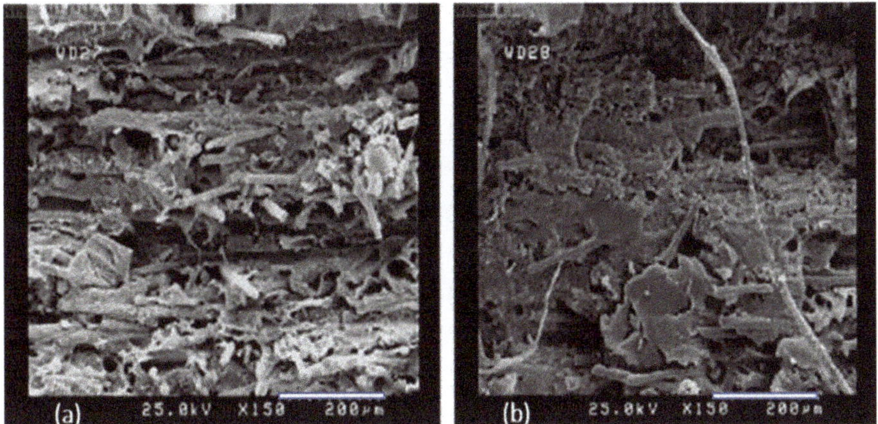

Figure 3. Scanning electron microscopy (SEM) images of the specimens manufactured using (**a**) 3D printing and (**b**) CM methods [33]. These images help to illustrate the differences in interlayer adhesion of 3D printed samples compared to that prepared using CM process.

A study by Sang et al. [35] investigated PLA composites reinforced with silane treated basalt fibre at three different fibre weight fractions (5, 10 and 20 wt %). The effects of fibre type, as well as its weight fraction, on the thermal properties, mechanical performance and rheological behaviour of PLA/BF composite filaments were investigated. The study included a comparison with composites fabricated using both compression-moulded and carbon fibre-reinforced counterparts with the same fibre weight fraction. The results of tensile tests indicated that 3D printed specimens exhibit similar tensile strength values to the compression moulded specimens, and this was attributed to the high alignment of fibre orientation in FFF 3D printed samples versus the random orientation of fibres in the moulded samples. A 33% increase in tensile strength was obtained for FFF samples reinforced with 20 wt % basalt fibre. Rheological results, as anticipated, demonstrated that the viscosity of the carbon fibre composites is higher than that of the PLA matrix, while that of the basalt fibre composite counterparts is close to the PLA only material. The high viscosity of the carbon fibre composite results in poor printing flowability and interlayer bonding defects, which causes stress concentration and failures in test samples. This study indicates the ease of processability of basalt compared to carbon fibre for the fabrication of FFF composites.

Applications of FFF Printed Short Fibre-Reinforced Composites

Some examples of the applications of fibre-reinforced polymer composites include high-temperature inlet guide vanes (IGV) for aerospace applications, printed using polyetherimides-Ultem 1000 mixed with 10% chopped carbon fibre [19,37]. Sang et al. [38] developed promising PLA-PCL/basalt fibre composite filaments, to be used as FFF feedstock for manufacturing honeycomb structures that exhibit elastic deformations with superior energy absorption under compressive loading, and which provide valuable means to obtain an excellent compressive mechanical performance in honeycomb structures (Figure 4).

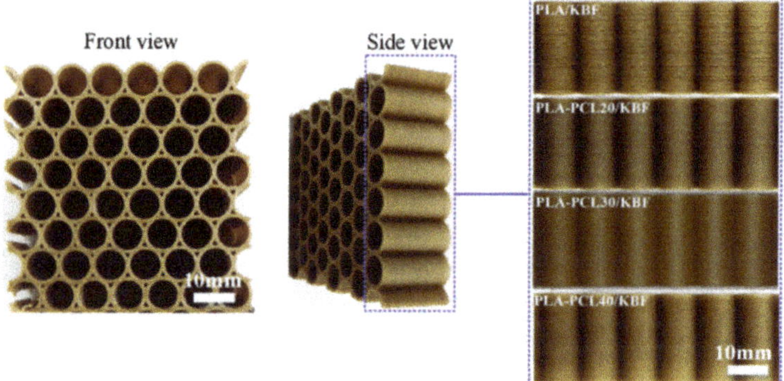

Figure 4. 3D printed circular honeycombs of PLA-PCL/KBF with varying ratios [38]. The PLA-PCL/KBF composite consists of polylactic acid (PLA) as the stiff matrix, polycaprolactone (PCL) as an elastomer phase and silane-treated basalt fibres (KBF) as the reinforcing filler.

3. 3D Printing of Continuous Fibre-Reinforced Composite

As discussed in Section 2, the incorporation of short fibre reinforcement can usually increase the stiffness of the resulting composite; however, the part strength is often only marginally increased. This is due to reliance on the matrix material to transfer loads between fibres. In contrast, a continuous fibre reinforcement transfers and retains primary loads within unbroken strands of fibre, and this results in a significantly lower load transfer through the polymer and allows for a load-bearing capacity orders of magnitude higher than that which short reinforcement is capable of achieving. In the case of continuous fibre composites, the polymer serves to transfer off-axis loads between fibres, such as shear forces. This protects the fibres, as high modulus reinforcement such as carbon and glass fibre exhibit poor mechanical properties under shear loading [39]. Continuous fibre-reinforced polymers are currently one of the largest areas of focus in 3D printing research [40]; this is due to their potential to match or exceed the mechanical performance of conventional composites. A number of authors have reported on modifications of the standard FFF process for the printing of continuous fibres; these include hardware changes, such as more wear-resistant nozzle materials, fibre cutting mechanisms, dual inlet hot-ends and fibre preheaters [41,42].

Two main categories of continuous fibre printing have been described in the literature, these being "in-situ fusion", and "ex-situ prepreg" [41,42]. The in-situ systems utilise two input materials, typically a dry fibre feedstock (the reinforcing fibre) and a neat polymer (the matrix polymer), which are combined together during the printing process. One of the most widely used techniques is known as "in-nozzle impregnation" [43]. In this process, the dry fibre is typically pre-threaded through the printer nozzle prior to printing, and the fibre is also preheated using a coil heater or IR lamps, so as not to excessively cool the molten polymer during printing. The polymer is fed by a motor-driven hobbed gear into the melt zone of the hot-end, and the preheated fibre and melted polymer converge in this melt zone where they are pushed together by the feeding polymer filament. The polymer continues to flow as long as the motor drives the filament, and the continuous fibre bundle is pulled through the nozzle by traction force as it is anchored to the build plate, as shown in Figure 5a. This method has the advantage of a single manufacturing step, which uses low-cost commercially available feedstocks, such as carbon fibre tow and FFF filaments. It also allows for real-time control over the local volume fraction of the part by altering the flowrate of polymer. This single-step printing approach, while rapid, has been reported to yield relatively poor-quality composite parts [44,45]. The short dwell time within the heated nozzle results in poor polymer infusion into the fibre bundles, and ultimately increases the porosity of the composite [43–45]. Tian et al. [45] reported that increasing the printing temperature led to a marked increase in polymer impregnation and decreased porosity, however a quantitative analysis was not performed. Despite these issues, these studies have demonstrated that significant strength increases can be achieved by the addition of the continuous fibre to polymer materials. Matsuzaki et al. [46] reported on a 3.4-fold improvement in the strength of carbon fibre-reinforced PLA, versus unreinforced polymer, when only 6.6 volume fibre percent (VF%) was used. Similarly, Bettini et al. [47] observed up to a six-fold improvement in strength with the incorporation of 8.6 VF% of aramid fibre in PLA. This aramid composite was shown to exhibit a lower porosity compared with that obtained for carbon fibre composites. Another "in-situ fusion" method involves first printing polymer parts using standard FFF 3D printing techniques; these are then sandwiched together with a composite reinforcement and bonded with the application of heat and/or pressure [48]. The fibres can be added during the printing process and overprinted or added after printing between layers of the printed polymer, and then placed into an oven to facilitate bonding (Figure 5b). Mori et al. [48] developed a method for overprinting and compared it with the in-nozzle impregnation method discussed earlier. Whilst overprinted carbon fibre/PLA samples led to a 180% increase in load before failure versus the unreinforced PLA, the in-nozzle impregnated composite withstood a force of 500%, which was obtained using PLA only. This superior performance was attributed to the better contact between the matrix and fibre, and as a result significantly reduced porosity. A case study by Brooks et al. [49] used a topologically optimised FFF polymer base structure, onto which large fibre tows were adhesively bonded, and the results were a lighter part that exhibited a 4000% higher strength, compared with the unreinforced equivalent. A potential weakness of these processes is the short amount of time the fibre spends in the molten polymer, as well as the low pressure applied to the polymer, which typically leads to poor infiltration and high porosity. The approach also necessitates that multiple manufacturing steps occur simultaneously, which makes optimisation difficult.

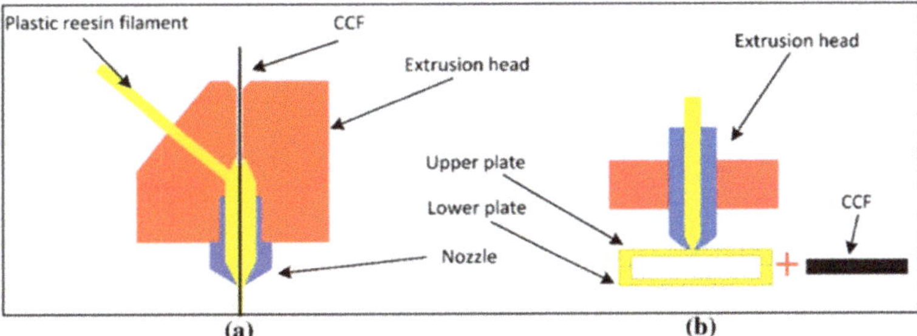

Figure 5. Schematics of in-situ fusion techniques: (**a**) in-nozzle impregnation with polymer and coaxial fibre extrusion, and (**b**) embedding of continuous carbon fibre (CCF) after 3D printing in a thermal bonding process. Images from [42].

In contrast with "in-situ fusion", the "ex-situ prepreg" systems separate the manufacturing of the filament and the printing of the composite into two separate steps. This allows for greater control over the individual processes. As with in-situ systems, the method utilises two input materials (a fibre tow and polymer); however, these are combined together prior to printing into a pre-impregnated filament (prepreg), via a separate extrusion process (Figure 6). The prepreg filament is then spooled and transferred to the printing system for deposition. Where in-nozzle impregnation requires a drive motor for the constant extrusion of the polymer, this method only requires a motor to feed the initial few millimetres of filament through the nozzle. After the filament is anchored to the plate, this motor disengages and is pulled through the nozzle by the continuous fibre reinforcement, which remains solid throughout the process. This simplifies the printing process significantly, and as well as allowing for superior fibre impregnation during the filament manufacture stage, the dedicated extrusion apparatus can exert more pressure on the polymer to fully infuse the fibre tow, and allows for higher manufacturing speeds.

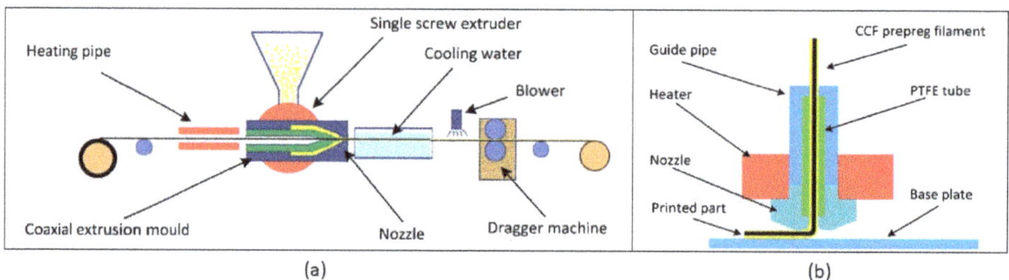

Figure 6. Schematic of ex-situ prepreg process: (**a**) The extrusion and cooling apparatus for production of the prepreg filament. (**b**) The printing process utilising the prepreg filament requires no drive gear as the fibre is pulled through the nozzle, extruding the polymer as it moves [42].

In 2014, the MIT spin-out company Markforged was the first to commercially offer the ex-situ prepreg composite printing system [50]. Their system could print using carbon, glass and Kevlar fibre, and used a cutting apparatus to deposit the correct amounts of fibre in specific locations in the part. The printer included a second FFF printhead for the printing of unreinforced nylon. Selected regions of a polymer part can be reinforced, rather than strengthening the entirety of the component. In doing so the components' fibre fractions are limited to approximately 34 VF% (slightly lower than the VF% of the prepreg filament used), and large areas of the parts remain unreinforced [41]. Compared with printed polymer only parts, the FFF of the composites exhibited significantly improved mechanical performances. For example, Blok et al. [51] reported tensile strengths as high as 725 MPa for carbon fibre/PA, compared with 84 MPa for the printed PA polymer. It is important to highlight, however, that the results from this study are based upon samples that were modified after printing. Several studies utilise an in-house designed filament or printer with a similar mechanism to that used by Markforged. Hu et al. [42], for example, produced a custom PLA/carbon fibre prepreg filament printed using a modified open source 3D printer, and these composites achieved flexural strengths five times higher than unreinforced PLA. However, the VF% was not provided for cross comparisons to be made. The air void content of these samples is typically lower than in the in-nozzle impregnated equivalents, primarily due to the initial impregnation step. As highlighted by Matsuzaki et al. [52], increasing the fibre count in a tow results in an increased air content, however after deposition this is typically reduced. This study also demonstrated that after printing under pressure (from the printing head) the nozzle serving to push the air out of the filament, a reduction in porosity from 33% to 4% was observed for some samples [52]. Whilst this result was reported for a single line of printed filament, the overall composite's void content increased due to the formation of air pockets between filaments and subsequent layers, as they are placed adjacent to and on top of one another.

Both "in-situ fusion" and "ex-situ prepreg" systems demonstrate that entrained air within the composite matrix is the primary challenge when 3D printing continuous fibre composites. Prepreg systems have the advantage of a dedicated manufacturing step, which can reduce filament air content to nearly 0%. However, some level of porosity remains after printing, and it is therefore evident that the printing process itself induces porosity and still requires optimisation. Goh et al. [41] observed that the overlapping of fibre bundles can reduce this porosity, however it could not be eliminated completely. The reduction in printed part porosity associated with the use of low pressure processing conditions during printing (1 Pa) has been successfully shown to increase the interlaminar shear strength (ILSS) of carbon, glass and Kevlar, by 33%, 22% and 12% respectively, compared to those materials printed under atmospheric pressure [53]. Another method of both decreasing the porosity of the FFF printed parts, as well as enhancing interlayer adhesion, is the use of atmospheric plasma surface activation treatments. This was demonstrated through the use of an in-line air atmospheric plasma jet treatment for the activation of sized basalt fibres, immediately prior to the application of polypropylene by extrusion coating, to form the polymer-coated filaments [25]. The flexural modulus and the maximum shear stress values of the resulting FFF composites were found to increase by 12% and 13%, respectively, compared with those obtained for composites fabricated using untreated fibres. It was concluded that this increased mechanical performance is likely due to the enhanced interfacial bond strength between the fibres and the polypropylene polymer, with an associated reduction in the level of air incorporation around the basalt filaments.

3.1. 3D Printed Composites—Mechanical Performance Comparison

Agarwal et al. [54] demonstrated how the 3D printed composites can outperform those produced using conventional approaches, particularly when the printing is carried out using optimised fibre orientations. Figure 7 provides an overview of the tensile properties of continuous reinforcement composites versus short, particle and unreinforced polymers, based on values reported by a number of authors in the literature. The wide range of tensile properties obtained for a given composite type is likely to be influenced by parameters such as fibre content, as well as composite processing technique used.

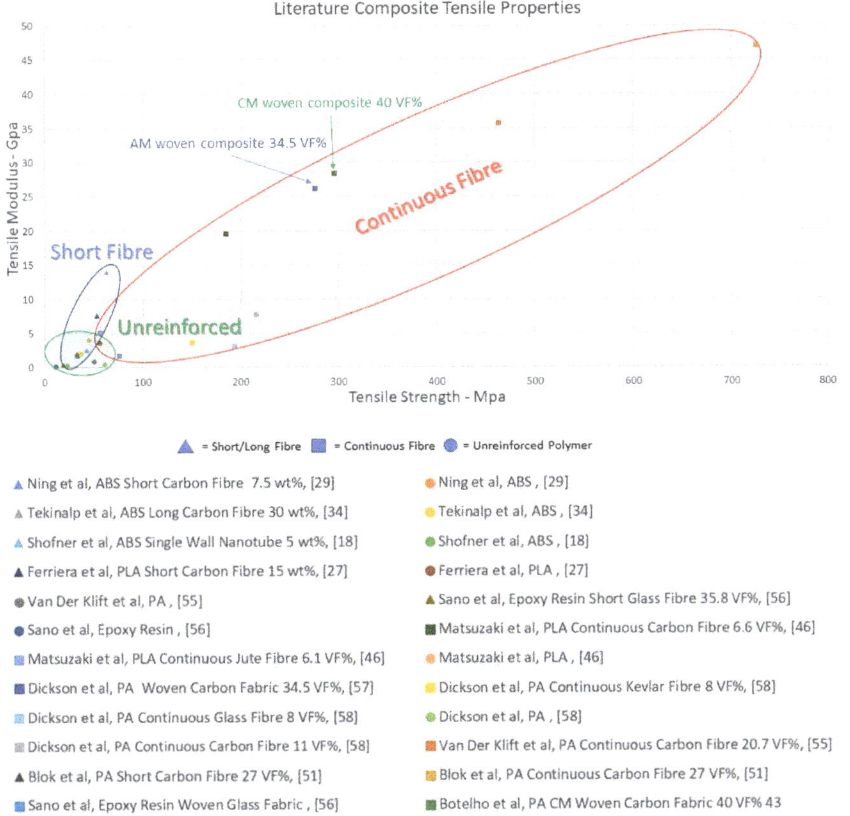

Figure 7. Literature values for tensile strength and modulus for short and continuous fibre-reinforced composites, as well as unreinforced polymers for comparison. A comparison between similar additive manufactured (AM) and compression moulded (CM) woven PA66/CF composites is highlighted, with tensile performance being comparable. Key: Author, matrix, reinforcement, fibre % [18,27,29,34,43,46,51,55–58].

3.2. Continuous Fibre Printing—Pathing Behaviour

3.2.1. Open Source Programme

Printing parameters such as temperature, speed, nozzle height, cornering radii and filament overlap can have an impact on the printed composite's mechanical performance. As the majority of composite printing systems under development are based upon a 3-Axis CNC (Computer numerical control) platform, they utilise Gcode to control movements. Toolpath logic and printing behaviour must accommodate the unique behaviour of a filament, containing a continuous unbroken reinforcement. As the fibre reinforcement remains in a solid state throughout the printing process, its mechanical properties should be unaffected by the printing process, however this assumes that no destructive printing behaviour has occurred during deposition. "Destructive printing behavior" refers to any movement or printing parameter that results in a reduction of the final properties of the composite, compared with those of the pre-deposition material (such as porosity or fibre discontinuity). FFF slicing software (GrabCAD, Ultimakers CURA, Slic3r, simplify3D etc.), has developed significantly over the last 30 years, since its invention in 1989 by Scott Crump, founder of Stratasys Ltd. [59]. The majority of studies performed on custom-built continuous composite systems use a simple raster deposition pattern, which is a back and forth movement separated by raster gaps [42,44–46], as shown in Figure 8a. As raster patterns can be easily programmed using existing FFF software, or with milling CNC software, they allow for the creation of rectangular specimens in a short period of time. This is likely the reason for many studies focusing on tensile and flexural testing, as samples can be manufactured without complex pathing software. The tight corners of 180° introduce major fibre damage and dislocation during printing, however the attachment of tabs during testing usually obscures this fact [52]. Another popular printing method is to use FFF software perimeter-following logic to form rings of material in a spiral like motion. This mode can mitigate most of the problems of a raster pattern by taking corners at reduced angles [47]. It can be used to make simple geometries, such as cylinders, wing cross-sections or dog bones, or any solid shape that contains no internal features (i.e., no hollow spaces) (see Figure 8b,c). The primary reason behind the use of these patterns is that the toolpath generated is continuous, as most custom-built systems do not include a cutting apparatus, which is necessary for the printing of more geometrically complex shapes.

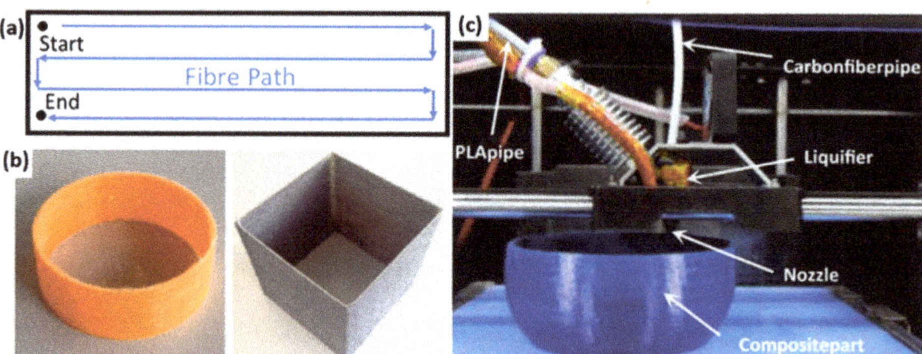

Figure 8. (a) Example of a raster pattern generated for printing tensile testing samples [41]. (b) Objects printed using "spiral" generated by an FFF slicer software [47]. (c) Printer producing a bowl-shaped component from PLA/CF [45].

3.2.2. Proprietary/Commercial Programs

In order to facilitate the commercialisation of the 3D printing technology for composite production, several companies have simplified the tool pathing software and aligned the method of operation more closely with that of FFF systems. Between 2014 and 2019, only one such software was available commercially; this was the Markforged slicer software "Eiger". This software allows for automated fibre placement based upon existing slicer logic, such as perimeter placement, but can only be utilised with Markforged Mark series composite printers [60]. As of 2019, additional companies have released open sourced equivalents of slicing software, with a specific focus on turning FFF printers into composite printers. For example, 9T labs of Zurich, Switzerland and Anisoprint of Moscow, Russia have both released open sourced slicer software for use on a wide range of 3-axis printing systems [61,62]. 9T labs is in early beta testing of its "CarbonKit" and accompanying slicing software; this system serves as a drop-in kit for existing FFF printers, expanding their capabilities into composite printing. The accompanying software "Fibrify" appears similar in function to Markforged "Eiger", with a greater emphasis on fibre optimisation. Anisoprint has followed a similar route to that of Markforged with its "Aura" slicer, which enables fibre printing with their desktop "Composer" series printers [61,62]. As with "Eiger" and "Fibrify", this slicer is fundamentally an FFF slicer, with the added functionality of fibre inclusion. These commercial systems cater to a model of "improving printed part performance" rather than "improving composite part performance". Figure 9 compares the Markforged "Eiger" and Anisoprint "Aura" slicer software for fibre composite printing. As demonstrated in this literature review, the inclusion of fibres will ultimately lead to an increase in the strength and stiffness of FFF parts, however the performance of the resulting composite part can be significantly enhanced by optimising the fibre placement.

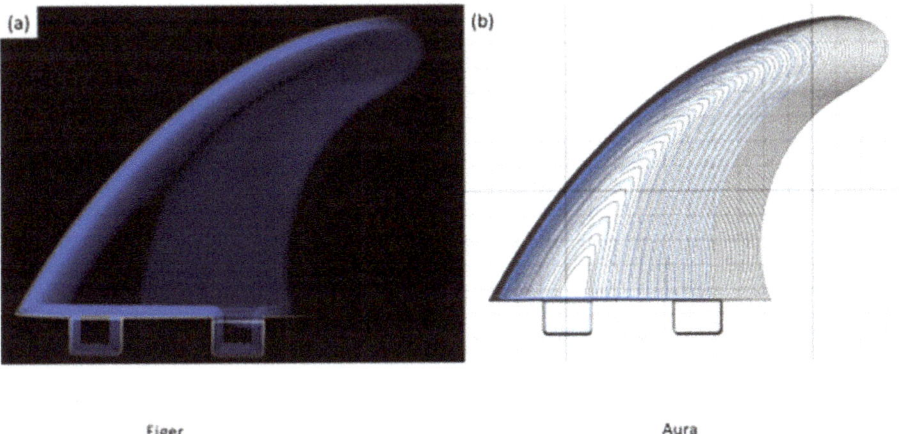

Figure 9. Comparison of (**a**) Markforged "Eiger" and (**b**) Anisoprint "Aura" slicer software for fibre composite printing. The Eiger software facilitates fibre placement in tighter spaces, with blue lines indicating fibre paths.

3.3. Geometrically Complex Composite Fabrication through 3D Printing

The majority of the studies to date on composites have focused on improving their mechanical properties. 3D printing systems, however, also facilitate greater design freedom than automated fibre placement (AFP), automated tape laying (ATL) and moulding techniques, particularly regarding fibre placement, localising volume fractions and part geometry.

Several studies have produced novel continuous fibre pathing programs to print unique structures. Hou et al., for example, printed a corrugated structure for use as a core material in composite sandwich paneling, by printing the panel sideways in the Z-direction of the printer [63]. These were reported to exhibit superior compression strengths and lower core densities to those of aluminum corrugated structures. Similarly, Sugiyama et al. printed sandwich structures from carbon fibre;

however, these were printed in the XY plane utilising fibre tension to print over the gaps in the core structure [64]. Utilising the continuous fibre filament under tension to form a bridge allows for printing with very little support material. A multitude of core patterns and shapes were tested in 3-point bending, with supporting abutment material being removed using a saw prior to testing. The printing was performed in a single continuous movement with no cutting apparatus present. These studies utilised a layer-by-layer approach for deposition, however a number of studies have taken this a step further to produce 3D curvilinear composites [65,66]. Liu et al. [65] produced a core structure for sandwich paneling, utilising a novel method of depositing small sections of composite in mid-air, without the need for a mould. In this study, the latticed core was printed and then adhesively bonded to epoxy composite face sheets. Compressive strengths were low, however the repeatability and filament placement error were significantly improved during the course of the study. To make a similar curvilinear component, Tse et al. [66] utilised a spring-loaded mechanism to reciprocate a heated nozzle over a 3D printed dissolvable mould. A number of contoured composite parts were produced, and in this case a 2D coordinate system was used to guide the head, with the spring mechanism following the mould in the Z-direction [66]. The exact method of achieving these printing paths is not disclosed in these studies, presumably to prevent replication, however it is likely that each study would have required bespoke programs/scripts to be written.

The use of 3D printing to minimise stress concentrators such as holes and notches has been explored in a number of modelling studies [67,68]. Stress concentrators are a major cause of the early/catastrophic failure of fibre composites. These can be associated, for example, with the machining of holes and notches in conventional composites. 3D printing has the potential to create holes within a part by reorienting fibres around the opening of a hole, rather than breaking the fibres through machining. Yamanaka et al. [67] produced a preliminary model of such a structure, indicating that by 3D printing the composite structure and preventing fibre breakage a significantly increased tensile strength could be achieved, compared with that obtained by the cutting of unidirectional reinforcements (Figure 10c) [67]. This study did not consider the width of a 3D printed fibre filament, and filaments are still cut at the point they reach the perimeter of the hole, meaning that fibre damage was also not accounted for at the point of cutting. Zhang et al. [68] performed a similar study; however, material properties were taken from the 3D printed composite literature, rather than the conventional composite values used by Yamanaka. Laminates were tested in single ply and cross ply configurations. In contrast to the procedure of Yamanaka, the fibres were not cut when the hole perimeter was reached, and instead formed a densified region on either side of the hole perimeter. This densified region prevented a major strain riser formation during testing, increasing the laminates strength and stiffness. Figure 10a,b represent the optimised and cut samples respectively.

Both studies are based upon the simulation of non-woven materials, with precise fibre placement, which are needed in order to reduce the effects of the induced off-axis forces generated by the stress concentration around the hole. Woven materials, however, typically exhibit superior through-thickness and off-axis properties [69]. It is therefore reasonable to conclude that woven laminates would be potentially favourable for resisting the forces developed around a notch.

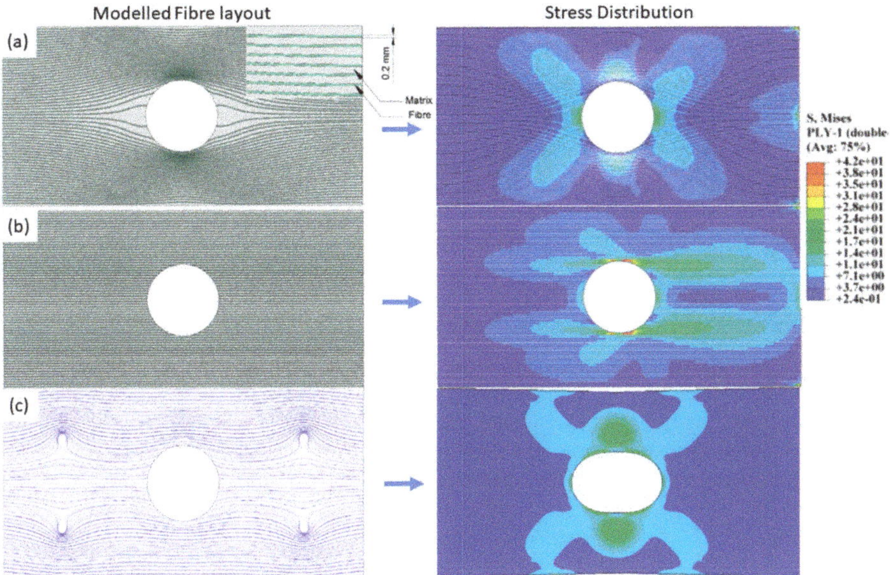

Figure 10. Modelled fibre layouts (left) and resulting stress heat maps (right). (**a**) represents an ideal fibre placement scenario with reduced strain concentration [68]. (**b**) represents a drilled/cut sample with discontinuous fibres resulting in large strain concentrators. (**c**) represents a cut sample containing fibre vorteces to reduce the strain concentration at the centre hole [67].

3.4. Six-Axis Robotics in the 3D Printing of Composites

In order to facilitate the wider adoption of 3D composite printing as a manufacturing process, automation, including robotics, is vital. The use of robotic techniques has been utilised for the layup process of composites for a number of years, and most are utilised in combination with ATL systems which, for example, are widely used in the aerospace industry [40]. A multi-axis articulating robot can be used to hold the ATL head, which is in turn connected to a gantry or rail. This can allow 5 to 10 degrees of freedom. This type of system is efficient for the depositing of large non-complex (flat or single curvature) parts; however, for non-flat structures with higher curvature, a second forming process is required after placement. This second forming step has a tendency to cause the warpage and buckling of the laminates, leading to fibre damage [70]. The tape width also precludes tight cornering, and results in machining being required for small feature inclusion.

Many companies have incorporated robotics as part of their composite printing activities, with a wider range of designs made possible by multi-axis printing. Arevo Labs, for example, reported on the printing of short fibre composites as early as 2016, and has since unveiled a new robotic printing system for the printing of PEEK/CF composites [71]. Arevo have also produced a demonstrator of their technology in the form of a 3D printed CF bicycle frame [71]; this frame is made of layered CF filaments that are deposited using a custom-built, laser-heated deposition head, the details of which are not disclosed (Figure 11). 9T labs have demonstrated a robotic system for the placing of CF/PA12 composite onto curved surfaces, however the details of this design have also not been disclosed publicly [62]. Atropos is a robotic technology demonstrator that originated at the Politechnico Milano +Lab, and features a unique thermosetting resin print head [72]. This system is reported to utilise carbon, glass, or basalt fibres, and can use a number of UV curable resins to produce complex geometric parts (Figure 11). The processing efficiencies obtained through the further automation of FFF composite printing are clearly key for the wider adoption of this 3D printing technology for the manufacture of individualised consumer products, such as sports bicycles, tennis rackets and golf clubs, as well as medical devices such as prostheses.

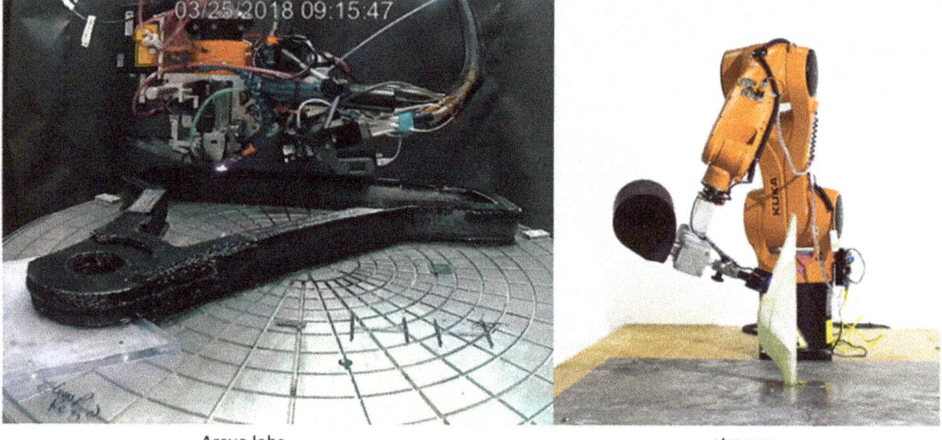

Figure 11. The Arevo labs robotic AM system printing a portion of a bicycle frame (**left**) [71]. The Atropos Robot system printing a glass/epoxy turbine blade without a mould for support (**right**) [72].

4. Summary and Conclusions

This paper provided an overview of the use of the fused filament fabrication (FFF) technique for the manufacture of both short and continuous fibre-reinforced polymeric materials, along with details of the mechanical properties of the resulting composites. The most widely reported short fibres in the literature are those of carbon and glass fibres, which is primarily due to the application focus within aerospace and automotive. Several other reinforcement fibres, including basalt, aramid and jute (and other natural fibres), are also reported, and have also been shown to improve the mechanical properties of polymer composites. The addition of short fibres into neat thermoplastics polymers can significantly improve their stiffness and strength. However, the maximum achievable properties of these composites are limited due to the presence of porosity in the printed parts. Despite the identified mechanical limitations, the FFF printing of short fibre-reinforced thermoplastic composites shows potential in 3D printing to produce some "end-use" components, such as moulds for tooling. Whilst short fibre composites have excellent utility, and can be processed through a standard FFF process, continuous fibre composites offer orders of magnitude higher strength and stiffness, versus neat polymers or short fibre composites. These materials contain continuous unbroken strands of reinforcing fibre, which allows for greater load capacity, but also requires specialist hardware for processing, such as cutting devices and multi-input printheads for proper polymer–fibre infusion. Two primary methods of continuous composite printing have been highlighted. "In-situ fusion" shows great promise for the rapid manufacturing of composite parts, with potential to produce variable volume fraction parts with a single manufacturing process. This technique, however, typically produces inferior quality parts due to entrained air contents and poor polymer permeation. "Ex-situ prepreg" provides superior quality parts with lower air contents and excellent polymer infusion, at the expense of a more complex multi-stage manufacturing process. A difficulty with the resulting 3D printed composites, however, is the presence of porosity, which can significantly impact on mechanical performance. Amongst the methods of addressing this are the application of pressure during printing, fibre bundle overlapping, the use of low-pressure printing conditions, as well as the use of atmospheric plasma pre-treatments.

The wider adoption of 3D printing for consumer products is facilitated through the use of robotic printing techniques. Developments in this area were reviewed and, when combined with the superior material properties of the high strength and low weight of the 3D printed composites, have the potential to produce a wide range of individualised parts, particularly for sectors such as sports goods and medical devices.

Funding: This work was funded by Science Foundation Ireland, through the I-Form Advanced Manufacturing Research Centre, grant number [16/RC/3872]. And the APC was funded by I-Form.

Acknowledgments: This work is partially supported by the Irish Centre for Composite Research (IComp) and the I-Form Advanced Manufacturing Research Centre (Grant Number 16/RC/3872).

Conflicts of Interest: The authors declare no conflict of interest. The funders had no role in the design of the study, in the collection or interpretation of data, in the writing of the manuscript, or in the decision to publish.

References

1. Sun, Q.; Rizvi, G.M.; Bellehumeur, C.T.; Gu, P. Effect of processing conditions on the bonding quality of FDM polymer filaments. *Rapid Prototyp. J.* **2008**, *14*, 72–80. [CrossRef]
2. Torrado, A.R.; Shemelya, C.M.; English, J.D.; Lin, Y.; Wicker, R.B.; Roberson, D.A. Characterizing the effect of additives to ABS on the mechanical property anisotropy of specimens fabricated by material extrusion 3D printing. *Addit. Manuf.* **2015**, *6*, 16–29. [CrossRef]
3. Abourayana, H.M.; Dobbyn, P.J.; Dowling, D.P. Enhancing the mechanical performance of additive manufactured polymer components using atmospheric plasma pre-treatments. *Plasma Process Polym.* **2018**. [CrossRef]
4. Torrado, A.R.; Roberson, D.A. Failure analysis and anisotropy evaluation of 3D-printed tensile test specimens of different geometries and print raster patterns. *J. Fail. Anal. Prev.* **2016**, *16*, 154–164. [CrossRef]
5. Bakarich, S.; Gorkin, R.; Panhuis, M.; Spinks, G. Three-dimensional printing fiber reinforced hydrogel composites. *ACS Appl. Mater. Interfaces* **2014**, *6*, 15998–16006. [CrossRef]
6. Wickramasinghe, S.; Do, T.; Tran, P. FDM based 3D printing of polymer and associated composite: A review on mechanical properties, defects and treatments. *Polymers* **2020**, *12*, 1529. [CrossRef]
7. Singh, S.; Ramakrishna, S.; Berto, F. 3D Printing of polymer composites: A short review. *Mat. Des. Process Comm.* **2020**, *2*, e97. [CrossRef]
8. O'Connor, H.J.; Dowling, D.P. Comparison between the properties of polyamide 12 and glass bead filled polyamide 12 using the multi jet fusion printing process. *Addit. Manuf.* **2020**, *31*, 100961. [CrossRef]
9. Kousiatza, C.; Karalekas, D. In-situ monitoring of strain and temperature distributions during fused deposition modelling process. *Mater. Des.* **2016**, *97*, 400–406. [CrossRef]
10. Wang, J.; Xie, H.; Weng, Z.; Senthil, T.; Wu, L. A novel approach to improve mechanical properties of parts fabricated by fused deposition modeling. *Mater. Des.* **2016**, *105*, 152–159. [CrossRef]
11. Le Duigou, A.; Castro, M.; Bevan, R.; Martin, N. 3D printing of wood fibre biocomposites: From mechanical to actuation functionality. *Mater. Des.* **2016**, *96*, 106–114. [CrossRef]
12. Alafaghani, A.; Qattawi, A.; Alrawi, B.; Guzman, A. Experimental optimization of fused deposition modelling processing parameters: A design-for-manufacturing approach. *Procedia Manuf.* **2017**, *10*, 791–803. [CrossRef]
13. Marcincinova, L.N.; Kuric, I. Basic and advanced materials for fused deposition modeling rapid prototyping technology. *Manuf. Ind. Eng.* **2012**, *11*, 24–27.
14. Cantrell, J.T.; Rohde, S.; Damiani, D.; Gurnani, R.; DiSandro, L.; Anton, J.; Young, A.; Jerez, A.; Steinbach, D.; Kroese, C.; et al. Experimental Characterization of the Mechanical Properties of 3D Printed ABS and Polycarbonate Parts. *Rapid Prototyp. J.* **2017**, *23*, 811–824. [CrossRef]
15. Chennakesava, P.; Narayan, Y.S. Fused Deposition Modeling-Insights. In Proceedings of the International Conference on Advances in Design & Manufacturing (ICAD&M'14), National Institute of Technology Tiruchirappalli, Trichy, India, 5–7 December 2014.
16. Nikzad, M.; Masood, S.H.; Sbarski, I. Thermo-mechanical properties of a highly filled polymeric composites for Fused Deposition Modeling. *Mater. Des.* **2011**, *32*, 3448–3456. [CrossRef]
17. Masood, S.H. Advanced in fused deposition modeling. *Compr. Mater. Process.* **2014**, *10*, 69–91. [CrossRef]
18. Shofner, M.L.; Lozano, K.; Rodríguez-Macías, F.J.; Barrera, E.V. Nanofiber-reinforced polymers prepared by fused deposition modeling. *J. Appl. Polym. Sci.* **2003**, *89*, 3081–3090. [CrossRef]
19. Wang, X.; Jiang, M.; Zhou, Z.; Gou, J.; Hui, D. 3D printing of polymer matrix composites: A review and prospective. *Compos Part B* **2017**, *110*, 442–458. [CrossRef]

20. Brenkena, B.; Barocio, E.; Favaloro, A.; Kunc, V.; Pipes, R.B. Fused filament fabrication of fiber reinforced polymer: A review. *Addit. Manuf.* **2018**, *21*, 1–16. [CrossRef]

21. Philip, S.M.; Raji, R. Composite Pressure Vessels and Nozzles. *Int. J. Eng. Res.* **2017**, *6*, 188–191. [CrossRef]

22. Fu, S.-Y.; Lauke, B.; Mäder, E.; Yue, C.-Y.; Hu, X. Tensile properties of short-glass-fiber- and short-carbon-fiber-reinforced polypropylene composites. *Compos. Part A* **2000**, *31*, 1117–1125. [CrossRef]

23. Fu, S.Y.; Lauke, B. An analytical characterization of the anisotropy of the elastic modulus of misaligned short-fiber-reinforced polymers. *Compos. Sci. Technol.* **1998**, *58*, 1961–1972. [CrossRef]

24. Blok, L.G.; Woods, B.K.S.; Yu, H.; Longana, M.L.; Potter, K.P. 3D printed composites Benchmarking the state-of-the-art. In Proceedings of the 21st International Conference on Composite Materials, Xi'an, China, 20–25 August 2017.

25. Dowling, D.P.; Abourayana, H.M.; Brantseva, T.; Antonov, A.; Dobbyn, P.J. Enhancing the mechanical performance of 3D-printed basalt fiber-reinforced composites using in-line atmospheric plasma pretreatments. *Plasma Process Polym.* **2020**, *17*, 1–8. [CrossRef]

26. Ferreira, R.T.L.; Amatte, I.C.; Dutra, T.A.; Bürger, D. Experimental characterization and micrography of 3D printed PLA and PLA reinforced with short carbon fibers. *Compos. Part B* **2017**, *124*, 88–100. [CrossRef]

27. Ning, F.; Cong, W.; Hu, Y.; Wang, H. Additive manufacturing of carbon fiber-reinforced plastic composites using fused deposition modeling: Effects of process parameters on tensile properties. *J. Compos. Mater.* **2017**, *51*, 451–462. [CrossRef]

28. Ning, F.; Cong, W.; Hu, Z.; Huang, K. Additive manufacturing of thermoplastic matrix composites using fused deposition modeling: A comparison of two reinforcements. *J. Compos. Mater.* **2017**, *51*, 3733–3742. [CrossRef]

29. Wang, P.; Zou, B.; Ding, S.; Hunag, C.; Shi, Z.; Ma, Y.; Yao, P. Preparation of short CF/GF reinforced PEEK composite filaments and their comprehensive properties evaluation for FDM-3D printing. *Compos. Part B* **2020**, *198*, 108175. [CrossRef]

30. Yu, S.; Hwang, Y.H.; Hwang, J.Y.; Hong, S.H. Analytical study on the 3D-printed structure and mechanical properties of basalt fiber-reinforced PLA composites using X-ray microscopy. *Compos. Sci. Technol.* **2019**, *175*, 18–27. [CrossRef]

31. Coughlin, N.; Drake, B.; Fjerstad, M.; Schuster, E.; Waege, T.; Weerakkody, A.; Letcher, T. Development and Mechanical Properties of Basalt Fiber-Reinforced Acrylonitrile Butadiene Styrene for In-Space Manufacturing Applications. *J. Compos. Sci.* **2019**, *3*, 89. [CrossRef]

32. Zhong, W.; Li, F.; Zhang, Z.; Song, L.; Li, Z. Short fiber reinforced composites for fused deposition modeling. *Mater. Sci. Eng.* **2001**, *301*, 125–130. [CrossRef]

33. Sodeifian, G.; Ghaseminejad, S.; Yousefi, A.A. Preparation of polypropylene/short glass fiber composite as Fused Deposition Modeling (FDM) filament. *Results Phys.* **2019**, *12*, 205–222. [CrossRef]

34. Tekinalp, H.L.; Kunc, V.; Velez-Garcia, G.M.; Duty, C.E.; Love, L.J.; Naskar, A.K.; Blue, C.A.; Ozcan, S. Highly oriented carbon fiber-polymer composites via additive manufacturing. *Compos. Sci. Technol.* **2014**, *105*, 144–150. [CrossRef]

35. Sang, L.; Han, S.; Li, Z.; Yang, X.; Hou, W. Development of short basalt fiber reinforced polylactide composites and their feasible evaluation for 3D printing applications. *Compos. Part B* **2019**, *164*, 629–639. [CrossRef]

36. Ning, F.; Cong, W.; Qiu, J.; Wei, J.; Wang, S. Additive manufacturing of carbon fiber reinforced thermoplastic composites using fused deposition modeling. *Compos. Part B* **2015**, *80*, 369–378. [CrossRef]

37. Grady, J.E.; Haller, W.J.; Poinsatte, P.E.; Halbing, M.C.; Schnulo, S.L. *A Fully Nonmetallic Gas Turbine Engine Enabled by Additive Manufacturing Part I: System Analysis, Component Identification, Additive Manufacturing, and Testing of Polymer Composites*; National Aeronautics and Space Administration: Washington, DC, USA, 2015.

38. Sang, L.; Han, S.; Peng, X.; Jian, X.; Wang, J. Development of 3D-printed basalt fiber reinforced thermoplastic honeycombs with enhanced compressive mechanical properties. *Compos. Part A* **2019**, *125*, 105518. [CrossRef]

39. Sugiyama, K.; Matsuzaki, R.; Malakhov, A.V.; Polilov, A.N.; Ueda, M.; Todoroki, A.; Hirano, Y. 3D printing of optimized composites with variable fiber volume fraction and stiffness using continuous fiber. *Compos. Sci. Technol.* **2020**, *186*, 107905. [CrossRef]

40. Frketic, J.; Dickens, T.; Ramakrishnan, S. Automated manufacturing and processing of fiber-reinforced polymer (FRP) composites: An additive review of contemporary and modern techniques for advanced materials manufacturing. *Addit. Manuf.* **2017**, *14*, 69–86. [CrossRef]

41. Goh, G.D.; Dikshit, V.; Nagalingam, A.P.; Goh, G.L.; Agarwala, S.; Sing, S.L.; Wei, J.; Yeong, W.Y. Characterization of mechanical properties and fracture mode of additively manufactured carbon fiber and glass fiber reinforced thermoplastics. *Mater. Des.* **2018**, *137*, 79–89. [CrossRef]

42. Hu, Q.; Duan, Y.; Zhang, H.; Liu, D.; Yan, B.; Peng, F. Manufacturing and 3D printing of continuous carbon fiber prepreg filament. *J. Mater. Sci.* **2018**, *53*, 1887–1898. [CrossRef]

43. Botelho, E.C.; Figiel, L.; Rezende, M.C.; Lauke, B. Mechanical behavior of carbon fiber reinforced polyamide composites. *Compos. Sci. Technol.* **2003**, *63*, 1843–1855. [CrossRef]

44. Namiki, M.; Ueda, M.; Todoroki, A.; Hirano, Y.; Matsuzaki, R. 3D printing of continuous fiber reinforced plastic. In Proceedings of the SAMPE, International SAMPE Symposium and Exhibition, Seattle, WA, USA, 2–5 June 2014.

45. Tian, X.; Liu, T.; Yang, C.; Wang, Q.; Li, D. Interface and performance of 3D printed continuous carbon fiber reinforced PLA composites. *Compos. Part A* **2016**, *88*, 198–205. [CrossRef]

46. Matsuzaki, R.; Ueda, M.; Namiki, M.; Jeong, T.K.; Asahara, H.; Horiguchi, K.; Nakamura, T.; Todoroki, A.; Hirano, Y. Three-dimensional printing of continuous-fiber composites by in-nozzle impregnation. *Sci. Rep.* **2016**, *6*, 1–7. [CrossRef]

47. Bettini, P.; Alitta, G.; Sala, G.; Di Landro, L. Fused Deposition Technique for Continuous Fiber Reinforced Thermoplastic. *J. Mater. Eng. Perform.* **2017**, *26*, 843–848. [CrossRef]

48. Mori, K.I.; Maeno, T.; Nakagawa, Y. Dieless forming of carbon fibre reinforced plastic parts using 3D printer. *Procedia Eng.* **2014**, *81*, 1595–1600. [CrossRef]

49. Brooks, H.; Tyas, D.; Molony, S. Tensile and fatigue failure of 3D printed parts with continuous fibre reinforcement. *Int. J. Rapid Manuf.* **2017**, *6*, 97–113. [CrossRef]

50. 3D Printer Manufacturer & Technology Company | Markforged. Available online: https://markforged.com/about/?mfa=gasearch&adg=55502082228&kw=markforged&device=c&gclid=EAIaIQobChMIuLze-pmM4gIVRLTtCh0oFg_jEAAYASAAEgJozvD_BwE (accessed on 21 September 2020).

51. Blok, L.G.; Longana, M.L.; Yu, H.; Woods, B.K.S. An investigation into 3D printing of fibre reinforced thermoplastic composites. *Addit. Manuf.* **2018**, *22*, 176–186. [CrossRef]

52. Matsuzaki, R.; Nakamura, T.; Sugiyama, K.; Ueda, M.; Todoroki, A.; Hirano, Y.; Yamagata, Y. Effects of Set Curvature and Fiber Bundle Size on the Printed Radius of Curvature by a Continuous Carbon Fiber Composite 3D Printer. *Addit. Manuf.* **2018**, *24*, 93–102. [CrossRef]

53. O'Connor, H.J.; Dowling, D.P. Low-pressure additive manufacturing of continuous fiber-reinforced polymer composites. *Polym. Compos.* **2019**, *40*, 4329–4339. [CrossRef]

54. Agarwal, K.; Kuchipudi, S.K.; Girard, B.; Houser, M. Mechanical properties of fiber reinforced polymer composites: A comparative study of conventional and additive manufacturing methods. *J. Compos. Mater.* **2018**, *52*, 3173–3181. [CrossRef]

55. Klift, F.V.D.; Koga, Y.; Todoroki, A.; Ueda, M.; Hirano, Y.; Matsuzaki, R. 3D Printing of Continuous Carbon Fibre Reinforced Thermo-Plastic (CFRTP) Tensile Test Specimens. *Open J. Compos. Mater.* **2016**, *6*, 18–27. [CrossRef]

56. Sano, Y.; Matsuzaki, R.; Ueda, M.; Todoroki, A.; Hirano, Y. 3D printing of discontinuous and continuous fibre composites using stereolithography. *Addit. Manuf.* **2018**, *24*, 521–527. [CrossRef]

57. Dickson, A.N.; Ross, K.-A.; Dowling, D.P. Additive Manufacturing of Woven Carbon Fibre Polymer Composites. *Compos. Struct.* **2018**, *206*, 637–643. [CrossRef]

58. Dickson, A.N.; Barry, J.N.; McDonnell, K.A.; Dowling, D.P. Fabrication of continuous carbon, glass and Kevlar fibre reinforced polymer composites using additive manufacturing. *Addit. Manuf.* **2017**, *16*, 146–152. [CrossRef]

59. 3D Printing Made Easy with GrabCAD Print Software. Available online: https://www.stratasys.com/software (accessed on 21 September 2020).

60. Eiger 3D Printing Software. Available online: https://markforged.com/eiger (accessed on 21 September 2020).

61. Continuous Carbon Fiber 3D Printing for Industrial-Grade Parts. Stronger, Lighter and Cheaper than Metal or Non-Optimal Composites. From Desktop to Industrial. Available online: http://anisoprint.com/ (accessed on 21 September 2020).

62. 9T Labs|Carbon Fiber 3D Printing. Available online: https://www.9tlabs.com/ (accessed on 21 September 2020).

63. Hou, Z.; Tian, X.; Zhang, J.; Li, D. 3D printed continuous fibre reinforced composite corrugated structure. *Compos. Struct.* **2018**, *184*, 1005–1010. [CrossRef]

64. Sugiyama, K.; Matsuzaki, R.; Ueda, M.; Todoroki, A.; Hirano, Y. 3D printing of composite sandwich structures using continuous carbon fiber and fiber tension. *Compos. Part A* **2018**, *113*, 114–121. [CrossRef]
65. Liu, S.; Li, Y.; Li, N. A novel free-hanging 3D printing method for continuous carbon fiber reinforced thermoplastic lattice truss core structures. *Mater. Des.* **2018**, *137*, 235–244. [CrossRef]
66. Tse, L.Y.L.; Kapila, S.; Barton, K. Contoured 3D printing of fiber reinforced polymers. In Proceedings of the 27th Solid Freeform Fabrication Symposium: An Additive Manufacturing Conference, Austin, TX, USA, 8–10 August 2016.
67. Yamanaka, Y.; Todoroki, A.; Ueda, M.; Hirano, Y.; Matsuzaki, R. Fiber Line Optimization in Single Ply for 3D Printed Composites. *Open J. Compos. Mater.* **2016**, *6*, 121–131. [CrossRef]
68. Zhang, H.; Yang, D.; Sheng, Y. Performance-driven 3D printing of continuous curved carbon fibre reinforced polymer composites: A preliminary numerical study. *Compos. Part B* **2018**, *151*, 256–264. [CrossRef]
69. Abot, J.L.; Daniel, I.M. Through-thickness Mechanical Characterization of Woven Fabric Composites. *J. Compos. Mater.* **2004**, *38*, 543–553. [CrossRef]
70. Grimshaw, M.N.; Grant, C.G.; Diaz, J.M.L. Advanced technology tape laying for affordable manufacturing of large composite structures. *Int. SAMPE Symp. Exhib Proc.* **2001**, *46*, 2484–2494.
71. AREVO Unveils First 3D-Printed Carbon-Fiber eBike Arevo. Available online: https://arevo.com/news_item/arevo-unveils-first-3d-printed-carbon-fiber-ebike/ (accessed on 21 September 2020).
72. This 6-Axis Robot Arm Can 3D Print Fiberglass Composites. Available online: https://www.archdaily.com/867696/atropos-this-6-axis-robot-arm-can-3d-print-fiberglass-composites (accessed on 21 September 2020).

Article

Novel Method for the Manufacture of Complex CFRP Parts Using FDM-Based Molds

Paul Bere [1],*, Calin Neamtu [2],* and Razvan Udroiu [3],*

[1]   Department of Manufacturing Engineering, Faculty of Machine Building,
    Technical University of Cluj-Napoca, Memorandumului 28, 400114 Cluj-Napoca, Romania
[2]   Department of Design Engineering and Robotics, Faculty of Machine Building,
    Technical University of Cluj-Napoca, Memorandumului 28, 400114 Cluj-Napoca, Romania
[3]   Department of Manufacturing Engineering, Faculty of Technological Engineering and Industrial
    Management, Transilvania University of Braşov, Mihai Viteazu 5, 500174 Braşov, Romania
*   Correspondence: Paul.Bere@tcm.utcluj.ro (P.B.); calin.neamtu@muri.utcluj.ro (C.N.);
    udroiu.r@unitbv.ro (R.U.); Tel.: +40-744-695-349 (P.B.); +40-740-258-225 (C.N.); +40-268-421-318 (R.U.)

Received: 8 September 2020; Accepted: 24 September 2020; Published: 27 September 2020

Abstract: Fibre-reinforced polymers (FRP) have attracted much interest within many industrial fields where the use of 3D printed molds can provide significant cost and time savings in the production of composite tooling. Within this paper, a novel method for the manufacture of complex-shaped FRP parts has been proposed. This paper features a new design of bike saddle, which was manufactured through the use of molds created by fused deposition modeling (FDM), of which two 3D printable materials were selected, polylactic acid (PLA) and acrylonitrile butadiene styrene (ABS), and these molds were then chemically and thermally treated. The novel bike saddles were fabricated using carbon fiber-reinforced polymer (CFRP), by vacuum bag technology and oven curing, utilizing additive manufactured (AM) molds. Following manufacture the molded parts were subjected to a quality inspection, using non-contact three-dimensional (3D) scanning techniques, where the results were then statistically analyzed. The statistically analyzed results state that the main deviations between the CAD model and the manufactured CFRP parts were within the range of ±1 mm. Additionally, the weight of the upper part of the saddles was found to be 42 grams. The novel method is primarily intended to be used for customized products using CFRPs.

Keywords: CFRP; PLA mold; additive manufacturing; fused deposition modeling; vacuum bag technology; 3D scanning; bike saddle

---

1. Introduction

Carbon fiber-reinforced polymers (CFRPs) often offer greater advantages than most other commonly used materials and are frequently used in many fields such as the aerospace, automotive, railway, naval, sports industry, medical, and civil construction industries [1,2]. Within the sports industry (CFRP bicycles, CFRP tennis rackets etc.) due to their excellent physical and mechanical properties, such as being lightweight, high strength, etc., CFRP products offer a good opportunity to improve the performance of the participants.

The uses of polymers are very popular within additive manufacturing (AM) processes having applications across many domains [3]. Fused deposition modeling (FDM) or fused filament fabrication (FFF) is defined as a material extrusion process within ISO/ASTM 52900:2015 standard [4]. Recently, due to the low production costs and the ability to feature a high degree of automation, the technology surrounding FDM/FFF has developed in a relatively rapid fashion [5–8]. This technology allows for the manufacture of complex parts, at a relatively low production cost, from various thermoplastic filaments including polylactic acid (PLA), acrylonitrile butadiene styrene (ABS), polycarbonates, nylons, etc. However, there are some limitations to the FDM process, for example the thermoplastic parts have poor mechanical properties when compared to traditional manufacturing methods, such as injection molding [9]. PLA is often treated with more importance in FDM, over other comparable materials, due to its relatively low production cost, good three-dimensional (3D) printability, and biodegradability; in addition to this, PLA has a low thermal tension, which reduces warping during the FDM process [10]. ABS material contains three monomers: acrylonitrile (A) in a proportion of 15%, butadiene, (B) 35% and styrene, (S) 60%. It should also be noted that ABS has good thermal and mechanical properties, resulting in ABS being used to produce jigs, fixtures, and tooling. The styrene from this material contributes to the achievement of a shiny surface. ABS is also compatible with soluble support materials for easy support removal, allowing more complex prints to occur.

Post-processes methods and variation of different printing parameters can enhance mechanical properties and surface finish of the part made by FDM. Porter et al. have investigated the optimal infill percentage of the beams from PLA that can maximised specific flexural rigidity, and obtained an infill percentage in the range of 10% to 20% [11].The post chemical treatment using acetone (99%) and dichloroethane (98%) solutions dramatically improves the surface finish and dimensional accuracy, and reduce the tensile strength of FDM specimens made of ABS [12]. Several researchers have analysed the effect of different annealing times and temperatures on thermoplastic materials which was 3D printed by FDM/FFF techniques [13–15]. The results showed that annealed PLA shape memory polymer at 75 °C increased the tensile and compressive strength (6% increases in ultimate compressive strength) compared to as-printed samples [13]. Wach et al., have printed PLA specimens by the FDM process and annealed they over its glass transition temperature determined an increase of 11–17% in the flexural stress of the parts [15].

There is a great diversity of available manufacturing methods and processes for the production of composite material (CM) parts. Parts with a complex geometry often require the use of molds, which can be manufactured using various methods and technologies (Figure 1). Traditionally, molds are made by computer numerical control (CNC) milling a block of raw material or of various epoxy resins, of which, this technological process is time and manpower intensive, usually with high associated costs. An additional method of making molds is by using CM, where it is necessary to manufacture a master model by a separate method, tailored to the composite material the mold is made from. This process involves the deposition of FRP layers on to the part model, followed by the polymerization of the manufactured part then extraction of the mold from the master model. This process has a relatively high level of accuracy, however, it is time intensive.

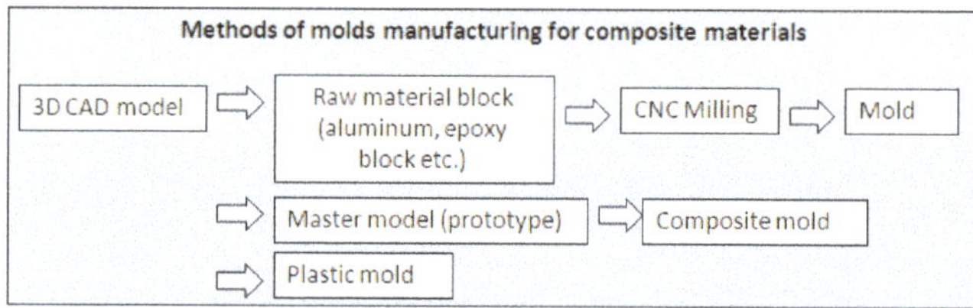

Figure 1. Methods of mold manufacturing for composite materials.

Rapid tooling (RT) [8,10,16,17] is a technology designed for short manufacturing runs that adopt rapid prototyping (RP) 3D printing techniques and apply the techniques to making soft or hard tooling, determined by the material used. RT can produce tooling and molds indirectly, through use of a master 3D printed model, or directly though AM. Thus, the use of AM allows for a direct and rapid way to produce hard tooling, jigs, and molds from different strong materials of the designer's choice (metals, resins). A mold was manufactured by Udroiu, using 3D inkjet printing techniques, fabricated out of the following materials: ZP 131 composite powder, ZB60 binder, and Z-max epoxy resin [16]. The mold was then used for the manufacturing of a part for an aircraft joystick, comprised of rubber. A research study of acrylic thermoplastic molds based on polymer jetting technology concluded that

3D printing is a rapid manufacturing method which can be used in a small series production of parts by using reaction injection molding (RIM) and casting [17]. Rodriguez et al. investigated the use of FDM technology to create molds using various types of material, for the thermoforming process [18], and found that the properties of the mold can transfer the surface defects of FDM technology to the thermoformed material sheets. Compression molds were 3D-printed using PLA and ABS filaments and then polished to produce personalized dental fillers [19]; this study demonstrated that 3D printed molds are able to be used in dental applications. Hay et al. used a FDM-printed mold of PLA to cast a customized PMMA (polymethyl methacrylate) implant for an external surface of a cranium [20]. Sieminsky et al. analyzed FDM/FFF technology used to manufacture mold inserts from PLA [21], to increase the life of the inserts and reduce the production time cycle, the inserts should be cooled using compressed air, after the removal of the molded part. Lusic et al. has analyzed, through the use of FEM (finite element method) simulations, the potential of CFRP-laminating molds to be used as rapid tooling through use of FDM [22]. It was concluded that thermoplastic molds comprising ULTEM 1010 material, manufactured by FDM, could be used for the manufacturing of CFRP components in autoclaves; however, they did not validate their FEM results. Additionally, Sudbury et al. investigated an industrial-grade approach to additive manufacturing, similar to extant FDM process used to produce molds used for hand lay-up of the fiberglass part [23]. The 3D-printed molds were machined by a CNC mill to improve the quality of the finished surface, and this paper concluded that 3D printed molds could be an effective approach for limited production runs of composite parts, although it must be noted that the molds used in this paper were not exposed to temperature or pressure.

The reinforcement of fibers on the thermoplastic filaments used within the 3D printing process can be created by combining short fibers and continuous fibers. The rapid manufacturing process of CFRP using FDM technology is treated in [24]. The thermoplastic matrix and the CFRP powder is used in this case and the process parameters and mechanical properties are investigated. Zhang et al. proposed a method for the 3D printing of continuous carbon fiber-reinforced plastics with a pressure roller, and found that there was an improvement in the strength of parts printed as compared to traditional FDM parts; however, this printer is considered in an experimental stage and thus cannot be used for 3D printing of molds at present [25]. Nanoparticles and short or long fibers in the thermoplastic filament can increase the specific desired mechanical properties of PLA, ABS, nylon etc. [26–28], although the 3D printing of continuous fiber-reinforced composites manufactured through an FDM process is a relatively new technique still in its infancy; thus there is a lack of experimental data on the mechanical performance of structures manufactured using this process [26]. The internal defects [29–31] resulting from the FDM process decrease the desired mechanical properties of the specified thermoplastic-reinforced fibers, and they cannot be compared with traditional FRP. Thus, traditional fiber-reinforced plastic materials with light weight and high strength are often the best option to manufacture parts to improve the performances of the athletes.

At present, the measurement of tools and molds through optical 3D scanning has gained much interest in the quality control process for different processes, e.g., forging [32,33], injection molding [34], and CFRP parts [35].

The use of advanced materials within performance cycling has led to outstanding results, and specialist bicycle components are now becoming lighter whilst retaining desired special mechanical properties. Liu et al. proposed to manufacture a bike frame comprising CFRP, using finite element analysis (FEA) the stacking sequence of the layers was investigated, however, they did not validate the results of the FEA [36]. The use of high-performance MC, especially CFRP, through the preferential arrangement of the layers can lead to both low total end mass and desired mechanical properties. Depending on the individual, some cyclists require more rigid parts, so that as much energy as possible is transmitted directly to the pedals. Alternatively, some solutions are centered around more elastic elements, in order to dampen vibrations encountered through use. It is accepted that elements made of CFRP can satisfy both of the aforementioned main requirements, through distinct approaches and the correct orientation of the material filaments for preferential reinforcement can lead to spectacular results. The position of the body on the bike during seated sprint cycling is studied in [36–39]. The optimization of lower limb joint cinematics of the participants under maximal intensity cycling was investigated. Also, a bicycle saddle 3D printed in colors can shown simulated pressure distribution of a rider [40]. A bicycle saddle with a special shape was tested by Piazza et al., where the effects on perineal compression, blood perfusion, genital sensation and sexual function of the cyclist were investigated [41]. A manufacturing method for a bicycle saddle was presented in [42]; this was created based upon a master 3D printed model of the bike saddle manufactured from ABS material. Using the ABS master model, the authors have applied a traditional technology by hand layup to manufacture the mold from glass fiber and epoxy material. The accuracy of the parts obtained and this traditional mold was not studied.

Traditional molds (metal or epoxy resin block, composite molds), used for composite materials are often very robust; however, they are only cost-effective when used in conjunction with mass production. For production of small batches or customized parts, there is a growing need for molds which can be produced quickly and at a relatively low cost. 3D printing through the use of FDM/FFF processes, utilizing low-price filaments, such as PLA or ABS, could be an approach that will significantly reduce production time and costs to produce molds for CM.

In this paper, the authors propose a novel method to obtain a mold comprising a thermoplastic material, used for laminating the CFRP parts, using an additive manufacturing technique. Two materials were selected for evaluation, PLA and ABS, which were used to create the mold(s) using FDM. The design of the novel mold and CFRP bike saddle were then presented within this paper, where the newly manufactured CFRP parts and mold(s) were then scanned using non-contact 3D scanning methods in order to perform dimensional evaluation of the results; these results were then presented and compared against the original CAD model. It was found that the manufacturing technology used to create the mold and CFRP parts allowed for good results to be achieved.

2. Materials and Methods

2.1. Design of a Novel Bike Saddle and its Mold

The proposed method consists of several steps as shown in Figure 2. The 3D CAD model of the mold is converted to a STL (standard triangulation language) file format. Within the second phase of the method, the mold is 3D printed by FDM using two different materials on two different 3D printers. The molds are prepared for vacuum bag forming methodology and oven curing, within the third step. Following the manufacture of CFRP, specimens are subjected to a quality inspection, using non-contact 3D scanning techniques, in order to perform the dimensional evaluation of the results. Three 3D scans were performed as follows for the 3D printed mold, the final mold, and the CFRP parts.

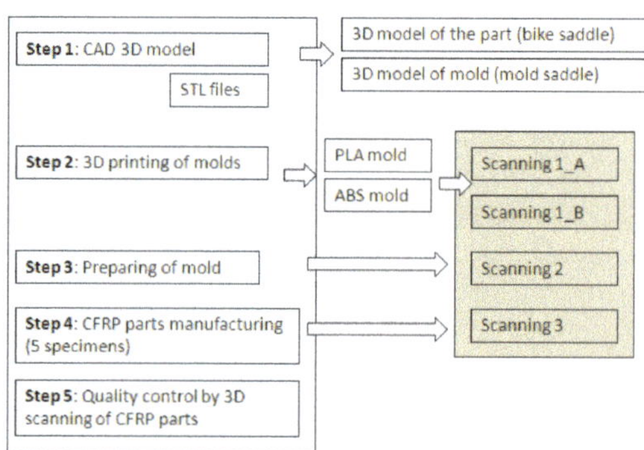

Figure 2. Novel method for manufacturing complex carbon fiber-reinforced polymer (CFRP) parts.

Saddles' dimensions used worldwide for performance bicycles are in the range of 110–170 mm in width and 240–280 mm in length, respectively, depending on the type of bike, kind of competition, and individual customization. Chen investigated the measurement of the external ischial tuberosity width and the anthropometric data using a group of men and women to determine the bicycle-seat sizes [43]. The proposed predictive model assists riders to determine their seats size for bicycling.

A saddle with a width of 145 mm and a length of 270 mm was taken into consideration in this research. The surface area of the bike saddle is 3300 mm^2. The design constraints of the saddle shape were established by analysis of the lateral and longitudinal position of the cyclist on the bicycle and by studying the anatomical particularities of the ischial bones (pelvic bones) and the perineum area of a group of cyclists (Figure 3). The authors cannot provide exact details on the anatomical dimensions of the analyzed individuals for reasons of protecting individual data. The geometric shape of the saddle (Figure 4) was determined taking into account the best possible contact of the pelvic bones with the saddle surface. A longitudinal channel designed on the central part of the saddle reduces the pressure applied in the perineum area on the prostate of male cyclists. Two CFRP bars of 8 mm in diameter are glued on the lower surface of the saddle, allowing the assembly to link with the bar inserted into the frame of the bike (Figure 5).

The 3D model of the saddle was designed based on non-uniform rational B-spline (NURBS) curves, and the multi-section surface technique within the CATIA V5 software (Dassault Systems, USA), as is shown in Figure 4a–c. A CAD model of the saddle mold with 4 mm wall thickness was obtained based on the 3D model of the saddle (Figure 6).

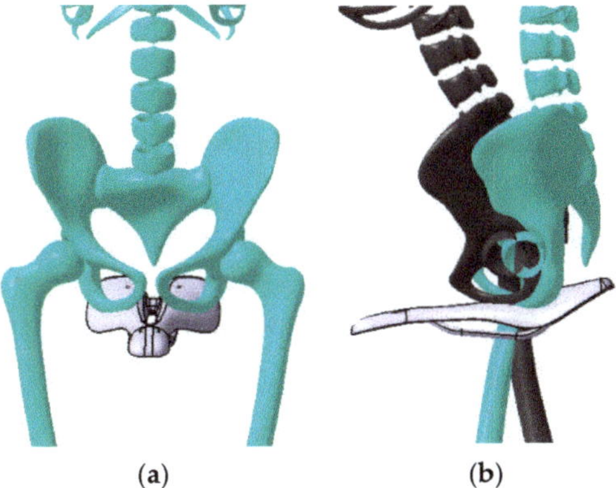

(a) (b)

Figure 3. Determination of the design curves based on the human bones position on the bike saddle: (a) front view; (b) side view.

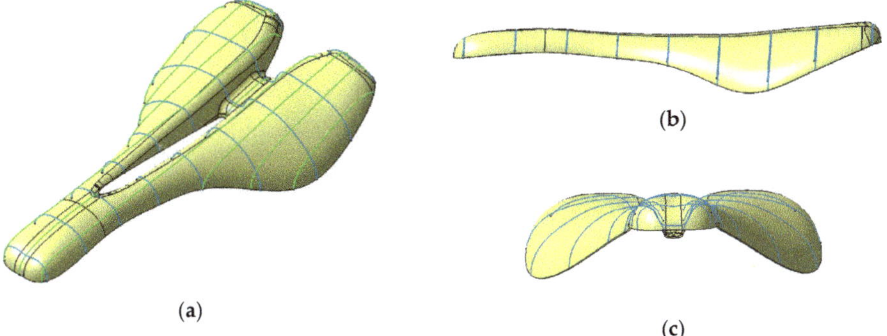

(b)

(a)

(c)

Figure 4. The CAD model of the novel saddle: (a) isometric view; (b) side view; (c) front view.

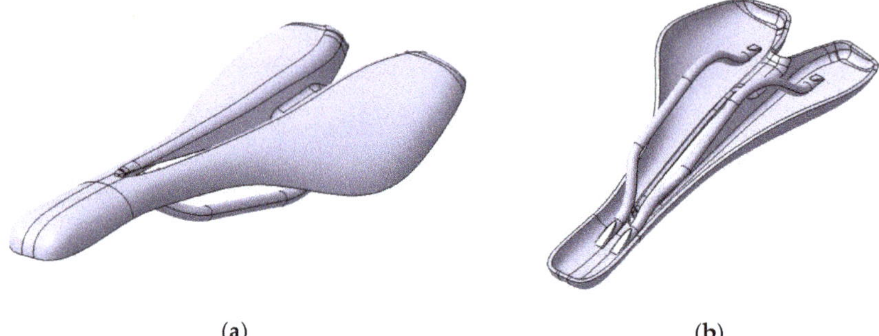

(a)   (b)

Figure 5. The 3D model of the assembled saddle: (**a**) upper side; (**b**) lower side.

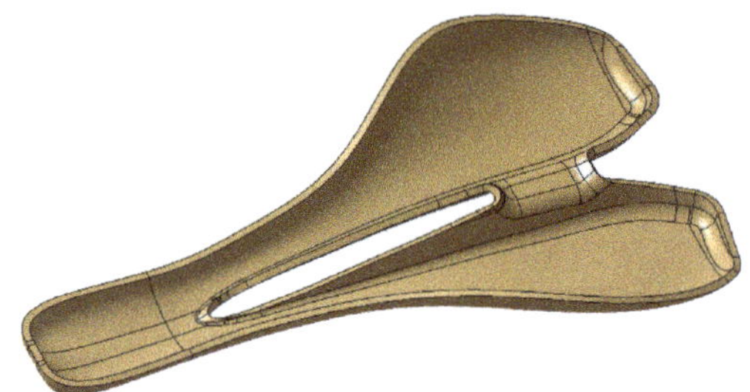

Figure 6. The 3D model of the mold saddle.

2.2. Manufacturing Method Using 3D Printing by Fused Deposition Modeling (FDM) Technology of the Mold Bike Saddle

Initially, a comparative study between two different mold manufacturing techniques was performed. Firstly, CNC milling using two different material blocks, aluminum and epoxy was considered. The processing time, the cost, and the masses of the mold were estimated by simulation. Two technological operations, roughing with a 6 mm end mill and finishing with a 3 mm ball-nosed tool were considered for manufacturing simulation within CATIA software. The mold manufacturing simulation by CNC milling is presented in Figure 7.

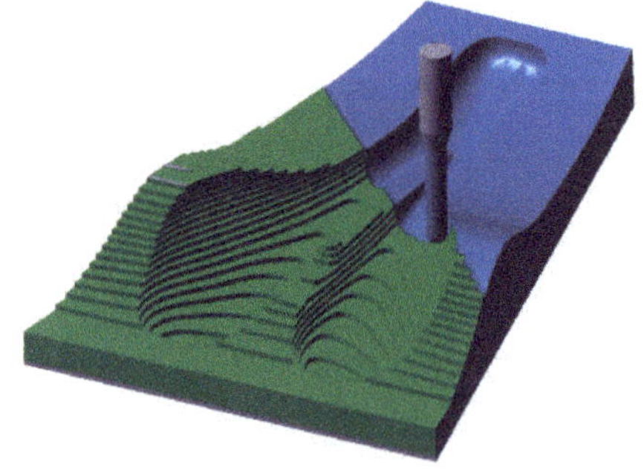

Figure 7. Mold manufacturing simulation by CNC milling.

The new method of manufacturing molds involves making them from plastic materials by 3D printing. Two thermoplastic molds were 3D printed using two different printers. Thus, the first mold marked with 'A' was 3D printed from SMARTFIL PLA filament (Smart Materials 3D, Alcalá la Real, Spain) on the Leapfrog Creatr XL printer (Leapfrog 3D Printers, Alphen aan den Rijn, The Netherlands) and the second one marked with 'B' from ABS-M30 (Stratasys, Eden Prairie, MN, USA) on a STRATASYS Fortus 380 mc dual extruder printer (Stratasys, Eden Prairie, MN, USA). The diameter of each type of filament was 1.75 ± 0.05 mm. The properties of printed materials provided by the suppliers are presented in Table 1 [44,45] and the processing parameters of the 3D printing are shown in Table 2. The mechanical properties of FDM parts from PLA and ABS were investigated by many authors in different printing conditions [11,14,15,24,27,28,46,47].

Table 1. Properties of the material used in 3D printing [44,45].

Properties	PLA	ABS-M30
Density	1.24 g/cm^3	1.04 g/cm^3 (ASTM D792)
Tensile strength at break	53 MPa	32 MPa (ASTM D638)
Tensile elongation	6%	7% (ASTM D638)
Flexural strength	83 MPa	60 MPa (ASTM D790)
Coefficient of thermal expansion	7×10^{-5} mm/mm/°C	8.82×10^{-5} mm/mm/°C

Table 2. The 3D printing process parameters.

Parameters	Value for 3D Printer Leapfrog Creatr XL	Value for 3D Printer Fortus 380 mc
Material	Smartfil PLA	ABS-M30 (model) SR30 (support)
Layer thickness	0.10 mm	0.178 mm
Infill	100%	100%
Print speed	70 mm/s	90 mm/s
Hotend	0.2 mm	-
Print temperature	215 °C	235 °C
Building plate temperature	70 °C	80 °C

The STL file of the mold, saved from CATIA V5 software, was imported to every specific software of the two printers. This was used to generate G-code and 3D print settings. In the first study, Ultimaker Cura software [48] was used to generate a G-code file for mold A. The mold A was positioned with the functional surface upward on the build plate (Figure 8a). The mold B was preprocessed using the Insight 3D printing software (Stratasys, Eden Prairie, MN, USA). The build orientation of the part was optimized in diagonal position with the functional surface downward (Figure 8b). The support structure of PLA-based mold was mechanically removed in the post-processing stage. The SR30 soluble support material used for 3D printing of the ABS-based mold was removed by dissolving it in a very mild NaOH solution.

The surface roughness of the mold has been measured using a Namicon TR-220 (Namicon Testing S.R.L., Otopeni, Romania) roughness tester, as per the ISO 4288 standard [49]. The roughness was measured perpendicular to the deposition direction, following a measurement procedure similar as in [50]. As is concluded in [51] the surface obtained by FDM technology is the surface composed of high peaks. The surface roughness of the FDM-based mold is not appropriate for laminating and polymerizing CFRP. The active surface of the mold requires mechanical processing and deposition of filler materials for its uniformity. Thus, layers of polyester gel coat were applied and then polished to obtain a glassy surface.

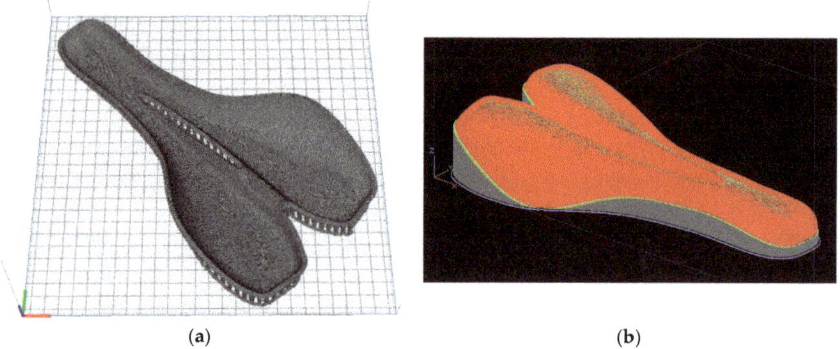

(a)   (b)

Figure 8. The 3D printing simulation of the molds: (**a**) mold 'A' made of PLA on Leapfrog Creatr XL; (**b**) mold 'B' made of ABS on STRATASYS Fortus 380 mc.

The PLA material does not make a very good chemical connection with the polyester gel coat. A chemical surface treatment or a deposition of an intermediate layer of another material would be suitable. The surface of the PLA-based mold was initially treated additionally with a layer of epoxy gel coat. Thus, OH 38 hardener and CH-3 epoxy gel coat (Ebalta Kunststoff GmbH, Rothenburg o.d.t. Germany) that contains aluminium (Al) powder was used, where the mixing ratio is 100: 35 parts by weight. The treated PLA mold was polymerized for 24 h at 20 °C, and then T35 polyester gel coat (Havel Composites CZ s.r.o., Czech Republic) layer was applied by spraying on both sides; 2% methyl ethyl ketone peroxide (MEKP) catalyst was used to polymerize it. The epoxy layer allows a better adhesion of the polyester layer to the surface of the mold, and creates a good interface between the T35 gel coat layer and the PLA. Also, Al powder creates high stiffness properties of the mold. The covered layers were applied on both sides of the molds.

The ABS mold was sprayed with T35 gel coat layers. Also, the styrene from ABS allows a good chemical adhesion with the styrene component from the polyester gel coat. A heat treatment in the oven at a temperature of 85 ± 2 °C for 24 h was applied for both molds in order to prepare them for the lamination of CFRP layers (Figure 9). Finally, the PLA mold was mechanically processed with abrasive paper up to 1000 grids and polished.

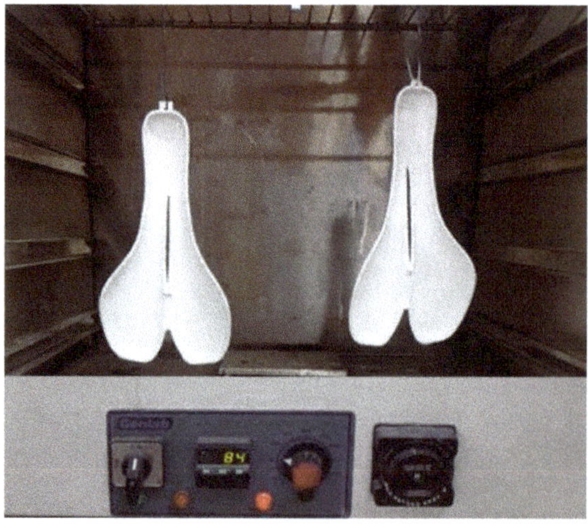

Figure 9. Heat treatment of the molds in the oven.

2.3. Manufacturing Method of Carbon Fiber-Reinforced Polymer (CFRP) Bike Saddle

The CFRP bike saddle specimens were manufactured by vacuum bag technology and oven curing. The used prepreg materials are codified as GG245T-DT806W-42 (Delta Tech S.p.A., Rifoglieto Italy) and consist of carbon fiber Twill prepreg fabric type 2 × 2 by 245 g/sq, (245T), 3K threads, and epoxy resin type DT806W. This prepreg material contains 42 wfr. epoxy resin with a curing range from 65 °C to 140 °C.

The mold was prepared by applying mold sealer layers, and mold release layers that prevent the adhesion of the CFRP material at the mold surface. A layer of mold sealer type S31 (Jost Chemicals, Laudenbach, Germany) was applied in order to close the material pores and create a sleek surface. Then, the surface was treated with five layers of liquid mold release type Frekote 770NC (Loctite, Düsseldorf, Germany). Each layer was applied at 20 min interval to allow evaporation and drying of the solvent.

Three CFRP prepreg layers were applied on the mold surface following a stacking sequence distributions of [0/90/±45/0/90] degrees (Figure 10a). The edges of the CFRPs prepreg were cut according to the PLA mold shape (Figure 10b), and the whole assembly was covered with release foil and the breather fabric (Figure 10c). The mold was inserted into a vacuum bag which was sealed on the separation plane using an electric welding machine. A uniform pressure of −0.09 MPa was applied for 30 min on the composite surface. Then, the vacuum pump was stopped in order to check the bag tightness. The composite material was cured in an oven at 80 ± 20 °C for 5 h, where the vacuum pressure was applied.

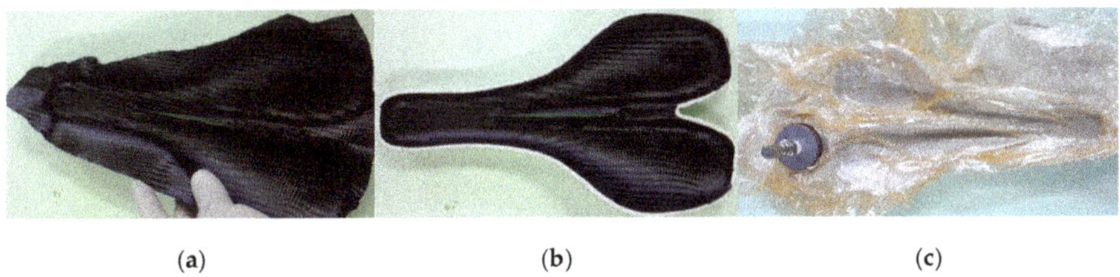

(a)(b)(c)

Figure 10. Application of CFRP prepreg layers on the PLA mold: (**a**) applying of a piece of CFRP prepreg; (**b**) cutting the edges of the CFRPs prepreg according to the PLA mold shape; (**c**) mold and composite material in the bag under vacuum pressure.

The mold was cooled down for 30 min. The auxiliary materials have been removed (vacuum bag, the breather fabric, and the release foil) and the CFRP bike saddle was released from the mold. The CFRP part was then sanded on the borders by glass paper and the central longitudinal channel of the part was cut.

Five specimens denoted by A1-A5 from CFRP and two specimens denoted by C1-C2 carbon-Kevlar fiber-reinforced polymer (CKFRP) were manufactured using the PLA mold. The mass of all samples was determined using a precision scale.

2.4. Quality Control by 3D Scanning Technique of Molds and CFRP Parts

The molds and parts were subjected to a quality inspection, using non-contact 3D scanning techniques. The Creaform GO!Scan 20 (Creaform, Québec, Canada) 3D optical scanning system used in this study integrated with VXelements software (Creaform, Québec, Canada) provides a fast acquisition speed at 550,000 points/second, and extensive capture of complex surfaces at 0.1 mm accuracy.

The parts sprayed with white powder were positioned on a rotary table on which positioning targets were placed (Figure 11). The scan was performed in one go resulting in a 3D mesh of points. VXelements software was used to acquire and optimize the 3D scanning data. In this case study, the 3D mesh of points was saved as an STL file and then transferred to the CATIAV5-6 2019 software to dimensionally compare the scanned 3D data and CAD model. The evaluation of the dimensional deviations was performed with the help of the deviation analysis tool within the CATIA software, where a quality control report was generated. Color comparison charts of 3D scanned model to CAD surface model which show the dimensional deviations were obtained. The dimensional deviation values of the 3D-printed molds, final molds, and CFRP specimens were determined. The sources of possible errors in the evaluation process can result from the scanning process, mesh processing, or the alignment procedure of the models being compared. The results of the dimensional deviations were statistically analyzed.

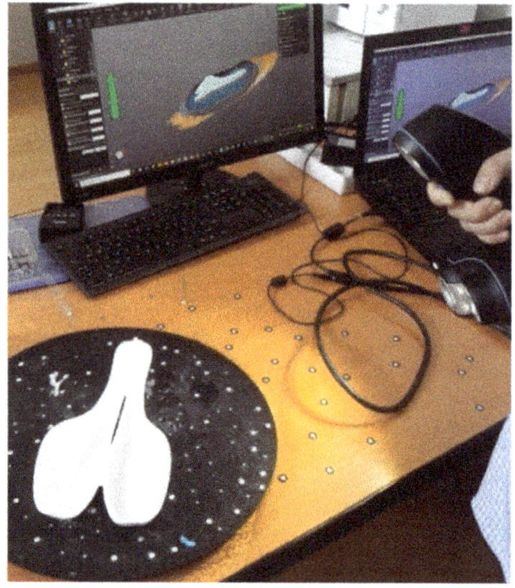

Figure 11. Positioning of the bike saddle on the rotary table of the scanning system

3. Results and Discussion

3.1. FDM-Based Molds Evaluation

Following the FDM process, two rigid molds of PLA and ABS were obtained as shown in Figure 12. The average roughness Ra of the upper mold surface was 25 ± 5 µm in both cases, as in the results obtained by Alsoufi et al. [52].

The surface aspect of treated molds was analyzed. The upper and lower surfaces aspect of the treated molds is shown in Figure 13. Some cracks were detected within the gel coat layer on the surface of the ABS mold after the heat treatment, as shown in Figure 13b. Thus, the results show that the treated ABS mold cannot be used for processing laminated composite materials. The unexpected behavior of the treated ABS mold can be attributed to the heat treatment and the expansion coefficient of ABS being different than that of the gel coat, although the deformations of 3D printed ABS parts are expected to appear at over 90 °C. The PLA mold had a different behavior than the ABS mold, the gel coat layer was adherent, and no cracks or mold deformation were detected.

The surface roughness (Ra) of the PLA mold was 0.03 µm after mechanical processing and polishing. The resulted surface of the PLA mold was smooth and continuous.

The results of the comparative study between different mold manufacturing techniques showed that the PLA mold weighed 180 g, the mold from epoxy 1200 g, and the mold of aluminum 2800 g, respectively. Thus, the molds obtained by FDM have a noticeably lower total mass than the molds conventionally used. The estimated cost of the molds and detail about the estimated manufacturing time of the molds can be found in Table 3.

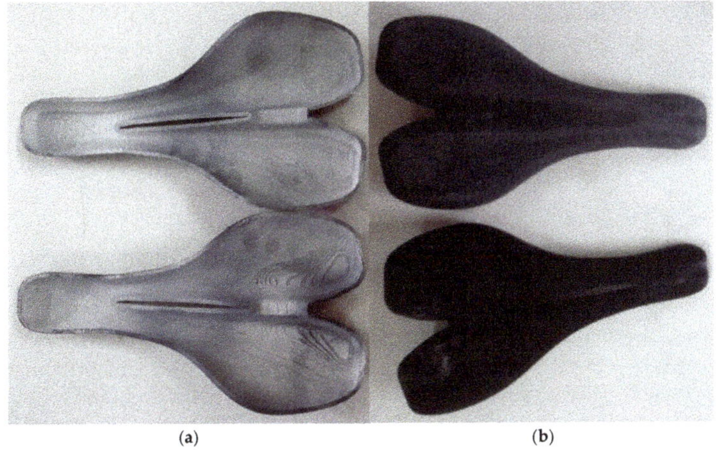

(a)   (b)

Figure 12. PLA mold (top) and ABS mold (down): (**a**) active surface of the molds (cavity); (**b**) back side of the molds.

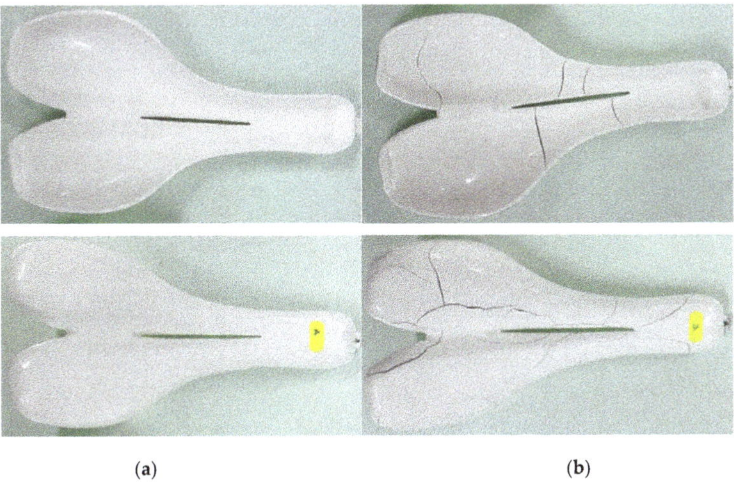

(a)   (b)

Figure 13. The molds after gel coat application and heat treatment: (**a**) PLA mold; (**b**) ABS mold.

Table 3. Estimate of mold cost, lead time, and the mold weight.

Manufacturing Method/Material Type	Cost	3D Print/Machining Time	Mold Weight/ Raw Material
CNC milling/Al block	1500 EUR	18 h	2800 g/6000 g
CNC milling/epoxy block	1300 EUR	14 h	1200 Kg/2600 kg
FDM process/PLA	150 EUR	18 h 37 min	180 g
FDM process/ABS	200 EUR	14 h 11 min	150 g

The estimated cost of the PLA mold was 150 euros. The PLA mold had a lower cost than the molds conventionally used, ≅10×, and ≅8.6× less in cost than CNC milled molds from the Al blocks and epoxy blocks, respectively.

The PLA molds can be heated up very easily at the same time as the laminated composite material. Due to the chosen solution, shell-type mold heating occurs rapidly. The mold surface has the same temperature as the CFRP material in the curing period time. This causes the first layer of the resin to fluidize and be pressed at the same time as the other layers. This removes the pores from the CFRP surface parts. This cannot be said of metal molds or epoxy blocks that are much harder to heat, where the outside layer is heated firstly and becomes fluid. Thus, the outside layer passes to the gel phase, seals the top layers, and the pores within the first deposited layer on the mold cannot be removed. They remain on the surface of the mold and CFRP part. It is undesirable that the surface of the CFRP specimen contains pores after the polymerization process. That generates important manual labor to cover pores and supplementary costs. From this point of view, the shell-type molds made of polymers have very good behavior.

It should also be noted that from the mechanical point of view they are quite fragile. For this reason, when the mold and CFRP are inserted into the vacuum bag and during vacuuming, the bag must be very well arranged. The bag must not strain the mold during the vacuum and curing process. It is important to leave as many as possible crimps of the bag on the surface of the mold. These can compensate for stresses generated by the vacuum pressure on both sides of the mold.

The mechanical properties of 3D printed materials by FDM (PLA and ABS) which have been post-processed by heat treatment (annealing) were studied by many authors [13–15]. Also, the mean ultimate strength at compression for annealed samples which were 3D printed with 100% infill was determined as 69.85 MPa [13]. The main stress of the mold during the vacuum bag-forming process and oven polymerization of CFRP is compression. The applied pressure during the CFRP part polymerization process is 0.09 MPa. In the most disadvantaged cases, this value is much lower than the compressive strength of the mold material (PLA in our case) which was 3D printed by FDM and thermal treatment.

3.2. Manufacturing of Bike Saddle Specimens of CFRP

The results have shown that the FDM-based mold made of PLA successfully withstands the manufacture of CFRP components. It has good behavior for the laminating of the composite materials. Heated and simultaneously exposed to a vacuum pressure of –0.9 bar, the mold remains rigid and can be used for the curing of CFRP prepreg in the oven.

Five samples from CFRP and two from CKFRP were manufactured as shown in Figure 14. A very well pressed and compact composite material of the saddle was the result. After manufacturing by oven vacuum bag curing technology of seven FRP specimens, no cracks of the gel coat layer were noticed, the mold successfully supporting the manufacture of the bike saddles.

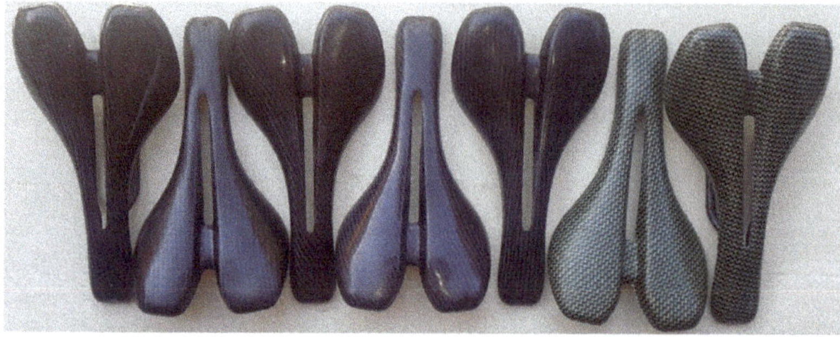

Figure 14. The CFRP and carbon-Kevlar fiber-reinforced polymer (CKFRP) bike saddle specimens.

The mass of the CFRP bike saddle specimens was 42 ± 2 g. The bike saddle assembly weight was 85 ± 2 g, where two fixing bars and the structural adhesive were included supplementary.

The proposed method that uses FDM-molds is suitable for obtaining, in a rapid way, CFRP parts in limited series. Customized CFRP parts can be manufactured at a low cost compared with traditional manufacturing. However, it should be mentioned that it must be undertaken with great care, especially when the edges of the CFRP part are cut at the same dimension as the borders of the mold. The borders of the mold can deteriorate easily or the release agent layer can be damaged.

The saddle assembled on a road bike (Figure 15) has resulted in a very comfortable biker position. The saddle position can be adjusted by changing the saddle tilt angle or by moving the saddle forward or backward.

(a)   (b)

Figure 15. The CFRP saddle prototype assembled on a road bike: (**a**) view about the bicycle; (**b**) detail of the saddle.

3.3. Results of the Quality Control by 3D Scanning Technique of Molds and CFRP Parts

The molds obtained using the FDM process from ABS and PLA were dimensionally evaluated. The results indicate a mean deviation of −0.94 mm and 86.14% of the scanned points are within the ±1 mm range, in the case of the ABS mold (Figure 16a). In the case of the PLA mold, the scanning results indicate a mean deviation of −0.219 mm, and 74% of the scanned points are within the ±1 mm range (Figure 16b). These negative values indicate a contraction of the materials for these two molds. In the case of the PLA mold, the deviations of the measurements indicated that the lower limit of deformations is within acceptable limits for such parts. The heat treatment can affect the dimensional tolerances of FDM-printed parts (PLA or ABS), as was mentioned by [14]. Therefore, it is necessary to take the shrinkage or expansion of the material into consideration when designing molds to be printed by FDM and subjected to heat treatment afterward. Thus, the shrinkage of the material can be reduced by scaling the 3D model of the mold in the design process; however, in this case study the model was not scaled to eliminate this issue. A scaling factor can be determined by dimensionally comparing the CAD model and the final part after post-treatment, as is investigated in [53].

Although the process has very good accuracy, after the deposition of the material we measured a small contraction of the real parts. It should be noted that this contraction can be amplified/decreased during the scanning process, as it is known that each measuring process has an uncertainty associated with the equipment and method. The appearance of the obtained pieces is glossy and black, due to which during the scanning process a series of errors can appear, and the points taken by mistake in the air must be removed. In some scans, there were areas where the scanned points are obviously erroneous due to the geometry (the part is very thin) and the appearance of the part. In Figure 16 it can be seen that a small percentage of points is marked with red, which are positioned on the side of the saddle that theoretically cannot be deformed only in that area.

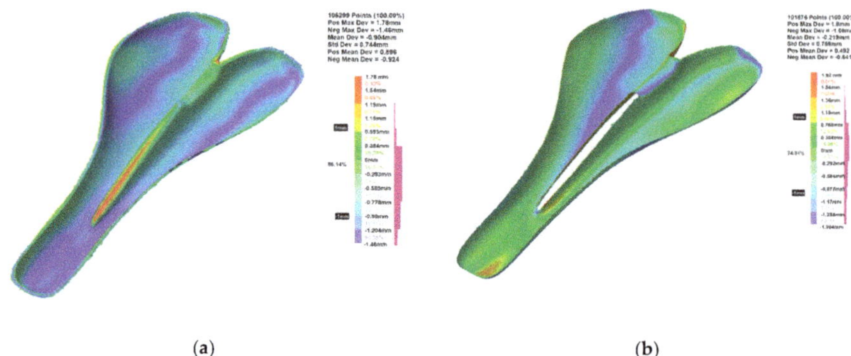

(a)   (b)

Figure 16. The comparison regarding the dimensional deviations from the CAD model and the FDM-based molds: (**a**) ABS mold; (**b**) PLA mold.

The dimensional deviations between the final PLA mold and the CAD model are shown in Figure 17. The mean deviation is −0.33 mm. The new contraction is due to the layers of gel coat applied on the mold. In this case, 87.45% of the scanned points are between ±1 mm. The value decreases from −0.21 mm to −0.33 mm regarding the mean deviation.

The deviations analysis between the CFRP saddle and the mold from PLA for A2 specimen is shown in Figure 18. It is found that 85.67% of the scanned points on the surface of the composite parts are between ±1 mm from the surface of the mold. In this case, the mean deviation is −0.233 mm. Similar analyzes were performed for all the specimens and the results are summarized in Table 4. The values of the mean deviation are in the range of −0.356 mm and −0.0296 mm.

A comparison between the designed CAD surface and the CFRP bike saddle is shown in Figure 19. It is found that 61% of the scanned points on the bike saddle's surface are in the range of ±1 mm from the surface of the CAD model. Some scanning deficiencies can be noticed here as well. Part of the scanned points is also in the area of the scanned surface and another part is on the edges of the workpiece. The mean deviation, in this case, is −0.102 from the CAD model.

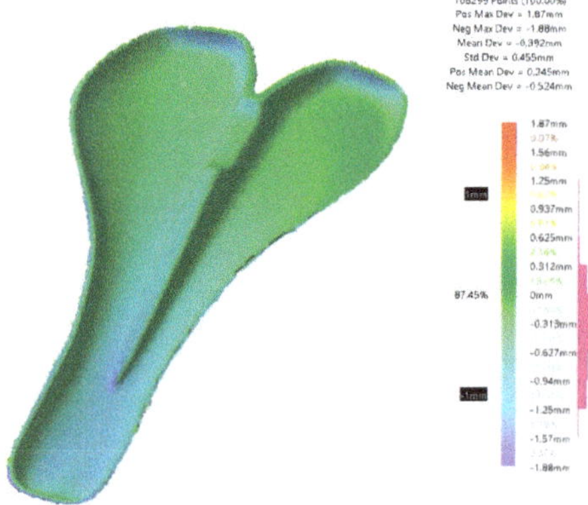

Figure 17. Dimensional deviation between the CAD model and the gel-coated PLA mold.

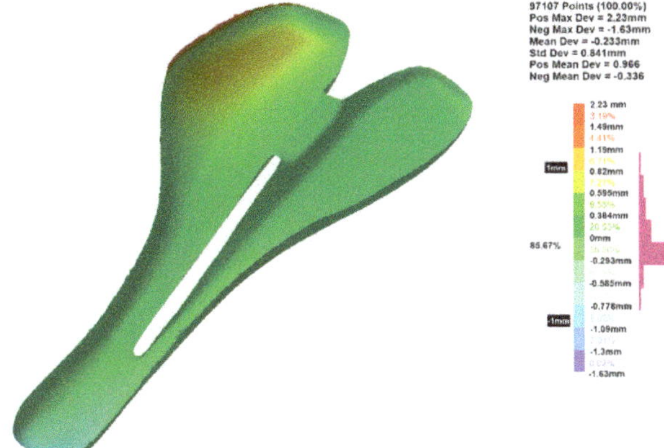

Figure 18. Dimensional deviation between a CFRP bike saddle specimen (A2) and the final PLA mold.

Table 4. Statistical parameters of comparison between the final PLA mold and CFRP parts.

Specimen	No. of Points	Mean Dev [mm]	Std Dev [mm]	Dev[−1,1] [%]
A1	99,412	−0.0838	0.92	78.84
A2	97,107	−0.233	0.841	85.67
A3	99,658	−0.0296	0.78	80.8
A4	100,860	−0.38	0.888	76.22
A5	101,345	−0.356	0.833	77.96

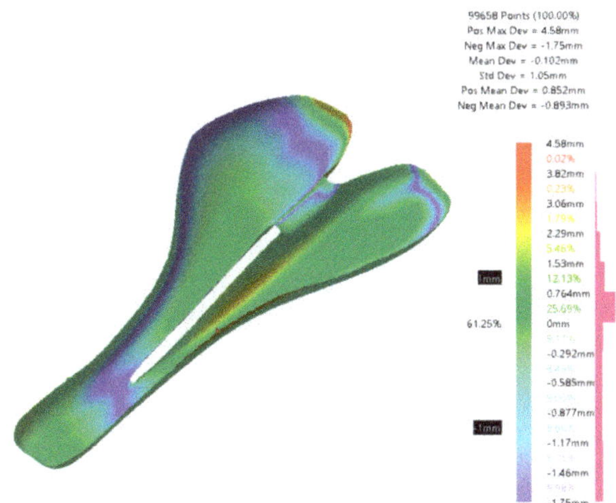

Figure 19. Comparison between the CAD model and specimen A3 obtained through CFRP.

Detailed results of the comparison between the CFRP bike saddle and the designed CAD model for all the specimens are shown in Table 5. The values of the mean deviation are in the range of −0.349 and −0.102 mm. The results of the statistical analysis for all the five specimens are shown in Table 5.

Table 5. Statistical parameters of comparison between mold CAD model and CFRP parts.

Specimen	No. of Points	Mean Dev [mm]	Std Dev [mm]	Dev[−1,1] [%]
A1	99,412	−0.349	0.848	66.61
A2	97,107	−0.302	0.896	64.74
A3	99,658	−0.102	0.854	61.25
A4	100,860	−0.142	0.96	64.07
A5	101,345	−0.123	0.984	62.36

Also, the interval plots of the main deviations in the range ±1 mm denoted as Dev[−1,1] of the active surface for all the samples is shown in Figure 20. Individual standard deviations were used to calculate the interval plot.

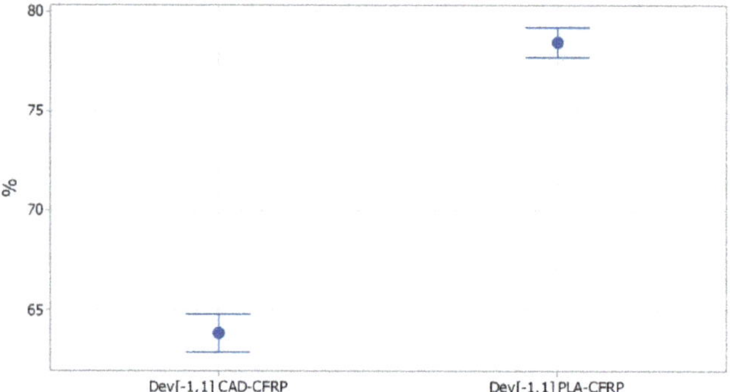

Figure 20. Interval plot of Dev[−1,1] CAD-CFRP and Dev[−1,1] PLA-CFRP; bars are one standard error from the mean.

The results show that the coefficient of variation is lower than 10% that assures the data heterogeneity and expresses the repeatability of the experiments, as shown in Table 6.

Table 6. Statistical parameters of percentage deviations Dev[−1,1] for PLA-CFRP and CAD-CFRP for five samples.

Dev[−1,1]	Mean [%]	Std Dev [%]	CV [%]
PLA mold to CFRP parts	78.468	1.650	2.10
CAD model to CFRP parts	63.806	2.087	3.27

The coefficient of variation (CV) is a measure of spread that describes the variation in the data relative to the mean.

4. Conclusions

This paper presented a novel method for the manufacturing of molds used to laminate complex geometry parts using CFRP prepreg. Specifically, the application this paper focused on was the production of a bike saddle for performance bicycles, where the design of a prototype bike saddle fabricated from CFRP was presented. The new mold and the CFRP prototype were created utilizing different technologies, where the mold was manufactured using additive manufacturing FDM processes and the CFRP fabricated part utilized vacuum bag-forming technologies and polymerization. This paper investigated the use of two materials (PLA and ABS) to create the molds through FDM.

The following conclusions can be drawn:

- The molds obtained by FDM from the thermoplastic polymers have a noticeably lower total mass than the molds conventionally used, ~15×, and ~6.6× less in mass than CNC milled molds from aluminum blocks and epoxy blocks, respectively.
- The weight of the manufactured CFRP bike saddle was measured to be 42 g and 85 g with the CFRP fixing bars fixed to the back of the saddle using adhesive. It was found that this allowed for a mass reduction of 3.52 times with the assumption that a typical saddle weighs around ~300 g [54].
- Due to the roughness of the active surface(s) of the mold(s) obtained through FDM, the active surface(s) were covered with additional layers of gel coat, as compared to the conventional mold(s). Following this, heat treatment of the mold(s) and the complete polymerization of the deposited layers were applied, and this contributes to the elimination of internal material stresses. The heat treatment comprised of exposing the specimens to 85 °C for 24 h. Subsequent to this, the ABS mold cracked, becoming unusable; it was noted that the thermal expansions of the two materials used, ABS and polyester gel coat, were different, which led to cracking of the polyester surface layer. However, it was observed that the PLA mold had very good behavior. Following the mechanical processing of the mold's active surface (grinding and polishing), it was measured that the PLA mold had a mean dimensional deviation from the CAD model of −0.392 mm.
- The PLA mold was used to make seven bike saddles, five comprising CFRP and two CKFRP. A dimensional evaluation of the CFRP bike saddle was carried out, which produced consistent results. Following the evaluated measurements of the new CFRP bike saddles, the mean deviations as compared to the PLA mold used to laminate were in the range of −0.0838 mm and −0.38 mm. After scanning and the processing of results, it was found that over 80% of the scanned points, from the surface of the CFRP saddles (101876 scans points), are within the range of ±1 mm, compared with PLA mold. With regard to differences between CAD model and the produced parts, the mean deviation was found to be −0.3 mm, where 65% percentage of the scanned points of the bike saddle are within the range ±1 mm from the original CAD model. However, it should be noted that this is a product with a complex geometry, and a maximum size of 270 mm long and 145 mm wide.
- When evaluating the measurements for the produced CFRP bike saddle, the observed extreme values were also taken into account for evaluation. These were determined to be scan errors, creating red points on the scanning diagrams or on areas that have unscanned red points. However, it should be noted that certain areas on the edges of the saddle should not be taken into account.
- The FDM-based mold made of PLA can be used in the manufacture of parts from CFRP prepreg, utilizing vacuum bag technologies, followed by curing in an oven. However, it is recommended that these molds are to be used only in the manufacture of limited series of prototypes or customized parts.
- An important aspect of the vacuum bag process is the distribution of the vacuum bag over the mold surface whilst a vacuum is being created, where a uniform pressure on both surfaces of the mold must be applied by the vacuum bag, as this will avoid deformations which can result from a non-uniform pressure being applied to the mold, which can deform the composite and the mold during the curing process. As a result of the dimensional evaluation of the CFRP parts, deformations can only be observed on a certain part of the mold. These deformations can be attributed to where the vacuum bag deformed the mold in the central channel area of the mold. Thus, the two parts of the saddle were contracted slightly. An important conclusion with regards to the design process of the mold(s) is that it is recommended to avoid drilling holes in the central areas of the mold, and for the mold to cover the longitudinal hole with solid material, where, the mold will consolidate and the two parts can no longer come together under the action of external pressure. Following the curing process of the CFRP parts, the channel can then be machined and eliminated.
- The manufacture of CFRP parts in 3D-printed molds is a variable process that depends on parameters such as the shape and structure of the printed mold, uniform distribution of

mold material, mold coating material and its thickness, constant thickness of the mold walls, uniform distribution of the vacuuming bag, vacuum pressure applied, oven baking temperature, cooling and demolding of the part.

- A future direction of research for these types of mold(s) would be to use them in the curing autoclave process of FRP. The same temperature can be used for the mold(s); however, various additional pressures that may rise up in the range of 4–5 bars will be needed. Another important study would be to investigate the design of these types of mold(s), so that they do not deform under the external pressures applied, whilst also retaining conditions resembling a low wall thickness. Additionally, studies would also need to investigate the use of different types of thermoset polymer on the mold surfaces in order to improve the stiffness and the thermal expansion qualities of the mold.

Author Contributions: Conceptualization, P.B., C.N. and R.U.; methodology, P.B., R.U. and C.N.; software, C.N.; formal analysis, R.U.; investigation, P.B. and C.N.; data curation, P.B., C.N. and R.U.; writing—original draft preparation, P.B., R.U. and C.N.; writing—review and editing, R.U. and P.B.; visualization, P.B., R.U. and C.N.; supervision P.B. and R.U.; project administration, P.B.; All authors have read and agreed to the published version of the manuscript.

Funding: This research received no external funding.

Acknowledgments: This article has benefited from the support of the project 345PED/2020 "Smart hospital bed - HoPE" funded by the Executive Agency for Higher Education, Research, Development and Innovation Funding (UEFISCDI), Romania. Also, the authors acknowledge of Technical University of Cluj-Napoca and Transilvania University of Braşov, for providing the infrastructure used in this work.

Conflicts of Interest: The authors declare no conflict of interest.

References

1. Henderson, L. *Carbon Fibers and Their Composite Materials*; MDPI AG: Basel, Switzerland, 2019; ISBN-13: 978-3039211029, ISBN-10: 3039211021.

2. Barbero, E.J. *Introduction to Composite Materials Design*, 3rd ed.; CRC Press: Boca Raton, FL, USA, 2018; ISBN-10: 1-138-19680-0.

3. Tofail, S.A.M.; Koumoulos, E.P.; Bandyopadhyay, A.; Bose, S.; O'Donoghue, L.; Charitidis, C. Additive manufacturing: Scientific and technological challenges, market uptake and opportunities. *Mater. Today* **2018**, *21*, 22–37. [CrossRef]

4. International Organization for Standardization. *Standard Terminology for Additive Manufacturing–General Principles–Terminology*; ISO/ASTM 52900-15; ISO/ASME International: Geneva, Switzerland, 2015.

5. Turner, B.N.; Strong, R.; Gold, S.A. A review of melt extrusion additive manufacturing processes: I. Process design and modeling. *Rapid Prototyp. J.* **2014**, *20*, 192–204. [CrossRef]

6. Gibson, I.; Rosen, D.W.; Stucker, B. *Additive Manufacturing Technologies*; Springer: New York, NY, USA, 2015.

7. Gerphard, A. *Understanding Additive Manufacturing, Rapid Prototyping-Rapid Tooling-Rapid Manufacturing*; Publisher Carl Hanser Verlag GmbH & Co. KG Munchen: Munich, Germany, 2011; ISBN-13: 978-1-56990-507-4. [CrossRef]

8. Gebhardt, A.; Kessler, J.; Thurn, L. *Basics of 3D Printing Technology*; Carl Hanser Verlag GmbH & Co. KG: Munich, Germany, 2018; ISBN 978-1-56990-702-3. [CrossRef]

9. Andreas, F.; Steve, R.; Thomas, B. New Fiber Matrix Process with 3D Fiber Printer—A Strategic In-process Integration of Endless Fibers Using Fused Deposition Modeling (FDM). In Proceedings of the IFIP International Conference on Digital Product and Process Development Systems, Dresden, Germany, 10–11 October 2013; pp. 167–175.

10. García Plaza, E.; Núñez López, P.J.; Caminero Torija, M.Á.; Chacón Muñoz, J.M. Analysis of PLA Geometric Properties Processed by FFF Additive Manufacturing: Effects of Process Parameters and Plate-Extruder Precision Motion. *Polymers* **2019**, *11*, 1581. [CrossRef] [PubMed]

11. Porter, J.H.; Cain, T.M.; Fox, S.L.; Harvey, P.S. Influence of infill properties on flexural rigidity of 3D-printed structural members. *Virtual Phys. Prototyp.* **2019**, *14*, 148–159. [CrossRef]

12. Jayanth, N.; Senthil, P.; Prakash, C. Effect of chemical treatment on tensile strength and surface roughness of 3D-printed ABS using the FDM process. *Virtual Phys. Prototyp.* **2018**, *13*, 155–163. [CrossRef]

13. Slavkovic, V.; Grujovic, N.; Dišic, A.; Radovanovic, A. Influence of annealing and printing directions on mechanical properties of PLA shape memory polymer produced by fused deposition modeling. In Proceedings of the 6th International Congress of Serbian Society of Mechanics Mountain Tara, Mountain Tara, Serbia, 19–21 June 2017; pp. 19–21.

14. Butt, J.; Bhaskar, R. Investigating the Efects of Annealing on the Mechanical Properties of FFF-Printed Thermoplastics. *J. Manuf. Mater. Process.* **2020**, *4*, 38. [CrossRef]

15. Wach, R.A.; Wolszczak, P.; Adamus-Wlodarczyk, A. Enhancement of mechanical properties of FDM-PLA parts via thermal annealing. *Macromol. Mater. Eng.* **2018**, *303*, 1800169. [CrossRef]

16. Udroiu, R. Rapid Tooling by Three Dimensional Printing (3DP). In Proceedings of the 3rd WSEAS International Conference on Manufacturing Engineering, Quality and Production Systems, Recent Researches in Manufacturing Engineering, Brasov, Romania, 11–13 April 2011; pp. 177–180.

17. Udroiu, R.; Braga, I.C. Polyjet technology applications for rapid tooling. In Proceedings of the IManE&E 2017 MATEC Web of Conferences, Iasi, Romania, 25–26 May 2017; pp. 1–6. [CrossRef]

18. Parada, L.R.; Mayuet, P.F.; Gámez, A.J. Industrial product design: Study of FDM technology for the manufacture of thermoformed prototypes. *Procedia Manuf.* **2019**, *41*, 587–593. [CrossRef]

19. Yanga, Y.; Lia, H.; Xua, Y.; Dongb, Y.; Shana, W.; Shenc, J. Fabrication and evaluation of dental fillers using customized molds via 3D printing technology. *Int. J. Pharm.* **2019**, *562*, 66–75. [CrossRef]

20. Hay, J.A.; Smayra, T.; Moussa, R. Customized Polymethylmethacrylate Cranioplasty Implants Using 3-Dimensional Printed Polylactic Acid Molds: Technical Note with 2 Illustrative Cases. *World Neurosurg.* **2017**, *105*, 971–979. [CrossRef]

21. Siemiński, P.; Szulc, B. Analysis of FDM/FFF additive manufacturing production mold inserts of injection molds. *Mechanik* **2018**, *1*, 53–55. [CrossRef]

22. Lušića, M.; Schneidera, K.; Hornfecka, R. A case study on the capability of rapid tooling thermoplastic laminating molds for manufacturing of CFRP components in autoclaves. *Procedia CIRP* **2016**, *50*, 390–395. [CrossRef]

23. Sudbury, T.Z.; Springfield, R.; Kunc, V.; Duty, C. An assessment of additivemanufacturedmolds for hand-laid fiber reinforced composites. *Int. J. Adv. Manuf. Technol.* **2017**, *90*, 1659–1664. [CrossRef]

24. Ning, F.D.; Cong, W.L.; Hu, Y.B.; Wang, H. Additive manufacturing of carbon fiber-reinforced plastic composites using fused deposition modeling: Effects of process parameters on tensile properties. *J. Compos. Mater.* **2017**, *51*, 451–462. [CrossRef]

25. Zhang, J.; Zhou, Z.; Zhang, F.; Tan, Y.; Tu, Y.; Yang, B. Performance of 3D-Printed Continuous-Carbon-Fiber-Reinforced Plastics with Pressure. *Materials* **2020**, *13*, 471. [CrossRef]

26. Reverte, J.M.; Caminero, M.Á.; Chacón, J.M.; García-Plaza, E.; Núñez, P.J.; Becar, J.P. Mechanical and Geometric Performance of PLA-Based Polymer Composites Processed by the Fused Filament Fabrication Additive Manufacturing Technique. *Materials* **2020**, *13*, 1924. [CrossRef]

27. Chacón, J.M.; Caminero, M.A.; Núñez, P.J.; García-Plaza, E.; Garcia-Moreno, I.; Reverte, J.M. Additive manufacturing of continuous fibre reinforced thermoplastic composites using fused deposition modelling: Effect of process parameters on mechanical properties. *Compos. Sci. Technol.* **2019**, *181*, 107688. [CrossRef]

28. Abeykoon, C.; Sri-Amphorn, P.; Fernando, A. Optimization of fused deposition modeling parameters for improved PLA and ABS 3D printed structures. *Int. J. Lightweight Mater. Manuf.* **2020**, *3*, 284–297. [CrossRef]

29. Caminero, M.Á.; Chacón, J.M.; García-Plaza, E.; Núñez, P.J.; Reverte, J.M.; Becar, J.P. Additive Manufacturing of PLA-Based Composites Using Fused Filament Fabrication: Effect of Graphene Nanoplatelet Reinforcement on Mechanical Properties, Dimensional Accuracy and Texture. *Polymers* **2019**, *11*, 799. [CrossRef]

30. Perez, A.R.T.; Roberson, D.A.; Wicker, R.B. Fracture Surface Analysis of 3D-Printed Tensile Specimens of Novel ABS-Based Materials. *J. Fail. Anal. Prevent.* **2014**, *14*, 343–353. [CrossRef]

31. Hauke, P.; Thomas, V. Design for Fiber-Reinforced Additive Manufacturing. *J. Mech. Des.* **2015**, *137*, 111409.

32. Hawryluk, M.; Ziemba, J.; Sadowski, P. A Review of Current and New Measurement Techniques Used in Hot Die Forging Processes. *Meas. Control* **2017**, *50*, 74–86. [CrossRef]

33. Hawryluk, M.; Gronostajski, Z.; Ziemba, J.; Janik, M.; Górski, P.; Lisowski, M. Support Possibilities for 3D Scanning of Forging Tools with Deep and Slim Impressions for an Evaluation of Wear by Means of Replication Methods. *Materials* **2020**, *13*, 1881. [CrossRef] [PubMed]

34. Nelson, J.W.; LaValle, J.J.; Kautzman, B.D.; Dworshak, J.K.; Johnson, E.M. Injection molding with an additive manufacturing tool study shows that 3D printed tools can create parts comparable to those made with P20 tools, at a much lower cost and lead time. *Plast. Eng. Conn.* **2017**, *73*, 60–66. [CrossRef]
35. Neamțu, C.; Bere, P. Methods for Checking the Symmetry of the Formula One Car Nose. *Appl. Mech. Mater.* **2014**, *657*, 785–789. [CrossRef]
36. Liu, T.J.C.; Wu, H.C. Fiber direction and stacking sequence design for bicycle frame made of carbon/epoxy composite laminate. *Mater. Des.* **2010**, *31*, 1971–1980. [CrossRef]
37. Bini, R.; Daly, L.; Kingsley, M. Changes in body position on the bike during seated sprint cycling: Applications to bike fitting. *Eur. J. Sport Sci.* **2019**, *20*, 35–42. [CrossRef]
38. Bini, R.; Hume, P.A.; Croft, J.L. Effects of Bicycle Saddle Height on Knee Injury Risk and Cycling Performance. *Sports Med.* **2011**, *41*, 463–476. [CrossRef]
39. Garcia-Lopez, J.; Rodriguez-Marroyo, J.A.; Juneau, C.E.; Peleteiro, J.; Martinez, A.C.; Villa, J.G. Reference values and improvement of aerodynamic drag in professional cyclists. *J. Sports Sci.* **2008**, *26*, 277–286. [CrossRef]
40. Thompson, M.K.; Moronib, G.; Vanekerc, T.; Fadel, G.; Campbell, R.I.; Gibson, I.; Bernard, A.; Schulz, J.; Grap, P.; Ahuja, B.; et al. Design for Additive Manufacturing: Trends, Opportunities, Considerations and Constraints. *CIRP Ann. Manuf. Technol.* **2016**, *65*, 737–760. [CrossRef]
41. Piazza, N.; Cerri, G.; Breda, G.; Paggiaro, A. The effect of a new geometric bicycle saddle on the genital-perineal vascular perfusion of female cyclists. *Sci. Sports* **2020**, *35*, 161–167. [CrossRef]
42. Bere, P.; Rozsos, R.; Dudescu, C.; Neamțu, C. Manufacturing method for bicycle saddle from carbon/epoxy composite materials. *Rom. J. Tech. Sci.* **2019**, *64*, 97–111.
43. Chen, Y.-L. Predicting external ischial tuberosity width for both sexes to determine their bicycle-seat sizes. *Int. J. Ind. Ergon.* **2018**, *64*, 118–121. [CrossRef]
44. Smart Materials 3D. Available online: https://www.smartmaterials3d.com/en/ (accessed on 10 August 2020).
45. Speed New Products to Market with ABS-M30. Available online: https://www.stratasys.com/materials/search/abs-m30 (accessed on 1 August 2020).
46. Akhoundi, B.; Behravesh, A.H.; Saed, A.B. Improving mechanical properties of continuous fiber-reinforced thermoplastic composites produced by FDM 3D printer. *J. Reinf. Plast. Compos.* **2019**, *38*, 99–116. [CrossRef]
47. Dudescu, C.; Racz, L. Effects of raster orientation, infill rate and infill pattern on the mechanical properties of 3D printed materials. *ACTA Uiversitatis Cibiniensis–Tech. Ser.* **2017**, *69*, 23–30. [CrossRef]
48. Ultimaker Cura: Advanced 3D Printing Software. Available online: https://ultimaker.com/en/products/ultimaker-cura-software (accessed on 15 May 2020).
49. International Organization for Standardization. *Geometrical Product Specifications (GPS)—Surface Texture: Profile Method—Rules and Procedures for the Assessment of Surface Texture*; ISO 4288:1996; ISO/ASME International: Geneva, Switzerland, 1996.
50. Udroiu, R.; Braga, I.C.; Nedelcu, A. Evaluating the Quality Surface Performance of Additive Manufacturing Systems: Methodology and a Material Jetting Case Study. *Materials* **2019**, *12*, 995. [CrossRef]
51. Krolczyk, G.; Raos, P.; Legutko, S. Experimental analysis of surface roughness and surface texture of machined and fused deposition modelled parts. *Teh. Vjesn.* **2014**, *21*, 217–221.
52. Alsoufi, M.S.; Elsayed, A.E. Surface Roughness Quality and Dimensional Accuracy—A Comprehensive Analysis of 100% Infill Printed Parts Fabricated by a Personal/Desktop Cost-Effective FDM 3D Printer. *Mater. Sci. Appl.* **2018**, *9*, 11–40. [CrossRef]
53. Ranganathan, R.; Ravi, T.; Pugalendhi, A. Analysis of Shrinkage Compensation Factor (SCF) of FDM uPrint SE for Accuracy Enhancement. *Int. J. Integr. Eng.* **2019**, *11*, 207–2016. [CrossRef]
54. Stevenson, J. 18 of the Best High-Performance Lightweight Saddles. Available online: https://road.cc/content/buyers-guide/18-best-high-performance-lightweight-saddles-195560 (accessed on 20 September 2020).

Article

Crystallization and Thermal Behaviors of Poly(ethylene terephthalate)/Bisphenols Complexes through Melt Post-Polycondensation

Shichang Chen [1,2], Shangdong Xie [3], Shanshan Guang [3], Jianna Bao [1,3,*], Xianming Zhang [1,3,*] and Wenxing Chen [3]

[1] Key Laboratory of Advanced Textile Materials and Manufacturing Technology, Ministry of Education, Zhejiang Sci-Tech University, Hangzhou 310018, China; scchen@zstu.edu.cn
[2] Zhejiang Provincial Key Laboratory of Fiber Materials and Manufacturing Technology, Zhejiang Sci-Tech University, Hangzhou 310018, China
[3] School of Materials Science and Engineering, Zhejiang Sci-Tech University, Hangzhou 310018, China; xshangdong96@163.com (S.X.); gssysb@163.com (S.G.); wxchen@zstu.edu.cn (W.C.)
* Correspondence: baojianna@zstu.edu.cn (J.B.); zhangxm@zstu.edu.cn (X.Z.); Tel.: +86-0571-8684-6663 (X.Z.)

Received: 30 September 2020; Accepted: 13 December 2020; Published: 19 December 2020

Abstract: Three kinds of modified poly(ethylene terephthalate) (PET) were prepared by solution blending combined with melt post-polycondensation, using 4,4′-thiodiphenol (TDP), 4,4′-oxydiphenol (ODP) and hydroquinone (HQ) as the bisphenols, respectively. The effects of TDP, ODP and HQ on melt post-polycondensation process and crystallization kinetics, melting behaviors, crystallinity and thermal stability of PET/bisphenols complexes were investigated in detail. Excellent chain growth of PET could be achieved by addition of 1 wt% bisphenols, but intrinsic viscosity of modified PET decreased with further bisphenols content. Intermolecular hydrogen bonding between carbonyl groups of PET and hydroxyl groups of bisphenols were verified by Fourier transform infrared spectroscopy. Compare to pure PET, both the crystallization rate and melting temperatures of PET/bisphenols complexes were reduced obviously, suggesting an impeded crystallization and reduced lamellar thickness. Moreover, the structural difference between TDP, ODP and HQ played an important role on crystallization kinetics. It was proposed that the crystallization rate of TDP modified PET was reduced significantly due to the larger amount of rigid benzene ring and larger polarity than that of PET with ODP or HQ. X-ray diffraction results showed that the crystalline structure of PET did not change from the incorporation of bisphenols, but crystallinity of PET decreased with increasing bisphenols content. Thermal stability of modified PET declined slightly, which was hardly affected by the molecular structure of bisphenols.

Keywords: poly(ethylene terephthalate); bisphenol; crystallization kinetics; thermal property; melt polycondensation

---

1. Introduction

Poly(ethylene terephthalate) (PET) is one of the most widely used linear semicrystalline thermoplastic polyester in many industrial and everyday applications because of its good mechanical and thermal properties, non-toxicity, processing low energy requirements and high chemical resistance [1–3]. Therefore, PET has been extensively used for the manufacture of beverage bottles, packing films, engineering plastics, industrial fibers and so on [4]. As a semicrystalline polymer, the crystallization and its degree of crystallinity play an important role on applications, which would highly affect physical and mechanical properties [5–12]. Generally speaking, the introduction of other polymers, nanofillers, comonomers and reactive functional groups are commonly used technique to modify the crystallization behaviors by means of copolymerization, blending, etc.

Tremendous effort has been ongoing to develop PET production methods by incorporation other polymers and adjusted crystallization kinetics of PET were characterized. It was found that in PET/PLA blends, the degree of crystallinity of PET was reduced by blending with polylactic acid (PLA) [13]. A more pronounced effect was observed in the crystallization rate of PET by the presence of PLA. It was proposed that PLA chains could act as barriers against movements of PET chains in diffusing through the surface of growing crystals [14]. In contrast, as in the polypropylene (PP)/PET blends, both of the crystallinity of PP and PET was enhanced [15]. With the addition of polyamide 56 (PA56) to PET, crystallization rate of PET was increased and increment was due to a nucleating effect by the early formed PA56 crystals [16]. Besides, both PA56 and PET exhibited smaller crystals compared large crystals with pure polymers. Several efforts have been made to regulate the crystallization rate of PET by inclusion of nanofillers [17,18]. Some authors found that SiO_2 nanoparticles [19,20], modified graphene oxide [21], multiwalled carbon nanotubes [22,23], polyhedral oligomeric silsesquioxanes [24,25] and calcium carbonate [26] could act as a heterogeneous nucleating agent and increase the crystallinity of PET.

PET is produced from petroleum-based monomers through a chemical process involving condensation reactions. In this sense, except for physical blending, PET can be copolymerized with various polymers and exhibit highly sensitive crystallization kinetic toward inclusion of other foreign monomers [27–29]. Related studies have revealed that addition of low content of other kinds of diols, substituted to the main raw diols of PET, could reduce crystallinity and crystallization rate. For example, it has been reported that incorporating low content of 1,3-propanediol, 2-methyl-1,3-propanediol and 2,2-dimethyl-1,3-propanediol into PET backbone could reduce the crystallinity due to the decreased macromolecular regularity of PET [30–32]. Zhou et al. added sodium-5-sulfo-bis-(hydroxyethyl)-isophthalate and 2,2-dimethyl-1,3-propanediol to improve the hydrophilicity and dyeability of PET [33]. Zhao et al. synthesized PET-based ionomers with flame retardant and antidripping by melt polycondensation using a phosphorus-containing ionic monomer as an end-capping agent into the PET chain end [34].

Bisphenol, endowing the same dihydroxyl-terminal structure as diols, can serve as a substitution of diol to react with diacid. As a result, it was often used to mediate the crystallization behaviors and physical properties of polyesters, such as polybutylene terephthalate (PBT) [35], polybutylene succinate (PBS) [36], polycaprolactone (PCL) [37] and so on. In previous study, the overall isothermal crystallization rate and growth rate of spherulites of PBS were depressed by bisphenol A [36]. In the case of PBT, the crystallinity and crystal size were reduced in the presence of TDP [35]. Besides, the thermal and mechanical properties of poly (L-lactic acid) PLLA and poly(3-hydroxybutyrate) [P(3HB)] were greatly influenced through blending with 4,4'-thiodiphenol (TDP) [38,39]. Despite the extensive literature published regarding the crystallization behavior of PET containing systems, very few studies have reported crystallization behaviors of PET/bisphenols complexes. In this study, several selected bisphenols with different molecular structures were incorporated into PET by solution blending combined with melt post-polycondensation. The main aim of this work is to evaluate the effect of a small amount of bisphenols on the intrinsic viscosity, crystallization kinetics, melting behavior, thermal property and crystallinity of PET. A variety of characterization methods were used including differential scanning calorimetry (DSC), thermogravimetric analysis (TGA) and two-dimensional X-ray diffraction (2D XRD). Furthermore, roles of the bisphenol structure and cooling rate of the nonisothermal crystallization process on the crystallization kinetics were investigated in detail. Using modified Avrami equation, the nonisothermal crystallization kinetic of the bisphenols-modified PET systems was explored.

2. Materials and Methods

2.1. Materials

PET chips with an intrinsic viscosity of 0.68 dL/g was kindly provided by Zhejiang Guxiandao Green Fiber Co., Ltd. (Hangzhou, China), TDP (AR grade), ODP (AR grade), HQ (AR grade) and 1,1,1,3,3,3-hexafluoroisopropyl alcohol (HFIP, LC grade) were purchased from Fluorochem (Derby, UK), molecular structures of four bisphenols were shown as Scheme 1. Phenol (AR grade) and 1,1,2,2-tetrachloroethane (AR grade) used as mixed solvent were bought from Shanghai Aladdin Bio-Chem Technology Co., Ltd. (Shanghai, China)

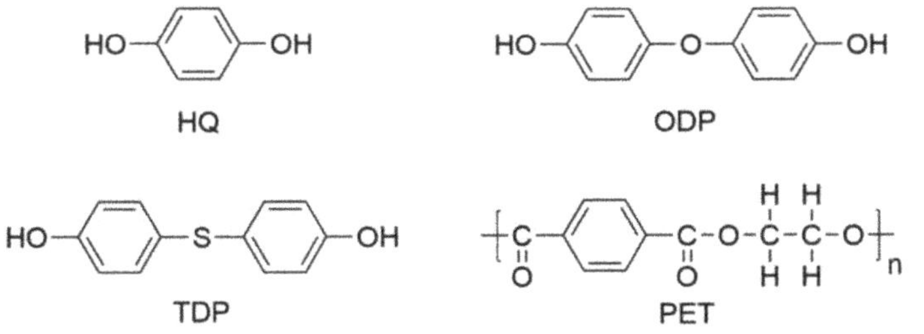

HQ

ODP

TDP

PET

Scheme 1. The chemical structure of bisphenols and PET.

2.2. Sample Preparation

In order to make the bisphenols and PET fully and evenly mixed, the solution blending method was utilized. Prior to solution blending, both PET chips and bisphenols were dried in a vacuum at 90 °C for 24 h to remove as much residual moisture as possible. Bisphenols were weighed accurately and dissolved into HFIP solutions of PET chips to obtain samples. Four weight concentrations of bisphenols in PET were selected: 0, 1, 2 and 4% wt/wt. To reach a uniform dispersion in polymer matrix, the solutions were stirred evenly and slowly to evaporate the solvent. Subsequently, the resulting mixtures were dried in a vacuum at 120 °C for 24 h. The blends were hot-pressed into the designed film thickness (1 mm) sheet using a press vulcanizer (S(X)LB-350 × 350-25) (Suyan, Changzhou, China). The melt post-polycondensation experiments were carried out in a glass drier at 270 °C and 200 Pa for 60 min. At the end of reaction, the materials were taken out and placed in water to let it cool down.

2.3. Characterization

Viscosified PET sample was dissolved into phenol-tetrachloroethane (50/50 wt/wt) solutions. After dissolution, the intrinsic viscosity (IV) of pure PET and PET/bisphenols complexes was measured with a VISCO 070 Ubbelohde viscometer (Julabo, Seelbach, Germany) at 25 °C and averaged by three times.

Fourier transform infrared (FTIR) analysis was carried out on a Nicolet 5700 FTIR spectrometer (Thermo Electron, Waltham, Massachusetts, USA, resolution: 0.09 cm^{-1}) to detect the functional groups of PET/TDP. The sample was prepared by KBr tablet and the spectra were recorded at room temperature.

Crystallization and melting behaviors of all samples were studied by means of a differential scanning calorimeter (DSC; Mettler Toledo, Zurich, Switzerland). Dry nitrogen was used as a purge gas at a rate of 45 mL/min during all measurements. For non-isothermal crystallization, samples of 6–8 mg encapsulated in aluminum pans were first heated from 25 to 280 °C and held for 5 min to eliminate any previous thermal history. Subsequently, the samples were cooled to 25 °C at different cooling rates of 5, 10, 20 and 30 °C/min, followed by a second heating to 280 °C at a constant cooling rate of 10 °C/min to evaluate their melting behaviors. Thermal stability analysis of all samples was performed on a thermal gravimetric analyzer (TGA, Mettler Toledo, Zurich, Switzerland), under nitrogen as DSC measurements. Samples of 7–8 mg were heated from 25 to 600 °C at a heating rate of 10 °C/min.

The crystalline structure and crystallinity of all specimens were analyzed by using two-dimensional wide-angle X-ray diffractions (WAXDs) with a D8 Discover diffractometer (Bruker, Karlsruhe, Germany). Before testing, samples were placed in a vacuum oven at 200 °C for 12 h crystallization. The wavelength of X-ray was 0.154 nm. The measurements were operated at a voltage of 40 kV and current of 40 mA. Coupled $2\theta/\theta$ was selected as a scanning mode and the test was divided into three steps ($2\theta = 20, 40, 60°$). The scan time per step was 70 s and total time was 210 s. The diffraction patterns were fitted using a deconvolution procedure to fit all crystalline peaks and amorphous background. On the basis of the WAXD data, the crystallinity was estimated by comparing the diffraction area of amorphous peak at total areas of crystalline peaks, i.e., $X_c = \sum A_c/(\sum A_c + \sum A_a)$ and the parameters for each crystalline peak (width) could be also obtained.

3. Results and Discussion

3.1. Melt Post-Polycondensation

The relationship between IV of melt-modified PET and content of bisphenols is shown in Figure 1. To investigate the growth ability of PET molecular chains by melt post-polycondensation in the presence of bisphenol compounds, the bisphenols/PET complexes have been tablet pressing into fixed thickness in advance, which will cause the IV of PET to be slightly lower than the original. It can be seen that the pure PET exhibited an IV of 1.048 dL/g. By reacting different bisphenols with PET, the IV of melt-modified PET first increased to around 1.30 dL/g by addition of 1 wt% bisphenols but decreased with further increase of bisphenols. PET with 4 wt% TDP exhibited the lowest IV of 0.778 dL/g. These results indicate that in the melt post-polycondensation process, the bisphenols were successfully reacted with PET. It is supposed that bisphenols can undergo chain growth with the active end groups of PET and degradation reactions with ester groups on the molecular chain simultaneously. When bisphenols content is low, the chain growth reaction occurred predominantly, which is manifested by a substantial increasement of IV. However, when the content of bisphenols further increased, the degradation reaction held the high ground, which led to a significant decrease in IV. The reaction of bisphenols with PET still needs further confirmation by means of analytical techniques other than IV measurements.

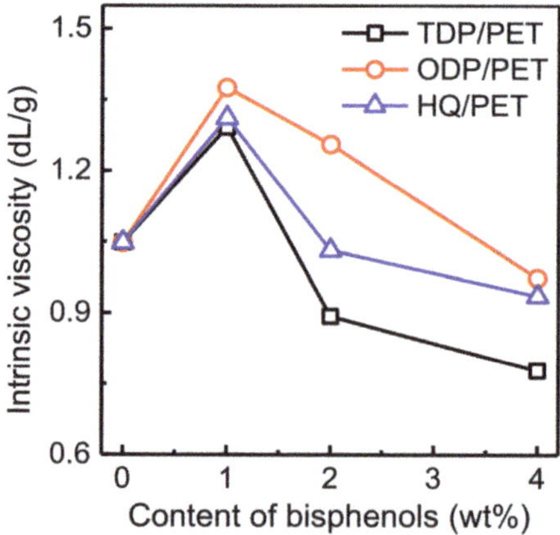

Figure 1. The intrinsic viscosity of PET modified by bisphenol compounds.

3.2. Nonisothermal Crystallization and Melting Behaviors

The processing forming of PET is usually carried out under non-isothermal conditions. Nonisothermal crystallization kinetics of all samples are investigated via DSC. Figure 2 displays DSC cooling curves of different complexes at the same cooling rate, and corresponding DSC curves of pure PET are also concluded for a better comparison. On the basis of these DSC results, the top of crystallization peaks (T_p) were evaluated and plotted as a function of the weight ratio of bisphenols (Figure 3). Crystallization kinetics of PET were significantly influenced by the presence of bisphenols. With the introduction of bisphenols, the crystallization peak became much broader and the crystallization peak shifted toward the low temperature direction compared to those of pure PET. The higher the bisphenols content is the greater the crystallization peak shifts. Besides, the value of melting onset temperature (T_{onset})$-T_p$ can characterize the overall crystallization rate [9]. After addition of bisphenols, $T_{onset}-T_p$ value increased, indicative of a lower crystallization rate. Based on T_p and $T_{onset}-T_p$ results, we could draw the conclusion that the introduction of bisphenols had a significant inhibitory effect on crystallization kinetics of PET. It is noticeable that low dosage bisphenols were used in modified PET, and considering the low reactivity between bisphenols and carboxyl end groups of PET chains, a combined effect of bisphenols referring to chain extension, transesterification and intermolecular hydrogen bonds are preferable to understand the results of a feeble inhibitory effect of PET crystallization.

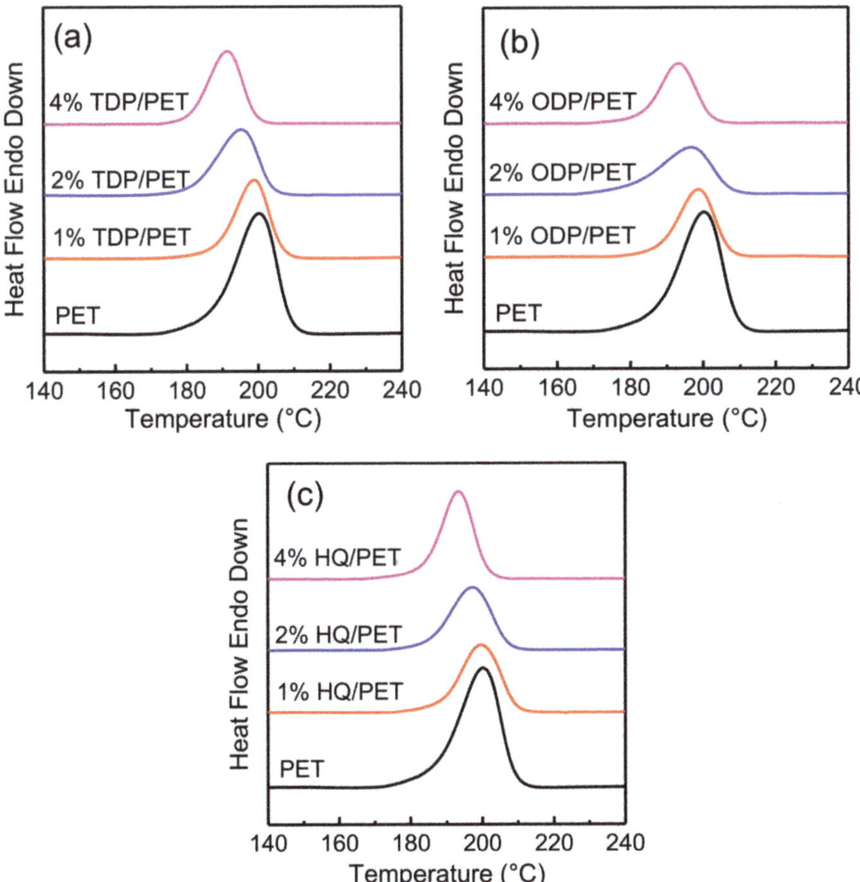

Figure 2. Differential scanning calorimeter (DSC) cooling curves obtained for PET and bisphenols/PET complexes at 10 °C/min: (**a**) TDP/PET, (**b**) ODP/PET and (**c**) HQ/PET.

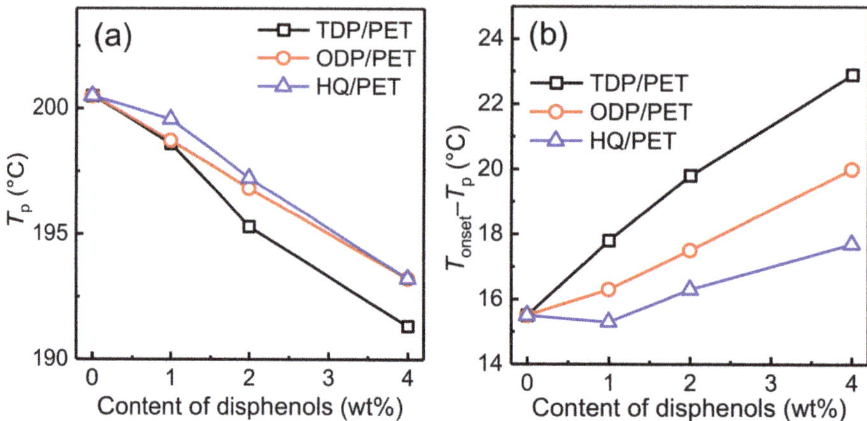

Figure 3. Plot of (**a**) T_p and (**b**) $T_{onset}-T_p$ as a function of content of bisphenols for PET and bisphenols/PET complexes.

Possible mechanisms of the depressed crystallization rate of PET by the presence of bisphenols were proposed. On one hand, as verified by FTIR measurements, bisphenol molecules are inclined to link to PET chains via hydrogen bonding and may serve as side groups or cross-linking agents. As a result, crystal growth of the polymers will be hindered due to the decreased the chain mobility by the formation of intermolecular hydrogen-bonding between the phenol group and carbonyl. Similar phenomena have been found in PBSA, PBT, PCL, PLLA, PHBV and so on [35,37–40]. For example, the T_p of PBSA was around 58.7 °C, and it shifted to 31.8 °C when incorporated 30 wt% of TDP into PBSA, demonstrating that the crystallization rate of PBSA was retarded by TDP [40]. By the presence of talc, the crystallization of PBT was inhibited because of the formation of chemical bonding interactions between talc and PBT [35]. On the other hand, another candidate reason may be related to the occurrence of transesterification between bisphenol and PET. Therefore, the density of rigid benzene ring structure on the molecular chain will increase to some degree of extent, resulting in reduced molecular chain flexibility. Therefore, PET chain is more difficult to be aligned in the crystallization process compared with pure PET.

With a careful comparison between the selected three bisphenols with different structures, it can be seen that degrees of the inhibition effect on the crystallization rate were different. The T_p of pure PET was around 200.5 °C, and it shifted to 191.3, 193.2 and 193.2 °C when incorporated 4 wt% of TDP, ODP and HQ into PEG, respectively. Besides, $T_{onset}-T_p$ for pure PET was 15.5 °C, and it changed slightly into 17.8, 16.3 and 15.8 °C with the addition of 1 wt% of TDP, ODP and HQ, which further increased to 22.9, 20.0 and 17.7 °C when the weight ratio of corresponding bisphenols reached 4 wt%. The highly coincident tendency towards T_p and $T_{onset}-T_p$ values suggest that the effect of TDP was much remarkable that that of ODP and HQ.

From the structural formula of the bisphenols compound in Scheme 1, it is noted that there was only one benzene ring in the HQ molecular structure and it was less than that of TDP and ODP. It is reasonable to expect that when the same amount of bisphenol is involved in the reaction, the increased rigidity of the PET molecular chain introduced by HQ is the least. Moreover, HQ is a non-polar compound, which exerts the least degree of inhibition on the crystallization process of PET. Although both ODP and TDP contain two benzene rings, the electron cloud distribution of the benzene ring in ODP is more uniform. In other words, the molecular chain polarity of TDP is relatively larger, resulting in the most obvious confinement effect on crystallization of PET. Additionally, the main difference between HQ and the other two compounds consists in the molecular size and shape, which affect the rotational motions along the PET-modified chains. Indeed, the two bent bisphenol units of ODP and

TDP limit the conformational degree of freedom of polyester chains much more than HQ, causing a slowdown of crystallization and higher decreases of the crystalline level with respect to the HQ units.

Figure 4 shows the effect of bisphenols on the melting behaviors of PET. Corresponding melting points are plotted as a function of content of bisphenols. The pure PET exhibited the highest top of melting peak (T_m) value among all samples. As the content of bisphenols increased, the T_m shifted to lower values. For example, T_m values for pure PET and TDP modified-PET with 1, 2 and 4 wt% are 252.7, and 251.7, 251.1 and 250.6 °C, respectively. When the amount was the same, the modified PET containing TDP exhibited the higher T_m value than those of ODP and HQ. This behavior suggests that the incorporation of bisphenols and hydrogen-bonding interaction suppressed the crystallization ability of PET and inhibited the formation of more perfect crystal structure with larger thickness.

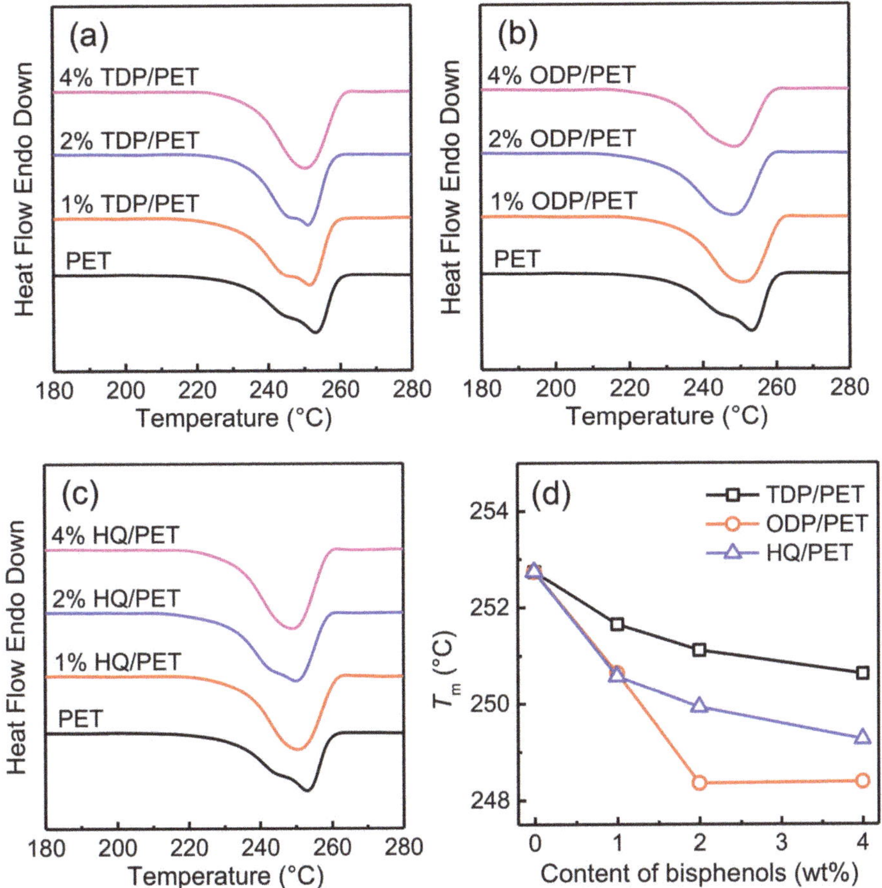

Figure 4. DSC heating curves obtained for PET and bisphenols/PET complexes at 10 °C/min: (**a**) TDP/PET, (**b**) ODP/PET and (**c**) HQ/PET. (**d**) Plots of T_m as a function of the content of bisphenols.

3.3. Effects of Cooling Rates

In general, the crystallization behaviors of semicrystalline polymers are functions of both processing and material parameters. Except for the bisphenols type and contents, cooling rate during the nonisothermal crystallization process plays an important role. Figure 5 presents the DSC cooling curve of melt-modified PET with 4 wt% TDPs at different cooling rates. When the cooling rate increased from

5 to 30 °C/min, the peak became broader and the T_p was shifted to lower values. It is clear that faster cooling rates provided less time for the molecular chain to regularly align and diffuse into the crystal lattice. Therefore, to complete the crystallization process, T_p was decreased and the crystallization temperature range was broadened. It is well-known that crystallization is an exothermal process, it is needed to remove heat from the growth front in order to allow advancement of crystallization and increases with decreasing temperature [41]. Therefore, the higher the cooling rate during crystallization, that is the lower the external temperature, the higher the crystallization rate and the crystalline level.

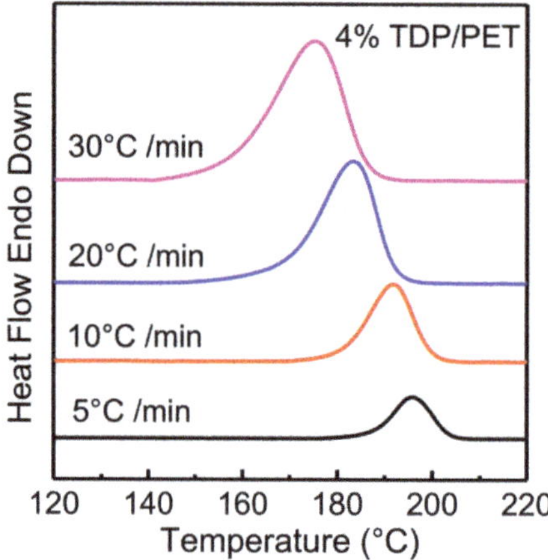

Figure 5. DSC cooling curves at various cooling rates for TDP/PET complexes.

In order to obtain important parameters extracted from DSC thermograms to further shed light on the effects of bisphenols and cooling rates on crystallization kinetics of the modified-PET systems. Non-isothermal crystallization data with various cooling rates (5–30 °C/min) of pure PET and its complexes are dealt with the Avrami model according to the following equations [42]:

$$1 - X_t = \exp(-Z_t t^n) \tag{1}$$

$$\lg[-\ln(1 - X_t)] = \lg Z_t + n\lg t \tag{2}$$

where X_t is the crystallinity at time t, Z_t is the rate parameter and n is the Avrami exponent obtained from the slope and intercept of the plot of $\lg[-\ln(1 - X_t)]$ versus $\lg t$. Z_t determines the crystallization rate and n indicates the mechanism of nucleation and crystal growth dimensions.

The modified form of the rate parameter characterizing the kinetics of the nonisothermal crystallization process by introducing a cooling rate (Φ) into the Avrami equation is given as follows [43]:

$$\lg Z_c = \frac{\lg Z_t}{\phi} \tag{3}$$

Finally, the main kinetic parameters of half crystallization time ($t_{1/2}$) can be obtained by Equation (4):

$$t_{1/2} = (\ln 2 / Z_c)^{1/n} \tag{4}$$

The plots of $\lg[-\ln(1 - X_t)]$ against $\lg t$ and corresponding fitting line for pure PET and modified-PET systems with 4 wt% bisphenols are shown in Figure 6. It is seen that a good linear relationship was obtained in the early stages of crystallization for all samples under various cooling rates, even with incorporation of 4 wt% TDP, ODP and HQ in PET matrixes. This result indicates that the presence of TDP, ODP and HQ exhibited a negligible effect on the induction periods. However, there are upward deviations in the later stages of crystallization at longer times, which is reported in previous literature [13]. The deviation is mainly attributed to the impingement and the nonlinear growth mode of the secondary crystallization in the later crystallization process. The curve shifted as the alternation of the cooling rate because the larger cooling rate would increase the degree of subcooling, thus accelerated the mobility of molecular chains.

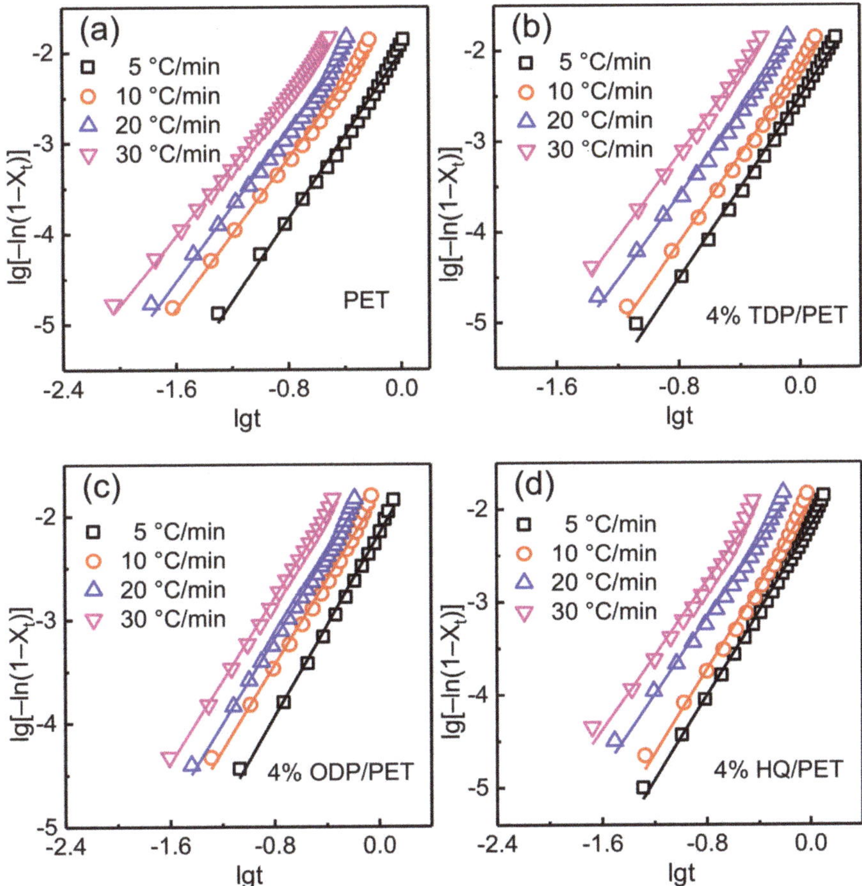

Figure 6. Plots of $\lg[-\ln(1-X_t)]$ against lgt for (**a**) PET, (**b**) TDP/PET, (**c**) ODP/PET and (**d**) HQ/PET at various cooling rates with a weight fraction of 4 wt%.

The values of $t_{1/2}$ and Z_c parameters for pure PET and modified-PET systems with 1 and 4 wt% bisphenols are plotted as a function of cooling rate (Figure 7). With the increase of the cooling rate, the values of Z_c increase and $t_{1/2}$ decrease significantly, which indicates that the crystallization rate of PET increased with the increase of the cooling rate. For pure PET, as the cooling rate rose from 5 to 30 °C/min, the values of Z_c and $t_{1/2}$ were changed from 0.40 $\min^{-n}$ and 1.27 min to 0.93 $\min^{-n}$ and 0.86 min, respectively, while those for and TDP modified-PET with a TDP content of 4 wt% were altered from 0.31 $\min^{-n}$ and 1.38 min to 0.9 $\min^{-n}$ and 0.89 min, respectively. Therefore, based on the results of Z_c and $t_{1/2}$, we could draw conclusion that the presence of bisphenols did not change dependence of crystallization kinetics on cooling rates.

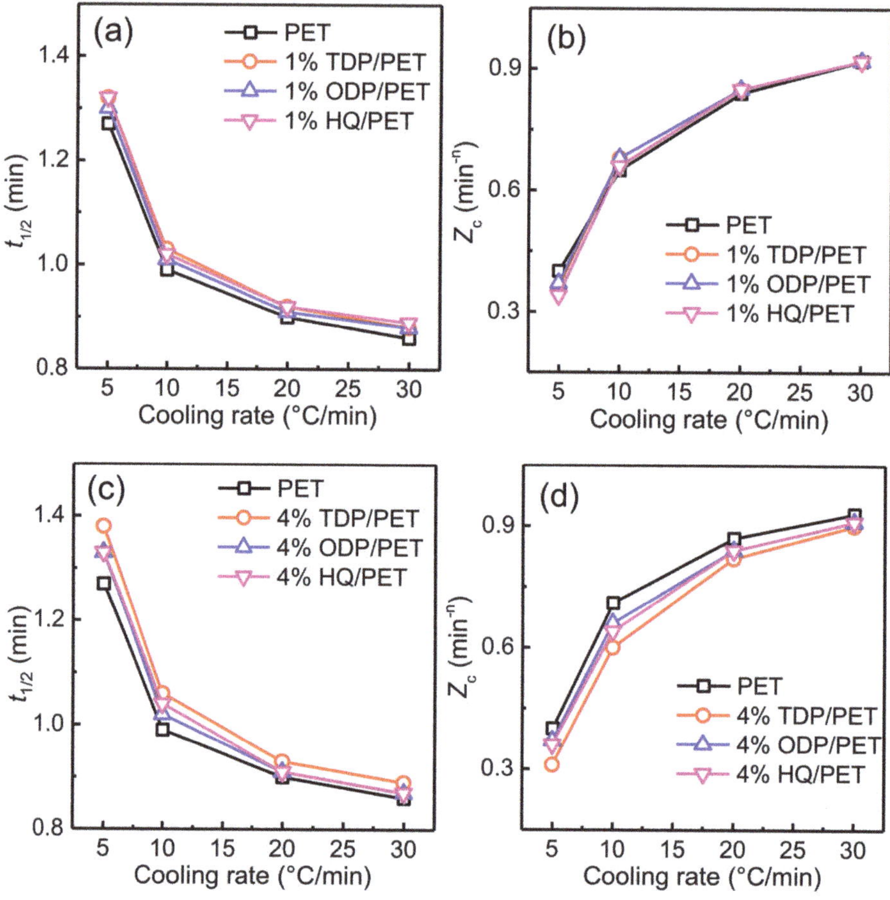

Figure 7. Plots of (**a,c**) $t_{1/2}$ and (**b,d**) Z_c as a function of the cooling rate for PET and bisphenols/PET complexes.

In comparison with pure PET under the same cooling rate, by addition of 1 wt% bisphenols, the crystallization rate did not change significantly as the Z_c and $t_{1/2}$ of PET/bisphenols complexes were comparable with those of pure PET, while lower values of Z_c and higher values of $t_{1/2}$ were obtained for all systems with 2 wt% bisphenols. By incorporation of 4 wt% of bisphenols to PET, the crystallization rate was depressed further due to decreased chain mobility by the higher interactions of hydrogen bonding and larger content of benzene rings in molecular chains. Besides, compared to ODP and HQ-modified PETs, $t_{1/2}$ and Z_c of the same amount of TDP-modified PET exhibited the largest differentials under various cooling rates due to the more significant effects on chain mobility, which was highly consistent with the data of T_p and $T_{onset}-T_p$. For instance, when the cooling rates were 5 °C/min, $t_{1/2}$ and Z_c of 4 wt% ODP and HQ-modified PETs were 8.2 and 7.9 min and 0.37 and 0.36 min, respectively, while they increased and decreased to 9.3 and 0.31 min in the case of TDP-modified PET. From the data in the Table 1, n value of pure PET ranged from 1.93 to 2.28, and that of modified PET with various contents range from 1.99 to 2.58 upon measured cooling rates. Generally, n is depending on both the nature of nucleation and growth geometry of crystals [44]. This indicates that the introduction of bisphenol had almost no significant effect on the nucleation and growth geometry of PET. Besides, it is observed that the variation of n parameter for TDP, OPD and HQ from pure PET was similar and had no distinctive dependence on the content of bisphenols.

Table 1. Values of n for PET and bisphenols/PET complexes.

Sample	Cooling Rate			
	5 °C/Min	10 °C/Min	20 °C/Min	30 °C/Min
PET	2.28	2.08	2.09	1.93
1% TDP/PET	2.35	2.20	2.26	2.21
2% TDP/PET	2.49	2.39	2.22	2.29
4% TDP/PET	2.48	2.41	2.30	2.27
1% ODP/PET	2.37	2.38	2.22	2.26
2% ODP/PET	2.58	2.52	2.48	2.31
4% ODP/PET	2.20	2.06	2.05	1.99
1% HQ/PET	2.58	2.57	2.32	2.35
2% HQ/PET	2.34	2.38	2.12	2.12
4% HQ/PET	2.27	2.26	2.06	2.00

3.4. Crystalline Structure and Crystallinity

2D WAXD patterns of pure PET and its bisphenols complexes after crystallization and the crystallinity are shown in Figure 8. The image was composed of a superposition of three circular diffraction patterns with a 2θ from 2.95 to 76°. There were three diffraction rings of PET in the pattern, where the three distinct bright rings were corresponding to (010), (−110) and (100) crystallographic plane of PET [45]. It can be seen that the brightness of diffraction ring became gradually darker as the content of bisphenols increased, suggesting a reduction of crystallinity. Besides, the variation in brightness of the diffraction ring become more distinctive for TDP than those for ODP and HQ when compared to pure PET, which may be related with their molecular structures.

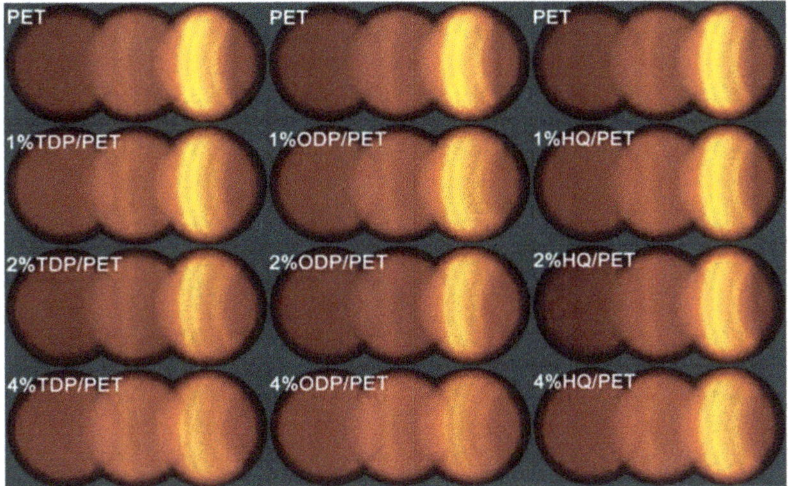

Figure 8. Selected 2D wide-angle X-ray diffraction (WAXD) patterns of PET and bisphenols/PET complexes.

Next, the 2D XRD patterns were converted into 1D profiles to obtain a peak fitting diagram. The crystallinity was calculated from the peak area and was plotted as a function of bisphenols content. After modification of PET, the main diffraction peaks were located at almost the same 2θ as pure PET, and no new diffraction peak is observed, which suggests that the inclusion of bisphenols did not alter the crystalline structure of PET. The sharpness and the intensity of diffraction peaks of (010), (−110) and (100) crystallographic planes decreased gradually with increasing bisphenol content

(Figure 9). At a content of 1 and 2 wt%, the PET shows low X_c by adding TDP, but ODP and HQ did not reduce the X_c as significant as TDP. Moreover, the X_c difference between TDP and ODP, HQ became much more significant as the contents increased. As shown in Figure 10, for pure PET, X_c was 34.8%. In contrast, the X_c decreased to 12.2%, 23.6% and 24.4% by increasing the TDP, ODP and HQ content to 4%. The proper reason was mentioned in previous nonisothermal crystallization analysis that TDP contained two benzene rings in one molecule and its polarity was relatively larger. Therefore, the inhibition of crystallization ability was more obvious due to the highest rigidity of PET molecular chain incorporated with TDP. Compared with the crystallinity measured by DSC, the decrease of crystallinity measured by 2D-WAXD was more obvious, mainly because the two methods use different principles to measure crystallinity. The crystallinity in 2D-WAXD was obtained by measuring the diffraction peak signal of the characteristic crystal surface and peak-differentiating and imitating process. The 2D WAXD patterns of PET could show the whole characteristic crystal surface if it had a relatively perfect crystallization. However, there was no in-depth report on the difference of the crystallinity measurement results of the two. It was still scientifically very appealing.

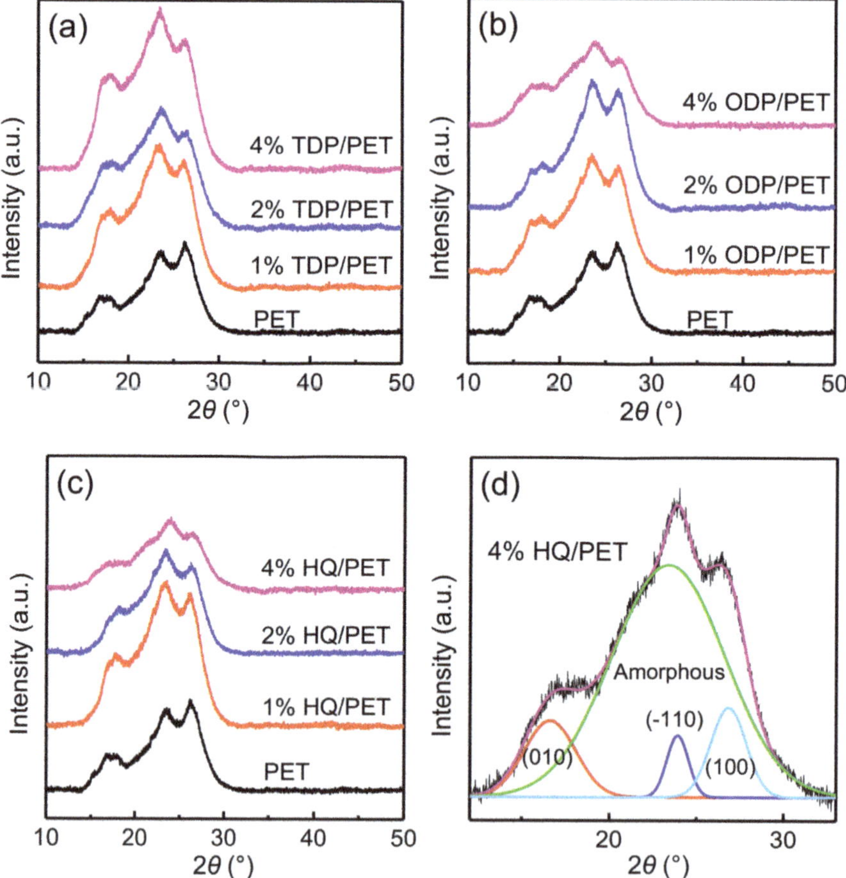

Figure 9. WAXD patterns for (**a**) TDP/PET, (**b**) ODP/PET, (**c**) HQ/PET and (**d**) typical fitting process with Pseudo-Voigt function.

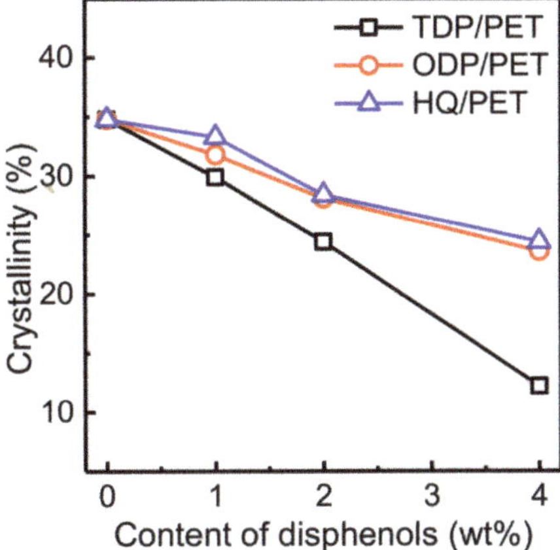

Figure 10. Composition dependence of X_c of PET and bisphenols/PET complexes.

3.5. Thermal Properties

Figure 11 presents the weight loss and derivative thermogravimetry curves of pure PET and bisphenol-modified PETs. The 5% ($T_{d5\%}$) mass loss of temperatures and the fastest decomposition rate (T_{dmax}) are listed in Table 2. It can be seen that the $T_{d5\%}$ and T_{dmax} of PET declined slightly with the addition of bisphenols. As bisphenol content increased, the thermal stability of PET demonstrated a bit decrease, which may be ascribed to the enhanced degradation reaction between the phenolic hydroxyl group and the PET chain. $T_{d5\%}$ for pure PET was 438.6 °C, and it dropped by the incorporation of TDP, ODP and QH modified-PET with a content of 4 wt%, but the reduction was within 5 °C. These results in Table 2 indicate that the addition of an appropriate amount of bisphenol would not have much effect on the thermal stability of PET.

Table 2. Thermal properties of PET and bisphenols/PET complexes.

Bisphenol Content (wt%)	/	TDP			ODP			HQ		
	0	1	2	4	1	2	4	1	2	4
T_{dmax} (°C)	438.6	437.2	435.2	434.2	436.7	436.0	434.0	438.3	438.1	436.7
$T_{d5\%}$ (°C)	402.6	399.1	397.1	394.9	399.1	398.4	396.8	401.1	398.4	397.9

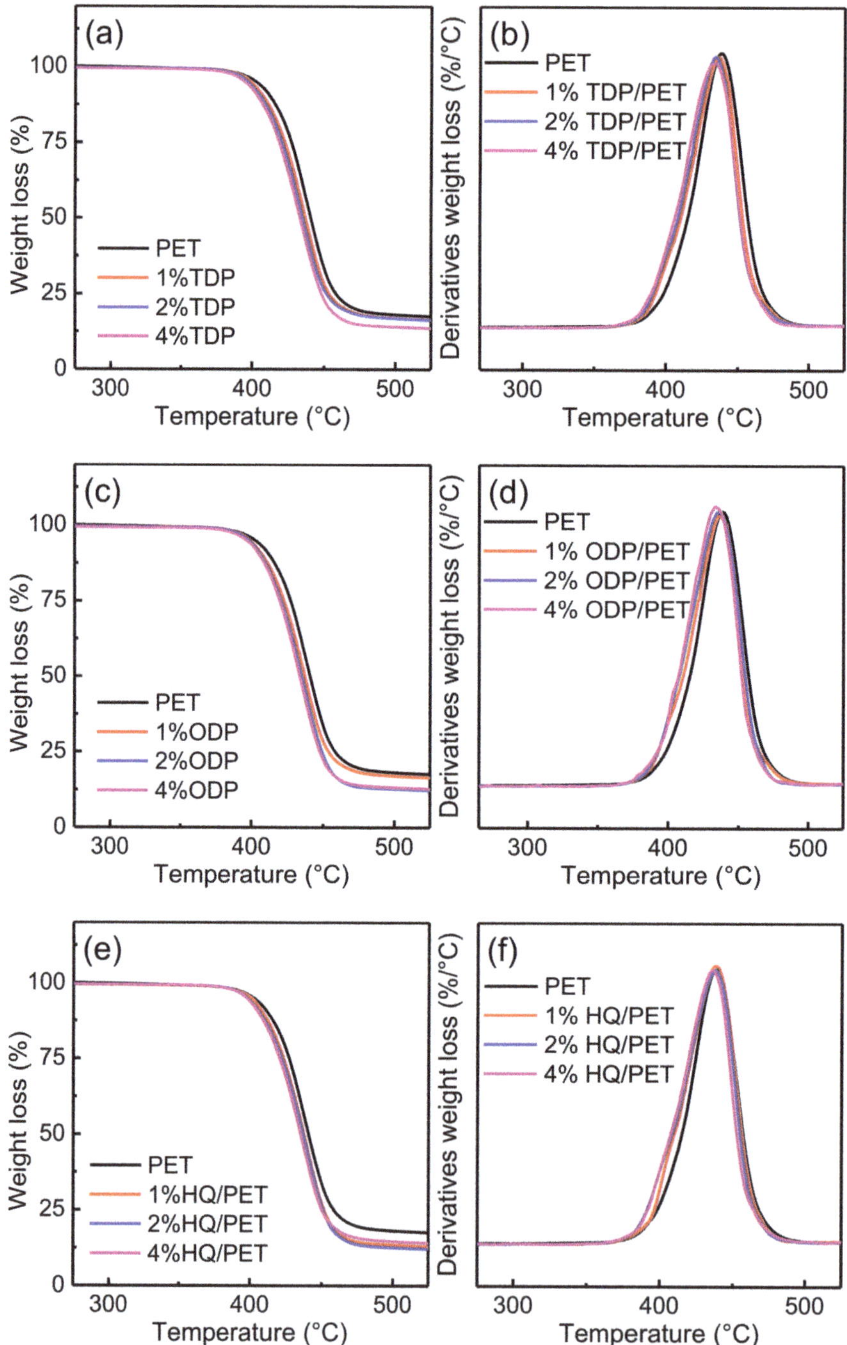

Figure 11. (**a,c,e**) Weight loss curves and (**b,d,f**) derivatives weight loss curves for PET and bisphenols/PET complexes.

4. Conclusions

In this work, three kinds of bisphenols were successfully introduced into PET by well-dissolved and blended, fixed-thickness compression and melt post-polycondensation. The effects of bisphenols on the melt post-polycondensation process, crystallization kinetic and thermal properties of PET were studied and quantitatively analyzed. With a small amount of bisphenols, the IV of PET was increased, suggesting a promotion on chain growth. However, it tended to decrease when the bisphenol content was further increased due to the dominant degradation reaction. It was found that the crystallization rate and the degree of crystallinity of PET were reduced by inclusion of foreign bisphenols due to the intermolecular hydrogen bonding and the decreased structure regularity. As the content of bisphenols increased, the crystallization rate and crystallinity of PET were further reduced. Besides, compared to ODP and HQ, the inhibition effect of TDP was more obvious, which might be ascribed to the more benzene ring in one molecule and larger polarity. As the cooling rate increased, the values of Z_c increased and $t_{1/2}$ decreased significantly, indicating that the crystallization rate of PET was increased. The petty differences of thermal decomposition temperature between PET and PET/bisphenols complexes demonstrated the introduction of bisphenols had little effect on thermal stability of PET, regardless of the kind of bisphenols.

Author Contributions: J.B. and X.Z. conceived and designed the experiments; S.C. contributed reagents/materials/analysis tools, he and S.G. performed the melt post-polycondensation experiments and thermal analysis testing; W.C. revised the experiments methods, S.X. and J.B. analyzed the data and wrote the paper. All authors have read and agreed to the published version of the manuscript.

Funding: This research was funded by National Natural Science Foundation of China (Grant No. 51803187, 51903221), Zhejiang Provincial Natural Science Foundation of China (No. LQ18E030011), Excellent Young Talent Training Fund of Key Laboratory of Advanced Textile Materials and Manufacturing Technology for Ministry of Education in Zhejiang Sci-Tech University (No. 2019QN03).

Conflicts of Interest: The authors declare no conflict of interest.

References

1. Benvenuta-Tapia, J.; Vivaldo-Lima, E.; Guerrero-Santos, R. Effect of copolymers synthesized by nitroxide-mediated polymerization as chain extenders of postconsumer poly(ethylene terephthalate) waste. *Polym. Eng. Sci.* **2019**, *59*, 2255–2264. [CrossRef]

2. Jafari, S.M.A.; Khajavi, R.; Goodarzi, V.; Kalaee, M.R.; Khonakdar, H.A. Nonisothermal crystallization kinetic studies on melt processed poly(ethylene terephthalate)/polylactic acid blends containing graphene oxide and exfoliated graphite nanoplatelets. *J. Appl. Polym. Sci.* **2019**, *136*, 47569. [CrossRef]

3. Antoniadis, G.; Paraskevopoulos, K.M.; Bikiaris, D.; Chrissafis, K. Non-isothermal crystallization kinetic of poly(ethylene terephthalate)/fumed silica (PET/SiO$_2$) prepared by in situ polymerization. *Thermochim. Acta* **2010**, *510*, 103–112. [CrossRef]

4. Zhang, Y.; Zhang, C.; Li, H.; Du, Z. Effect of cyclotrimerization of bisphenol-A dicyanate monomer on poly(ethylene terephthalate) chain extension. *Polym. Eng. Sci.* **2011**, *51*, 1791–1796. [CrossRef]

5. Bao, J.; Guo, G.; Lu, W.; Zhang, X.; Mao, H.; Dong, X.; Chen, S.; Lu, W.; Chen, W. Thermally induced physical gelation and phase transition of stereocomplexable poly(lactic acid)/poly(ethylene glycol) copolymers: Effects of hydrophilic homopolymers. *Polymer* **2020**, *208*, 122965. [CrossRef]

6. Bao, J.; Dong, X.; Chen, S.; Lu, W.; Zhang, X.; Chen, W. Fractionated crystallization and fractionated melting behaviors of poly(ethylene glycol) induced by poly(lactide) stereocomplex in their block copolymers and blends. *Polymer* **2020**, *190*, 122189. [CrossRef]

7. Tong, Z.; Zhou, J.; Wang, R.-Y.; Xu, J.-T. Interplay of microphase separation, crystallization and liquid crystalline ordering in crystalline/liquid crystalline block copolymers. *Polymer* **2017**, *130*, 1–9. [CrossRef]

8. Ma, J.; Yu, L.; Chen, S.; Chen, W.; Wang, Y.; Guang, S.; Zhang, X.; Lu, W.; Wang, Y.; Bao, J. Structure–Property Evolution of Poly(ethylene terephthalate) Fibers in Industrialized Process under Complex Coupling of Stress and Temperature Field. *Macromolecules* **2018**, *52*, 565–574. [CrossRef]

9. Gaonkar, A.A.; Murudkar, V.V.; Deshpande, V.D. Comparison of crystallization kinetics of polyethylene terephthalate (PET) and reorganized PET. *Thermochim. Acta* **2020**, *683*, 178472. [CrossRef]

10. Liu, Y.; Wirasaputra, A.; Jiang, Z.; Liu, S.; Zhao, J.; Fu, Y. Fabrication of improved overall properties of poly (ethylene terephthalate) by simultaneous chain extension and crystallization promotion. *J. Therm. Anal. Calorim.* **2018**, *133*, 1447–1454. [CrossRef]

11. Hassan, M.K.; Cakmak, M. Strain-Induced Crystallization during Relaxation Following Biaxial Stretching of PET Films: A Real-Time Mechano-Optical Study. *Macromolecules* **2015**, *48*, 4657–4668. [CrossRef]

12. Zekriardehani, S.; Jabarin, S.A.; Gidley, D.R.; Coleman, M.R. Effect of Chain Dynamics, Crystallinity, and Free Volume on the Barrier Properties of Poly(ethylene terephthalate) Biaxially Oriented Films. *Macromolecules* **2017**, *50*, 2845–2855. [CrossRef]

13. Chen, H.; Pyda, M.; Cebe, P. Non-isothermal crystallization of PET/PLA blends. *Thermochim. Acta* **2009**, *492*, 61–66. [CrossRef]

14. Topkanlo, H.A.; Ahmadi, Z.; Taromi, F.A. An in-depth study on crystallization kinetics of PET/PLA blends. *Iran. Polym. J.* **2018**, *27*, 13–22. [CrossRef]

15. Li, Z.-M.; Yang, W.; Li, L.-B.; Xie, B.-H.; Huang, R.; Yang, M.-B. Morphology and nonisothermal crystallization of in situ microfibrillar poly(ethylene terephthalate)/polypropylene blend fabricated through slit-extrusion, hot-stretch quenching. *J. Polym. Sci. Part B Polym. Phys.* **2004**, *42*, 374–385. [CrossRef]

16. Yan, Y.; Gooneie, A.; Ye, H.; Deng, L.; Qiu, Z.; Reifler, F.A.; Hufenus, R. Morphology and Crystallization of Biobased Polyamide 56 Blended with Polyethylene Terephthalate. *Macromol. Mater. Eng.* **2018**, *303*, 1800214. [CrossRef]

17. Anoop Anand, K.; Agarwal, U.S.; Joseph, R. Carbon nanotubes induced crystallization of poly(ethylene terephthalate). *Polymer* **2006**, *47*, 3976–3980. [CrossRef]

18. Yang, Y.; Gu, H. Preparation and properties of deep dye fibers from poly(ethylene terephthalate)/SiO_2 nanocomposites by in situ polymerization. *J. Appl. Polym. Sci.* **2007**, *105*, 2363–2369. [CrossRef]

19. Zhang, Y.; Wang, Y.; Li, H.; Gong, X.; Liu, J.; Huang, L.; Wang, W.; Wang, Y.; Zhao, Z.; Belfiore, L.A.; et al. Fluorescent $SiO_2@Tb^{3+}(PET-TEG)_3$Phen Hybrids as Nucleating Additive for Enhancement of Crystallinity of PET. *Polymers* **2020**, *12*, 568. [CrossRef]

20. Han, Z.; Wang, Y.; Wang, J.; Wang, S.; Zhuang, H.; Liu, J.; Huang, L.; Wang, Y.; Wang, W.; Belfiore, L.A.; et al. Preparation of Hybrid Nanoparticle Nucleating Agents and Their Effects on the Crystallization Behavior of Poly(ethylene terephthalate). *Materials* **2018**, *11*, 587. [CrossRef]

21. Tong, Z.; Zhuo, W.; Zhou, J.; Huang, R.; Jiang, G. Crystallization behavior and enhanced toughness of poly(ethylene terephthalate) composite with noncovalent modified graphene functionalized by pyrene-terminated molecules: A comparative study. *J. Mater. Sci.* **2017**, *52*, 10567–10580. [CrossRef]

22. Heeley, E.L.; Hughes, D.J.; Crabb, E.M.; Bowen, J.; Bikondoa, O.; Mayoral, B.; Leung, S.; McNally, T. The formation of a nanohybrid shish-kebab (NHSK) structure in melt-processed composites of poly(ethylene terephthalate) (PET) and multi-walled carbon nanotubes (MWCNTs). *Polymer* **2017**, *117*, 208–219. [CrossRef]

23. Pan, J.; Zhang, D.; Wu, M.; Ruan, S.; Castro, J.M.; Lee, L.J.; Chen, F. Impacts of Carbonaceous Particulates on Extrudate Semicrystalline Polyethylene Terephthalate Foams: Nonisothermal Crystallization, Rheology, and Infrared Attenuation Studies. *Ind. Eng. Chem. Res.* **2020**, *59*, 15586–15597. [CrossRef]

24. Lee, A.S.; Jeon, H.; Choi, S.-S.; Park, J.; Hwang, S.Y.; Jegal, J.; Oh, D.X.; Kim, B.C.; Hwang, S.S. Crystallization derivation of amine functionalized T12 polyhedral oligomeric silsesquioxane-conjugated poly(ethylene terephthalate). *Compos. Sci. Technol.* **2017**, *146*, 42–48. [CrossRef]

25. Sirin, H.; Turan, D.; Ozkoc, G.; Gurdag, S. POSS reinforced PET based composite fibers: "Effect of POSS type and loading level". *Compos. Part B Eng.* **2013**, *53*, 395–403. [CrossRef]

26. Lorenzo, M.L.D.; Errico, M.E.; Avella, M. Thermal and morphological characterization of poly(ethylene terephthalate)/calcium carbonate nanocomposites. *J. Mater. Sci.* **2002**, *37*, 2351–2358. [CrossRef]

27. Li, W.; Kong, X.; Zhou, E.; Ma, D. Isothermal crystallization kinetics of poly(ethylene terephthalate)–poly(ethylene oxide) segmented copolymer with two crystallizing blocks. *Polymer* **2005**, *46*, 11655–11663. [CrossRef]

28. Flores, I.; Etxeberria, A.; Irusta, L.; Calafel, I.; Vega, J.F.; Martínez-Salazar, J.; Sardon, H.; Müller, A.J. PET-ran-PLA Partially Degradable Random Copolymers Prepared by Organocatalysis: Effect of Poly(l-lactic acid) Incorporation on Crystallization and Morphology. *ACS Sustain. Chem. Eng.* **2019**, *7*, 8647–8659. [CrossRef]

29. Flores, I.; Basterretxea, A.; Etxeberria, A.; González, A.; Ocando, C.; Vega, J.F.; Martínez-Salazar, J.; Sardon, H.; Müller, A.J. Organocatalyzed Polymerization of PET-mb-poly(oxyhexane) Copolymers and Their Self-Assembly into Double Crystalline Superstructures. *Macromolecules* **2019**, *52*, 6834–6848. [CrossRef]
30. Wei, G.; Wang, L.; Chen, G.; Gu, L. Synthesis and characterization of poly(ethylene-co-trimethylene terephthalate)s. *J. Appl. Polym. Sci.* **2006**, *100*, 1511–1521. [CrossRef]
31. Lewis, C.L.; Spruiell, J.E. Crystallization of 2-methyl-1,3-propanediol substituted poly(ethylene terephthalate). I. Thermal behavior and isothermal crystallization. *J. Appl. Polym. Sci.* **2006**, *100*, 2592–2603. [CrossRef]
32. Kiyotsukuri, T.; Masuda, T.; Tsutsumi, N. Preparation and properties of poly(ethylene terephthalate) copolymers with 2,2-dialkyl-1,3-propanediols. *Polymer* **1994**, *35*, 1274–1279. [CrossRef]
33. Zhou, R.; Wang, X.; Huang, L.; Li, F.; Liu, S.; Yu, J. Poly(ethylene terephthalate) copolyesters and fibers modified with NPG and SIPE for improved hydrophilicity and dyeability. *J. Text. Inst.* **2017**, *108*, 1949–1956. [CrossRef]
34. Zhao, H.B.; Wang, Y.Z. Design and Synthesis of PET-Based Copolyesters with Flame-Retardant and Antidripping Performance. *Macromol. Rapid Commun.* **2017**, *38*, 1700451. [CrossRef] [PubMed]
35. Shen, Z.; Luo, F.; Bai, H.; Si, P.; Lei, X.; Ding, S.; Ji, L. A study on mediating the crystallization behavior of PBT through intermolecular hydrogen-bonding. *RSC Adv.* **2016**, *6*, 17510–17518. [CrossRef]
36. Luo, F.-L.; Luo, F.-H.; Xing, Q.; Zhang, X.-Q.; Jiao, H.-Q.; Yao, M.; Luo, C.-T.; Wang, D.-J. Hydrogen-bonding induced change of crystallization behavior of poly(butylene succinate) in its mixtures with bisphenol A. *Chin. J. Polym. Sci.* **2013**, *31*, 1685–1696. [CrossRef]
37. Li, J.; He, Y.; Inoue, Y. Thermal and infrared spectroscopic studies on hydrogen-bonding interactions between poly-(ε-caprolactone) and some dihydric phenols. *J. Polym. Sci. Part B Polym. Phys.* **2001**, *39*, 2108–2117. [CrossRef]
38. He, Y.; Asakawa, N.; Li, J.; Inoue, Y. Effects of low molecular weight compounds with hydroxyl groups on properties of poly(L-lactic acid). *J. Appl. Polym. Sci.* **2001**, *82*, 640–649. [CrossRef]
39. He, Y.; Asakawa, N.; Inoue, Y. Blends of poly(3-hydroxybutyrate)/4,4′-thiodiphenol and poly(3-hydroxybutyrate-co-3-hydroxyvalerate)/ 4,4′-thiodiphenol: Specific interaction and properties. *J. Polym. Sci. Part B Polym. Phys.* **2000**, *38*, 2891–2900. [CrossRef]
40. Si, P.; Luo, F. Hydrogen bonding interaction and crystallization behavior of poly (butylene succinate-co-butylene adipate)/thiodiphenol complexes. *Polym. Adv. Technol.* **2016**, *27*, 1413–1421. [CrossRef]
41. Raimo, M.; Lotti, E. Rebuilding growth mechanisms through visual observations. *ChemTexts* **2016**, *2*, 14. [CrossRef]
42. Tao, Y.; Mai, K. Non-isothermal crystallization and melting behavior of compatibilized polypropylene/recycled poly(ethylene terephthalate) blends. *Eur. Polym. J.* **2007**, *43*, 3538–3549. [CrossRef]
43. Jeziorny, A. Parameters characterizing the kinetics of the non-isothermal crystallization of poly(ethylene terephthalate) determined by d.s.c. *Polymer* **1978**, *19*, 1142–1144. [CrossRef]
44. Qiu, Z.; Fujinami, S.; Komura, M.; Nakajima, K.; Ikehara, T.; Nishi, T. Nonisothermal Crystallization Kinetics of Poly(butylene succinate) and Poly(ethylene succinate). *Polym. J.* **2004**, *36*, 642–646. [CrossRef]
45. Qiu, Z.; Yang, W. Nonisothermal crystallization kinetics of biodegradable poly(butylene succinate)/poly(vinyl phenol) blend. *J. Appl. Polym. Sci.* **2007**, *104*, 972–978. [CrossRef]

Publisher's Note: MDPI stays neutral with regard to jurisdictional claims in published maps and institutional affiliations.

Article

PVDF-BaTiO₃ Nanocomposite Inkjet Inks with Enhanced β-Phase Crystallinity for Printed Electronics

Hamed Abdolmaleki and Shweta Agarwala *

Department of Engineering, Aarhus University, 8200 Aarhus, Denmark; hamedabdolmaleki@eng.au.dk
* Correspondence: shweta@eng.au.dk

Received: 18 September 2020; Accepted: 19 October 2020; Published: 21 October 2020

Abstract: Polyvinylidene difluoride (PVDF) and its copolymers are promising electroactive polymers showing outstanding ferroelectric, piezoelectric, and pyroelectric properties in comparison with other organic materials. They have shown promise for applications in flexible sensors, energy-harvesting transducers, electronic skins, and flexible memories due to their biocompatibility, high chemical stability, bending and stretching abilities. PVDF can crystallize at five different phases of α, β, γ, δ, and ε; however, ferro-, piezo-, and pyroelectric properties of this polymer only originate from polar phases of β and γ. In this research, we reported fabrication of PVDF inkjet inks with enhanced β-phase crystallinity by incorporating barium titanate nanoparticles ($BaTiO_3$). $BaTiO_3$ not only acts as a nucleating agent to induce β-phase crystallinity, but it also improves the electric properties of PVDF through synergistic a ferroelectric polarization effect. PVDF-$BaTiO_3$ nanocomposite inkjet inks with different $BaTiO_3$ concentrations were prepared by wet ball milling coupled with bath ultrasonication. It was observed that the sample with 5 w% of $BaTiO_3$ had the highest β-phase crystallinity, while in higher ratios overall crystallinity deteriorated progressively, leading to more amorphous structures.

Keywords: piezoelectric; PVDF; barium titanate; nanocomposites; printed electronics; inkjet printing; nanomaterial ink

---

1. Introduction

Direct printing with nanomaterial inks has gained much interest lately due to its potential in fabricating flexible electronic components such as sensors [1], actuators [2], batteries [3], supercapacitors [4], transistors [5], etc. Generally, printing techniques can be classified into two categories, namely "contact" and "non-contact" printing. In contact printing, patterns are formed inside the printer (on engraved rollers or a stencil) and transferred directly onto the substrate; however, in non-contact techniques, the patterns are deposited onto the substrate through one or a series of computer-controlled nozzles. Flexography, gravure, offset, and screen printing are the most well-known contact printing techniques, while inkjet, aerosol jet, and electrohydrodynamic jet printing are the most prominent non-contact methods.

Nanomaterial inks used in inkjet printing play a crucial role in imparting functionality to the printed pattern. Hence, a lot of impetus has been put on building a library of new nanomaterial inks for flexible electronics. For satisfactory inkjet printing, inks have to meet certain fluid mechanical requirements, otherwise printing may suffer from satellite drops, splashing, the coffee-ring effect, and nozzle clogging. To quantify fluid mechanical requirements, Fromm et al. introduced a dimensionless Z value, which is defined by dividing the Reynold number (Re) number by the square root of the Weber (We) number [6].

According to Reis and Derby, the interval of $1 < Z < 10$ is the inkjet-printable window [7], where the lower limit demonstrates the minimum Z value below which the drops cannot be ejected from the nozzles and the upper limit is the starting point of satellite drop formation. In other research, Jang et al.

reported the interval of $4 < Z < 14$ as the inkjet-printable range [8], which was more consistent with the results achieved in this research. Much focus has been put on synthesizing conducting inks that can replace interconnects and electrodes in devices. However, multifunctional devices require multilayers of semiconducting, conducting, insulating, and piezoelectric materials. Thus, there is a need to develop novel functional inks for printing functional devices.

This work was a step forward in preparing piezoelectric inks for inkjet printing employed for fabricating sensing devices. Polyvinylidene difluoride (PVDF), a thermoplastic polymer, was the material of choice due to its high chemical and mechanical stability and outstanding ferroelectric, piezoelectric, and pyroelectric properties in comparison with other organic materials [9]. These interesting properties lead to its wide application in electronic devices such as sensors, actuators, and capacitors [10,11]. PVDF can crystallize at five different phases of α, β, γ, δ, and ε based on the processing method; however, ferro-, piezo-, and pyroelectric properties of this polymer only stem from polar phases of β and γ. Several methods have been reported to increase the ratio of β-phase crystallinity to other non-polar phases, such as annealing [12], mechanical stretching [13], electrical poling [14], electrospinning [15], solvent casting [16], and addition of nucleating fillers [17]. So far, several fillers have been used to enhance the performance of PVDF films. Maity et al. [18] introduced molybdenum disulfide (MoS_2) into PVDF using a polyaniline (PANI) interlinker. They reported that the incorporation of 10% filler led to 86% of β-phase crystallinity. Li et al. [19] enhanced the performance of PVDF-based nanogenerators by using ZnO nanorods as filler, coupled with electrospinning. The obtained membrane demonstrated promising open circuit voltage of ~85 V and short circuit current of ~2.2 µA. Pariy et al. [20] reported the incorporation of 0.1 w% reduced graphene oxide to enhance the piezoelectric constant (d_{33}) of PVDF to 87 pm/V.

In this paper, we reported a method that uses $BaTiO_3$ as nucleating filler to enhance β-phase crystallinity of PVDF for inkjet printing. Typical inkjet ink comprises four main components, namely (I) solvent, (II) functional particles, (III) binder, and (IV) additives. Solvent is used for dissolving binders and additives and tuning the final viscosity of ink. Furthermore, the solvent has crucial influence on the drying behavior of deposited droplets. The coffee-ring effect and slow/fast drying are typical problems associated with the improper selection of solvent. In this work, DMF (*N,N*-dimethylformamide) was used as the solvent of choice as it showed good solubility towards PVDF (with a dipole moment of 3.86 D) and promising surface wettability ($\gamma = 37.1$ mN/m) on a variety of substrates, including Kapton polyimide films. In the majority of inks, functional particles are the most crucial part of the ink, since they determine the final properties of printed patterns. In graphical inks, pigments are used as functional particles to endow different colors to printed images, while in functional inks the desired electrical properties such as conductivity, semi-conductivity, and piezoelectricity are provided by the particles (metal, carbon, ceramic, etc.). Here, $BaTiO_3$ nanoparticles were used as functional particles to endow the ink with piezoelectric properties. $BaTiO_3$ could also enhance polar-phase crystallinity of PVDF by acting as a nucleating agent, which resulted in enhancing the final piezoelectric performance of the obtained film. Polymeric binder is the other component of inkjet ink that brings about stable dispersion of functional particles and prevents their aggregation. PVDF is a polymer with good binding ability and demonstrates promising piezoelectric properties. Unlike metal inks, which require a sintering step to remove the binders, fabricated piezoelectric ink does not require any post-processing steps, since both the binder and the particles are piezoelectric, thus enabling printed devices on substrates with low thermal resistivity.

2. Experimental Section

2.1. Materials

PVDF pellets with average $M_w = 275,000$ and $M_n = 107,000$, DMF with purity of 99.8%, and $BaTiO_3$ nanoparticles with an average size of 50 nm and purity of 99.9% were purchased from Sigma–Aldrich (St. Louis, MO, USA). All chemicals were used without further purifications.

2.2. Ink Preparation

To prepare PVDF-BaTiO$_3$ nanocomposite inks, 5 g of PVDF was added to 95 g of DMF and magnetically stirred at 65 °C for 2 h until fully dissolved. Next, 1 g of BaTiO$_3$ was added to 24 g of DMF and ball-milled for 1 h in a Retch planetary ball mill PM 100 (Haan, Germany) at the speed of 400 rpm with 50 g of 0.5 mm zirconia balls. To prepare inks with different concentrations, suitable amounts of PVDF solution and BaTiO$_3$ dispersion were mixed and, if necessary, diluted with a suitable amount of DMF. Afterwards, the mixtures were placed in an ultrasonic bath for 15 min and then ball-milled at the speed of 300 rpm for 10 min.

2.3. Characterization

A Dimatix material printer DMP-2850, equipped with a drop-watcher and a fiducial camera, was used to print and study ejection behavior of the prepared inks. The fabricated inks were passed through Teflon filters with 2 μm pores and then injected into 1pL inkjet cartridges with a nozzle diameter of 9 μm. A capillary Cannon–Fenske (State College, PA, USA) viscometer tube was used to measure the static viscosity of the prepared inks at ambient temperature. A Theta Flex Optical Tensiometer (Gothenburg, Sweden) was used to measure the contact angle of the inks on microscope glass slides. Surface tension of the inks was calculated with contact angle data using Young's equation:

$$\gamma_{sv} = \gamma_{sl} + \gamma_{lv} \cdot \cos\theta \tag{1}$$

γ_{sl} is solid–liquid interfacial surface energy, γ_{lv} is liquid surface energy, and θ is the contact angle. Scanning electron microscopy (Zeiss Evo 10, Jena, Germany) was used to study the morphology and distribution of fillers in the printed films. To determine the crystalline structure of PVDF films, a powder X-ray diffraction spectroscopy technique (STOE fixed-stage powder diffractometer with a curved IP detector) was used. PerkinElmer (Waltham, MA, USA) Fourier transform infrared spectroscopy (FTIR) was used to determine the chemical structure and crystallinity of the printed films. A PerkinElmer DSC 8500 (Waltham, MA, USA) with an autosampler was used to investigate thermal behavior and the degree of crystallinity of the nanocomposite, according to the equation below:

$$x_c = \frac{1}{x} \frac{\Delta H_m}{\Delta H_{\%100}} \times 100 \tag{2}$$

where x_c is the degree of crystallinity, x is the mass fraction of PVDF in the nanocomposite, ΔH_m is the melting enthalpy, and $\Delta H_{\%100}$ is the melting enthalpy of a pure crystalline polymer, which is 104.6 J g^{-1} for PVDF [21]. Both heating and cooling measurements were carried out in the temperature range of 20–250 °C with a rate of 10 K min^{-1}.

3. Results and Discussion

3.1. Determining Inkjet Printable Region

Drop ejection in a DOD inkjet printer is the result of consecutive deformation of piezoelectric ceramics in the printhead, which generates pressure pulses to push ink droplets out of the nozzles. Fluid mechanical properties determine jetting behavior and printability of inkjet inks. With fluids that are too viscous, the piezoelectric pulses cannot overcome viscous dissipation and the energy associated with forming a new surface, which leads to no ejection [8]. At the other end, with liquids that are too dilute, the pushing forces result in formation of two or more droplets (satellite drops), which can have deleterious effects on resolution and printability. In order to define the inkjet-printable region for the PVDF-BaTiO$_3$ nanocomposite, six inks with a constant PVDF:BaTiO$_3$ ratio but different concentrations were prepared, and their jetting behavior was observed at different cartridge temperatures. Table 1 summarizes the fluid mechanical properties of the fabricated inks and their jetting behavior at a constant trigger voltage of 30 V.

Table 1. Properties of fabricated inks and their jetting behavior at a constant trigger voltage of 30 V.

Sample Name	PVDF Concentration (mg mL^{-1})	BaTiO$_3$ Concentration (mg mL^{-1})	Density (kg m^{-3})	Viscosity (cP) (at Room Temperature)	Surface Tension (mN m^{-1})	Printhead Temperature(s) (°C)	Z Value	Jetting Behavior
PB1	40	8.0	972	13.6	30.2	30, 40, 50	1.17	No ejection
PB2	32	6.4	967	9.7	31.7	30, 40, 50	1.72	No ejection
PB3	24	4.8	963	6.0	32.4	30, 40, 50	2.79	Chaotic jet
PB4	16	3.2	959	3.7	33.5	30	4.59	Chaotic jet
PB4	16	3.2	959	3.7	33.5	40, 50	4.59	Ideal jet
PB5	8	1.6	955	2.1	34.8	30, 40, 50	8.23	Ideal jet
PB6	1	0.2	959	1.3	36.0	30, 40, 50	13.56	Satellite drop

PB1 and PB2 inks with viscosities of 13.6 and 9.7 cP, surface tensions of 30.2 and 32.7 mN m^{-1}, and corresponding Z values of 1.17 and 1.72, respectively, could not be ejected from the nozzle at any cartridge temperature (Figure 1a), indicating that the pushing pulses in the printhead could not overcome their viscous dissipation and supply sufficient energy to form a new surface in the form of a droplet. In ink PB3 with Z value of 2.79, it was observed that ink droplets were ejected from some nozzles in a completely haphazard way, yet other nozzles had no ejection. This ejection mode was named chaotic ejection and can be observed in Figure 1b. PB4 ink with Z value of 4.59 showed different ejection behaviors with changing cartridge temperature. At cartridge temperature of 30 °C the ejection was chaotic; however, when the temperature increased to 40 and 50 °C, ideal inkjet ejection was observed. Ideal inkjet ejection was characterized by consecutive and simultaneous ejection of droplets from nozzles in the form of singular drops (Figure 1c). Ideal ejection was also observed in PB5 ink with Z value of 8.23 at all three cartridge temperatures. In ink PB6 with Z value of 13.56, the ejection mode was in the form of satellite drops (Figure 1d). Satellite drops, also known as secondary drops, are the extra droplets formed during the detaching of the main droplet from the nozzle. This ejection behavior is observed when the ink is too dilute and piezoelectric pulses in the printhead can produce more separate surfaces from bulk ink. In summary, the interval of 4.59 < Z < 13.56 was obtained as the inkjet-printable range for PVDF-BaTiO$_3$ nanocomposite inks. Inks with lower Z value could not be ejected from nozzles, and inks with Z value above this range formed satellite drops during ejection.

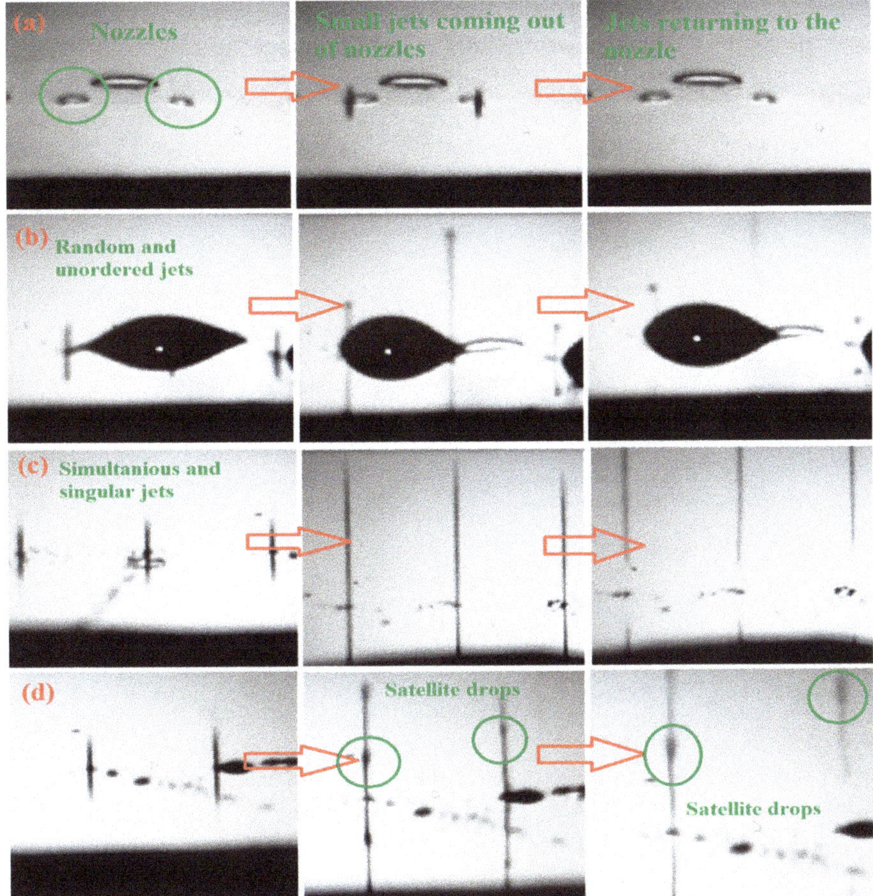

Figure 1. Jetting behavior of fabricated inks, (**a**) no ejection, (**b**) chaotic jet, (**c**) ideal jet, (**d**) satellite drops.

In order to confirm the chemical structure of the printed films, FT-IR and XRD spectroscopy were employed. Figure 2a demonstrates the FT-IR spectra of PVDF, $BaTiO_3$, and their printed nanocomposite. In PVDF, the absorption peaked at 1231, 1266, and 1401 cm^{-1}, which was attributed to the wagging vibration of CH_2; the peak at 836 cm^{-1} originated from C–C–C asymmetric stretching, and absorption bands at 874 cm^{-1} were attributed to C–F vibrations [22–24]. Different crystalline phases of PVDF could also be determined using the FT-IR technique. Peaks around 480, 532, 760, 795, 1149, 1209, and 1423 cm^{-1} could be assigned to α-phase, while peaks attributed to β-phase appeared at 473, 840, and 1266 cm^{-1} [25–28]. $BaTiO_3$ was characterized by two main absorption peaks, one at around 492 cm^{-1}, which was attributed to Ti–O vibrations, and the other at 1440 cm^{-1}, related to Ba–Ti–O stretching vibrations [29]. As seen in Figure 2a, the FT-IR spectrum of PVDF-$BaTiO_3$ nanocomposite possessed all the characteristic peaks of both materials. Although the characteristic peaks of PVDF and $BaTiO_3$ overlapped at 490 and 1440 cm^{-1}, the broad absorption bands of $BaTiO_3$ were distinguishable within the sharp peaks of PVDF.

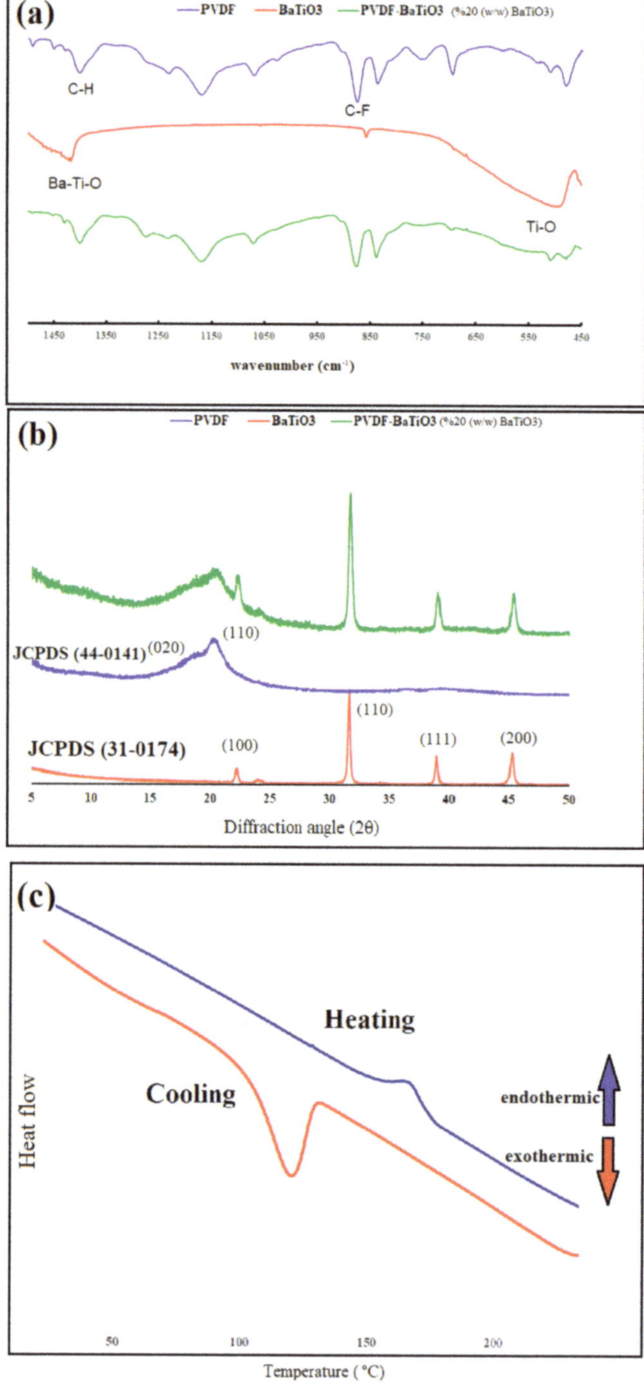

Figure 2. (**a**) FT-IR and (**b**) XRD spectra of polyvinylidene difluoride (PVDF), BaTiO3, and PVDF-BaTiO$_3$ nanocomposite, (**c**) DSC thermogram of the nanocomposite with 20% BaTiO$_3$ content.

To further confirm the chemical structure of the printed films, XRD measurements were carried out on a pure PVDF film, BaTiO$_3$, and the nanocomposite (Figure 2b). PVDF showed a broad XRD peak starting at 2θ = 15.5° and finishing at 2θ = 21.7°. This broad peak was composed of two separate peaks: one at around 18.7°, corresponding to the α crystalline phase, and the other at 20.5°, attributed to the β-phase crystalline structure of PVDF. The XRD pattern of BaTiO$_3$ showed four sharp peaks in the diffraction angle interval of 5° < 2θ < 50°. The peaks were located at diffraction angles of 22.2°, 31.5°, 38.9°, and 45.3°, which were indexed (100), (110), (111), and (200), respectively. No split in the peak of (200) at 2θ = 45.3° indicated cubic crystalline structure of nano-BaTiO$_3$ powders. The XRD spectrum of the printed PVDF-BaTiO$_3$ nanocomposite possessed all the characteristic peaks of both products, confirming the predicted structure of the obtained film. Crystallite size of PVDF could also be calculated from the XRD spectra using Scherrer´s equation [30]. β-phase crystallite size of the nanocomposite was 63.6 nm, which was smaller than that of pure PVDF (76.3 nm). Heating and cooling DSC measurements were also carried out to investigate melting temperature and degree of crystallinity of the nanocomposite with 20% of BaTiO$_3$. In the endothermic process, the melting peak appeared around 166 °C, and the degree of crystallinity was 43.7%, according to Equation (2). The glass transition temperature (T_g) of PVDF was around −35 °C, so the corresponding peak did not appear in the obtained thermogram. SEM was used to investigate the morphology of the obtained nanocomposite. Figure 3a demonstrates the surface structure of PVDF, which consisted of both amorphous regions and spherulite phases. Figure 3b shows BaTiO$_3$ before the wet milling process, which indicated highly aggregated particles with average diameter of around 5 μm. The printed PVDF-BaTiO$_3$ film (Figure 3c,d) was homogeneous and showed uniform morphology. The nanocomposite film did not show agglomerated particles, as evident before milling. Furthermore, no sign of voids and micro cracks was observed in the surface of the nanocomposite, confirming that inkjet printing could be employed for preparation of uniform PVDF-BaTiO$_3$ films. Figure 3e is the image of patterns printed on a Kapton substrate.

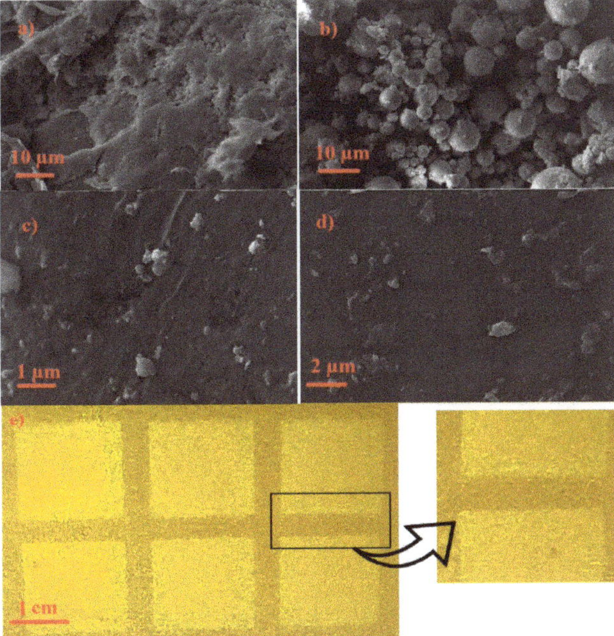

Figure 3. SEM images of (**a**) PVDF, (**b**) BaTiO$_3$, (**c**) and (**d**) PVDF-BaTiO$_3$ (%20 (*w/w*) BaTiO$_3$) nanocomposite at different magnifications. (**e**) Image of the printed PVDF-BaTiO$_3$ (%20 (*w/w*) BaTiO$_3$) inkjet ink on a Kapton substrate.

3.2. Influence of BaTiO₃ Content on β-Phase Crystallinity

The unit cell of α-phase PVDF consists of chains with trans-gauche-trans-gauche (TGTG) conformation and total lattice energy of -25.23 kJ mol^{-1}, while β-phase structure had all-trans conformation (TTTT) with total lattice energy of -23.97 kJ mol^{-1} [31]. As nature prefers lower energy states, α-phase should be the dominant polymorph of PVDF. However, there are several strategies to enhance the ratio of β- to α-phase. For instance, when transforming bulk PVDF to thin films through printing or solution casting methods, polymer chains tend to attain more parallel structures, which results in increasing the β-phase ratio.

Addition of nucleating fillers was the approach taken in this work to enhance the ratio of polar to non-polar phases. To investigate the influence of BaTiO₃ nanofillers on β-phase crystallinity of PVDF, inks with different concentrations of BaTiO₃ (0%, 5%, 10%, 15%, 20%, 25%, 30%) were prepared and inkjet-printed. Figure 4a shows the XRD spectra of the printed films in the diffraction angle interval of $15° < 2\theta < 25°$. The spectra showed two distinct peaks at 18.7°, corresponding to the (020) reflection of α-phase, and 20.6° representing the (110) plane diffraction peak of the β-phase lattice. According to Figure 4a, the film containing 5% BaTiO₃ had higher (110) and lower (020) intensity of peaks in comparison with those of pure PVDF, thus confirming β-phase enhancement through α-to-β transformation. By increasing the ratio of BaTiO₃, the intensity of both (110) and (020) peaks decreased progressively, indicating that more filler content deteriorated the overall crystallinity of PVDF and led to more amorphous structures. FT-IR spectroscopy was used to further corroborate the obtained XRD results. Characteristic absorption bands of the α-phase appeared around 410, 489, 532, 614, 763, 795, 854, 975, 1149, 1209, 1383, and 1423 cm^{-1}, whereas characteristic bands of β-phase were around 445, 473, 840, 1275, and 1431 cm^{-1} [28]. According to Figure 4b, the intensity of the peak at 763 cm^{-1}, representing the α-phase, decreased remarkably in the spectrum of the sample with 5% BaTiO₃, in comparison with that pure of PVDF, while the absorption band at 840 cm^{-1}, corresponding to the β-phase, was strengthened. Figure 5 shows how BaTiO₃ nanoparticles worked as nucleating spots for growing the β crystalline phase in PVDF. In summary, crystallization occurs in two stages: nucleation and growth. The free energy of crystallization is the sum of (I) free energy for formation of a stable nucleus embryo ($\Delta G°$) and (II) free energy of polymer chain diffusion to join the growing crystals (ΔG_η) [32]. The presence of nucleating fillers significantly decreases $\Delta G°$, leading to more overall crystallinity. The domination of the β-phase in comparison to other polymorphs can be attributed to its all-trans (TTTT) configuration, where negative dipoles (C–F bonds) are in one direction and positive dipoles (C–H bonds) on the other side. As BaTiO₃ particles possess positive surface charge, they tend to absorb negative dipoles; hence, β-phase crystals, which have the highest dipole moment among all the phases, tend to grow and dominate on the surface [33].

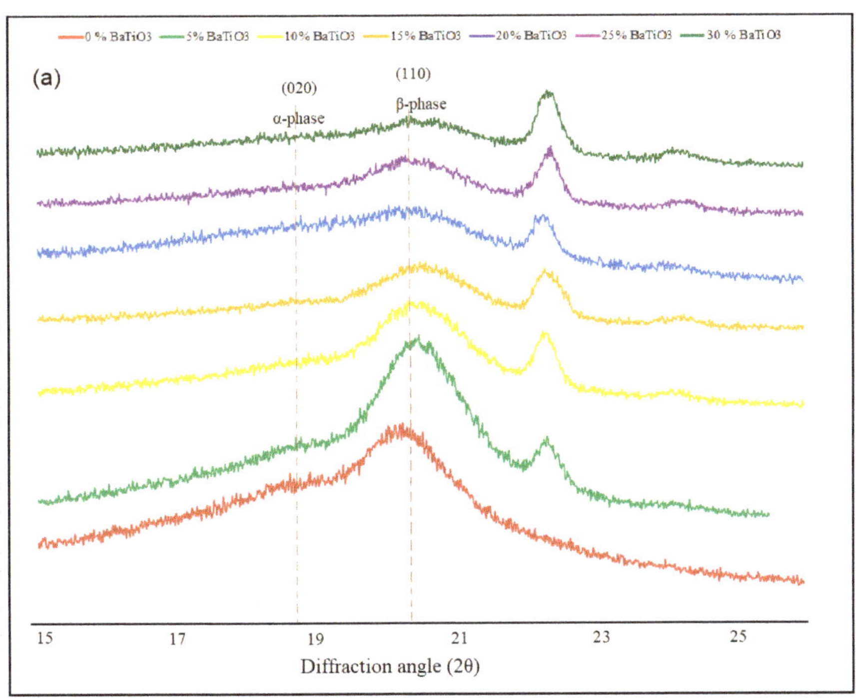

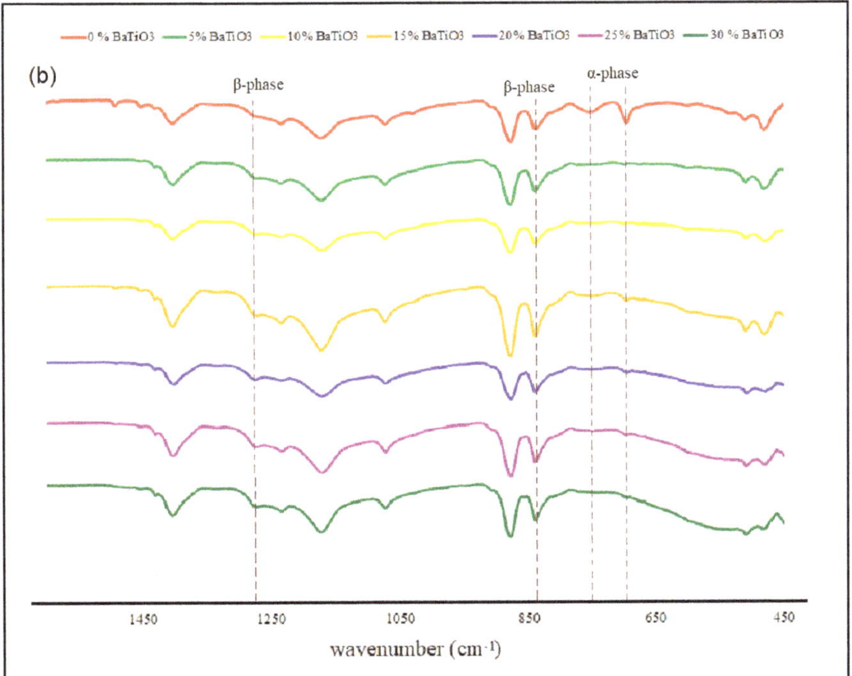

Figure 4. (**a**) XRD and (**b**) FT-IR spectra of PVDF nanocomposites with different $BaTiO_3$ content.

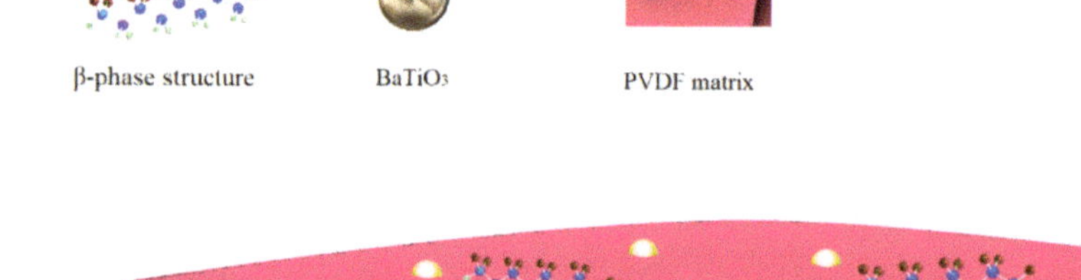

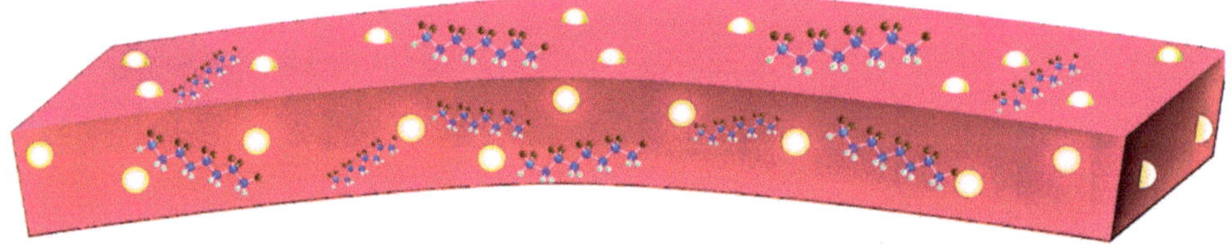

Figure 5. Schematic of the nucleating effect of BaTiO$_3$ nanoparticles in PVDF.

4. Conclusions

PVDF and its copolymers are promising electroactive organic materials, which demonstrate remarkable piezoelectric properties. PVDF can crystallize at five different phases of α, β, γ, δ, and ε, while piezoelectric properties only originate from the polar phases of β and γ. This work employed the addition of nucleating fillers to enhance the ratio of polar to non-polar phases in PVDF. We developed inkjet-printable PVDF-BaTiO$_3$ inks with enhanced β-phase crystalline structure through the incorporation of 5 w% BaTiO$_3$. The ink was printed successfully and made a homogeneous thin film without observable voids and microcracks. The showcased work is a step forward in building a library of functional materials for printed electronic applications.

Author Contributions: H.A. carried out the experimental works and drafted the manuscript. S.A. designed and supervised the research. All authors have read and agreed to the published version of the manuscript.

Funding: This research received no external funding.

Acknowledgments: This work was supported by the Dean Start-up grant, Science and Technology, Aarhus University, Denmark. The authors would also like to thank Youssif Merhi for his help in designing the schematic of Figure 5.

Conflicts of Interest: The authors declare no conflict of interest.

References

1. Xu, Y.; Wu, X.; Guo, X.; Kong, B.; Zhang, M.; Qian, X.; Mi, S.; Sun, W. The boom in 3D-printed sensor technology. *Sensors* **2017**, *17*, 1166. [CrossRef] [PubMed]
2. Zolfagharian, A.; Kouzani, A.Z.; Khoo, S.Y.; Moghadam, A.A.A.; Gibson, I.; Kaynak, A. Evolution of 3D printed soft actuators. *Sens. Actuators A Phys.* **2016**, *250*, 258–272. [CrossRef]
3. Pang, Y.; Cao, Y.; Chu, Y.; Liu, M.; Snyder, K.; MacKenzie, D.; Cao, C. Additive manufacturing of batteries. *Adv. Funct. Mater.* **2020**, *30*, 1906244. [CrossRef]
4. Li, H.; Liang, J. Recent development of printed micro-supercapacitors: Printable materials, printing technologies and perspectives. *Adv. Mater.* **2020**, *32*, 1805864. [CrossRef]
5. Conti, S.; Pimpolari, L.; Calabrese, G.; Worsley, R.; Majee, S.; Polyushkin, D.K.; Paur, M.; Pace, S.; Keum, D.H.; Fabbri, F.; et al. Low-voltage 2D materials-based printed field-effect transistors for integrated digital and analog electronics on paper. *Nat. Commun.* **2020**, *11*, 1–9. [CrossRef]
6. Fromm, J.E. Numerical calculation of the fluid dynamics of drop-on-demand jets. *IBM J. Res. Dev.* **1984**, *28*, 322–333. [CrossRef]
7. Reis, N.; Derby, B. Ink jet deposition of ceramic suspensions: Modeling and experiments of droplet formation. *Mater. Res. Soc.* **2000**, *625*, 117–122. [CrossRef]
8. Jang, D.; Kim, D.; Moon, J. Influence of fluid physical properties on ink-jet printability. *Langmuir* **2009**, *25*, 2629–2635. [CrossRef]

9.   Ruan, L.; Yao, X.; Chang, Y.; Zhou, L.; Qin, G.; Zhang, X. Properties and applications of the β Phase Poly (vinylidene fluoride). *Polymers* **2018**, *10*, 228. [CrossRef]

10.   Luo, H.; Zhou, X.; Ellingford, C.; Zhang, Y.; Chen, S.; Zhou, K.; Zhang, D.; Bowen, C.R.; Wan, C. Interface design for high energy density polymer nanocomposites. *Chem. Soc. Rev.* **2019**, *48*, 4424–4465. [CrossRef]

11.   Chen, S.; Yan, X.; Liu, W.; Qiao, R.; Chen, S.; Luo, H.; Zhang, D. Polymer-based dielectric nanocomposites with high energy density via using natural sepiolite nanofibers. *Chem. Eng. J.* **2020**, *401*, 126095. [CrossRef]

12.   Shaik, H.; Rachith, S.N.; Rudresh, K.J.; Sheik, A.S.; Thulasi Raman, K.H.; Kondaiah, P.; Mohan Rao, G. Towards β-phase formation probability in spin coated PVDF thin films. *J. Polym. Res.* **2017**, *24*, 35. [CrossRef]

13.   Li, L.; Zhang, M.; Rong, M.; Ruan, W. Studies on the transformation process of PVDF from α to β phase by stretching. *RSC Adv.* **2014**, *4*, 3938–3943. [CrossRef]

14.   Senthil Kumar, R.; Sarathi, T.; Venkataraman, K.K.; Bhattacharyya, A. Enhanced piezoelectric properties of polyvinylidene fluoride nanofibers using carbon nanofiber and electrical poling. *Mater. Lett.* **2019**, *255*, 126515. [CrossRef]

15.   Lei, T.; Zhu, P.; Cai, X.; Yang, L.; Yang, F. Electrospinning of PVDF nanofibrous membranes with controllable crystalline phases. *Appl. Phys. A* **2015**, *120*, 5–10. [CrossRef]

16.   Mahato, P.; Seal, A.; Garain, S.; Sen, S. Effect of fabrication technique on the crystalline phase and electrical properties of PVDF films. *Mater. Sci. Pol.* **2015**, *33*, 157–162. [CrossRef]

17.   Tienne, L.G.P.; de Abreu, T.B.; Gondim, F.F.; da Cruz, B.D.S.M.; Martins, G.R.; Simão, R.A.; Marques, M.D.F.V. Low contents of graphite improving general properties of poly(vinylidene fluoride). *Polym. Test.* **2020**, *91*, 106790. [CrossRef]

18.   Maity, N.; Mandal, A.; Nandi, A.K. High dielectric poly(vinylidene fluoride) nanocomposite films with MoS2 using polyaniline interlinker via interfacial interaction. *J. Mater. Chem. C* **2017**, *5*, 12121–12133. [CrossRef]

19.   Li, J.; Chen, S.; Liu, W.; Fu, R.; Tu, S.; Zhao, Y.; Dong, L.; Yan, B.; Gu, Y. High performance piezoelectric nanogenerators based on electrospun ZnO Nanorods/Poly(vinylidene fluoride) composite membranes. *J. Phys. Chem. C* **2019**, *123*, 11378–11387. [CrossRef]

20.   Pariy, I.O.; Ivanova, A.A.; Shvartsman, V.V.; Lupascu, D.C.; Sukhorukov, G.B.; Ludwig, T.; Bartasyte, A.; Mathur, S.; Surmeneva, M.A.; Surmenev, R.A. Piezoelectric response in hybrid micropillar arrays of poly (vinylidene fluoride) and reduced graphene oxide. *Polymers* **2019**, *11*, 1065. [CrossRef]

21.   Gong, X.; Chen, Y.; Tang, C.Y.; Law, W.C.; Chen, L.; Wu, C.; Hu, T.; Tsui, G.C.P. Crystallinity and morphology of barium titanate filled poly (vinylidene fluoride) nanocomposites. *J. Appl. Polym. Sci.* **2018**, *135*, 46877. [CrossRef]

22.   Bai, H.; Wang, X.; Zhou, Y.; Zhang, L. Preparation and characterization of poly(vinylidene fluoride) composite membranes blended with nano-crystalline cellulose. *Prog. Nat. Sci. Mater. Int.* **2012**, *22*, 250–257. [CrossRef]

23.   Rahimpour, A.; Madaeni, S.S.; Zereshki, S.; Mansourpanah, Y. Preparation and characterization of modified nano-porous PVDF membrane with high antifouling property using UV photo-grafting. *Appl. Surf. Sci.* **2009**, *255*, 7455–7461. [CrossRef]

24.   Gu, S.; He, G.; Wu, X.; Hu, Z.; Wang, L.; Xiao, G.; Peng, L. Preparation and characterization of poly(vinylidene fluoride)/sulfonated poly(phthalazinone ether sulfone ketone) blends for proton exchange membrane. *J. Appl. Polym. Sci.* **2010**, *116*, 852–860. [CrossRef]

25.   Gomes, J.; Nunes, J.S.; Sencadas, V.; Lanceros-Méndez, S. Influence of the β-phase content and degree of crystallinity on the piezo-and ferroelectric properties of poly (vinylidene fluoride). *Smart Mater. Struct.* **2010**, *19*, 065010. [CrossRef]

26.   Salimi, A.; Yousefi, A.A. FTIR studies of b-phase crystal formation in stretched PVDF films. *Polym. Test.* **2003**, *22*, 699–704. [CrossRef]

27.   Ren, Y.; Wang, Y.; Zhang, W.; Yan, X.; Huang, B. Improved battery performance contributed by the optimized phase ratio of β and α of PVDF. *RSC Adv.* **2019**, *9*, 29760–29764. [CrossRef]

28.   Cai, X.; Lei, T.; Sun, D.; Lin, L. A critical analysis of the α, β and γ phases in poly(vinylidene fluoride) using FTIR. *RSC Adv.* **2017**, *7*, 15382–15389. [CrossRef]

29.   Kappadan, S.; Gebreab, T.W.; Thomas, S.; Kalarikkal, N. Tetragonal BaTiO$_3$ nanoparticles: An efficient photocatalyst for the degradation of organic pollutants. *Mater. Sci. Semicond. Process.* **2016**, *51*, 42–47. [CrossRef]

30.   Muniz, F.T.L.; Miranda, M.A.R.; Morilla dos Santos, C.; Sasaki, J.M. The Scherrer equation and the dynamical theory of X-ray diffraction. *Acta Crystallogr. Sect. A Found. Adv.* **2016**, *72*, 385–390. [CrossRef]

31. Chen, C.; Cai, F.; Zhu, Y.; Liao, L.; Qian, J.; Yuan, F.-G.; Zhang, N. 3D printing of electroactive PVDF thin films with high β-phase content. *Smart Mater. Struct.* **2019**, *28*, 065017. [CrossRef]

32. Shanks, R.A.; Tiganis, B.E. Nucleating agents for thermoplastics. In *Plastics Additives: An A–Z Reference*; Pritchard, G., Ed.; Springer: Dordrecht, The Netherlands, 1998; pp. 464–471.

33. Sebastian, M.; Larrea, A.; Gonçalves, R.; Alejo, T.; Vilas, J.; Sebastian, V.; Martins, P.; Lanceros-Mendez, S. Understanding nucleation of the electroactive β-phase of poly (vinylidene fluoride) by nanostructures. *RSC Adv.* **2016**, *6*, 113007–113015. [CrossRef]

Publisher's Note: MDPI stays neutral with regard to jurisdictional claims in published maps and institutional affiliations.

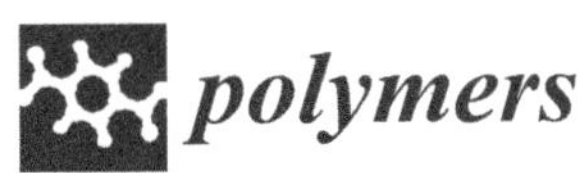 

Article

Rheology-Assisted Microstructure Control for Printing Magnetic Composites—Material and Process Development

Balakrishnan Nagarajan [1], Martin A.W. Schoen [2], Simon Trudel [2], Ahmed Jawad Qureshi [1] and Pierre Mertiny [1,*]

[1] Department of Mechanical Engineering, University of Alberta, 9211-116 St., NW Edmonton, AB T6G 1H9, Canada; bnagaraj@ualberta.ca (B.N.); ajquresh@ualberta.ca (A.J.Q.)
[2] Department of Chemistry, University of Calgary, 2500 University Dr. NW, Calgary, AB T2N 1N4, Canada; martin.schon@ucalgary.ca (M.A.W.S.); trudels@ucalgary.ca (S.T.)
* Correspondence: pmertiny@ualberta.ca

Received: 11 August 2020; Accepted: 18 September 2020; Published: 20 September 2020

Abstract: Magnetic composites play a significant role in various electrical and electronic devices. Properties of such magnetic composites depend on the particle microstructural distribution within the polymer matrix. In this study, a methodology to manufacture magnetic composites with isotropic and anisotropic particle distribution was introduced using engineered material formulations and manufacturing methods. An in-house developed material jetting 3D printer with particle alignment capability was utilized to dispense a UV curable resin formulation to the desired computer aided design (CAD) geometry. Formulations engineered using additives enabled controlling the rheological properties and the microstructure at different manufacturing process stages. Incorporating rheological additives rendered the formulation with thixotropic properties suitable for material jetting processes. Particle alignment was accomplished using a magnetic field generated using a pair of permanent magnets. Microstructure control in printed composites was observed to depend on both the developed material formulations and the manufacturing process. The rheological behavior of filler-modified polymers was characterized using rheometry, and the formulation properties were derived using mathematical models. Experimental observations were correlated with the observed mechanical behavior changes in the polymers. It was additionally observed that higher additive content controlled particle aggregation but reduced the degree of particle alignment in polymers. Directionality analysis of optical micrographs was utilized as a tool to quantify the degree of filler orientation in printed composites. Characterization of in-plane and out-of-plane magnetic properties using a superconducting quantum interference device (SQUID) magnetometer exhibited enhanced magnetic characteristics along the direction of field structuring. Results expressed in this fundamental research serve as building blocks to construct magnetic composites through material jetting-based additive manufacturing processes.

Keywords: additive manufacturing; magnetic composites; ferrite composites; field structuring; microstructure control; rheological modifications

1. Introduction

Material jetting is an additive manufacturing process where a three-dimensional solid part is manufactured by dispensing polymeric material from a print head and, subsequently, solidifying it utilizing an ultraviolet light or thermal curing methodology. Material jetting employing photopolymerization is similar to stereolithography, where acrylate-type photopolymers are deposited and exposed to ultraviolet light [1]. Promising material systems for future applications involve resins modified with carbon-based fillers, ceramics, and metals [2–6]. Factors that mutually influence the material jetting process are machine and material parameters, including liquid material viscosity, shear-thinning effects and surface tension, print head nozzle design, print speed, and droplet velocity and frequency [7]. The development of fabrication processes includes fine-tuning of material and machine parameters in order to achieve a robust and effective material jetting-based additive manufacturing process [8,9].

Field-structured magnetic composites are manufactured by aligning magnetic particles within a polymer matrix by applying an external magnetic field. The concept of magnetic field-induced particle structuring is illustrated in Figure 1. Within the polymer matrix, the magnetic particles are randomly distributed after the dispersion process, i.e., random distribution of the crystallographic easy axis of magnetization. When this suspension is subjected to an external magnetic field, magnetic moments are induced along the easy axis, producing particle chaining that enhances magnetic properties like remanence and susceptibility along the direction of field structuring [10–13]. Such field-structured magnetic materials are of significant interest in applications like magnetic sensors and data storage devices [14,15].

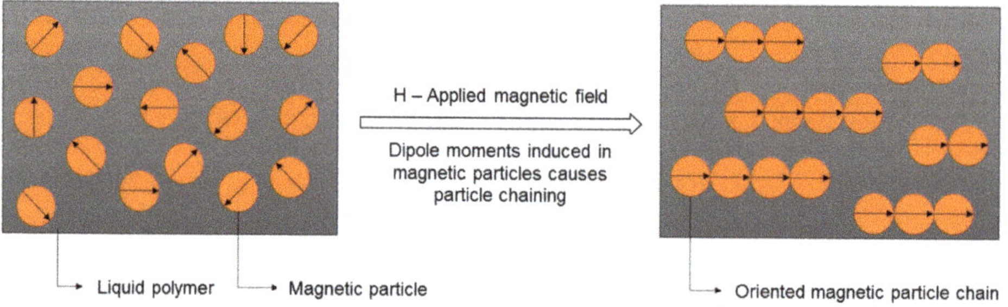

Figure 1. Schematic of magnetic field-induced particle structuring in magnetic polymer composite.

Anisotropy in thermal conductivity has been achieved by orienting ferromagnetic particles in epoxy resin, applying an external magnetic field where the chain-like microstructures serve as enhanced heat flow paths [16]. Inkjet-printed one-dimensional arrays of monodisperse Fe_3O_4 nanoparticles with high anisotropic magnetization with possible applications for magnetic field sensing has been demonstrated in the technical literature [17]. Composites 3D printed and poled, applying an external magnetic field, utilizing strontium ferrite particles dispersed in SU8 photoresist, have been observed to be suitable for onboard integration of magnetic components in millimeter-wave circuits [18]. Among the available magnetic materials, ferrite-based magnetic materials have become particularly important for a multitude of applications. Strontium ferrite ($SrFe_{12}O_{19}$) is one such material, having a hexagonal structure like magnetoplumbite [19]. Under the influence of an external magnetic field, dipole moments induced in the particles along the crystallographic *c*-axis orient the particles along the direction of the applied magnetic field. Additive manufacturing of magnetorheological fluid-dispersed photopolymers using magnetic field-assisted stereolithography has been studied and reported in the technical literature [20]. The influence of applied magnetic flux density on the degree of particle alignment and orientation behavior at multiple angles has already been reported in previous work by the present authors [21]. The finite element method magnetics (FEMM) simulation for a two-cube magnet particle alignment system has indicated an exponential reduction in magnetic flux density with the increasing separation distance between the magnet faces. Simulations additionally have shown that the magnetic flux density is enhanced by 0.21 Tesla in a magnetic array type particle alignment system compared to a two-cube permanent magnet system [22]. The developed particle alignment systems have been integrated with an in-house developed material jetting 3D printer, which, in addition to material deposition, enables particle structuring during the manufacturing process [23]. Moreover, the effectiveness of utilizing additives to mitigate particle settling in polymers has been reported in the technical literature [24].

The present study investigated the capabilities of polymer formulations engineered with magnetic fillers and additives for the material jetting additive manufacturing process. First, the developed formulations were characterized for their rheological behavior, and mathematical models were employed to derive suspension properties. Derived properties were utilized to correlate different aspects of material and process behavior observed at various manufacturing process stages. The role of additives toward controlling particle aggregation and enabling particle alignment was evaluated using optical microscopy. Optical microscopy, coupled with directionality analysis using image processing, enabled quantifying particle alignment within the dispensed photopolymers. The fundamental understanding thus obtained for developed materials and process scenarios permitted the fabrication of field-structured composites using a suspension engineered with 10 wt% magnetic particle loading. Magnetic characterization of field-structured composites was conducted using a SQUID (superconducting quantum interference device) magnetometer. The goal of this work was to scientifically rationalize material behavior and utilize this knowledge to develop magnetic composite structures with controllable microstructures. Ultimately, this research sought to provide an in-depth understanding of the role of material formulations, magnetic alignment setup, and manufacturing process methods in order to further evolve processes to produce 3D magnetic solids with in-situ microstructure control using the material jetting-based additive manufacturing process.

2. Experimental Procedures

2.1. Materials

For this study, strontium ferrite ($SrFe_{12}O_{19}$, abbreviated as SrFeO) powder with an average particle size of 1.41 μm, density 3.41 g/cm^3 was purchased from DOWA Electronics Materials Co. Ltd. (Tokyo, Japan). Photosensitive polymer resin PR-48 (UV curable acrylate) was purchased from Colorado Photopolymer Solutions (Boulder, CO, USA), and rheological additive BYK 7410ET was obtained from BYK Additives & Instruments (Wesel, Germany). All chemicals were used as received without any further purification.

2.2. Magnetic Filler Dispersion Methodology

The dispersion of magnetic fillers in the UV curable prepolymer was accomplished using a combination of mechanical mixing and sonication. Strontium ferrite powder was added in the desired quantity to the PR-48 resin, and the resultant combination was agitated using an impellor agitator from Calframo Ltd. (Georgian Bluffs, ON, Canada). After mechanical agitation, ultrasonic mixing was initiated using a Branson model S-75 sonicator (Branson Ultrasonics Corporation, Danbury, CT, USA) adopting 15 s p7ulsed on/off mode for 15 min. For experiments involving magnetic particle settling mitigation, the rheological additive was added after the sonication step using mechanical mixing with an impeller. Suspensions were allowed to rest for one day to allow the additive to form a network structure [25]. The suspensions were again mechanically agitated to enhance the efficiency of the rheological additive. Any air that was entrapped during the mixing processes was degassed in a vacuum.

3. Characterization Methods

3.1. Rheological Behavior—Viscosity and Flow Curve Analysis

Rheological analysis of magnetically loaded prepolymer suspensions containing ferromagnetic particles and rheological additive materials was performed using a rotational rheometer (Rheolab QC, Anton Parr GmbH, Graz, Austria) equipped with double gap type measuring system. Table 1 lists the material formulations that were characterized by their rheological properties.

Table 1. Materials characterized by rheological properties.

Material Code	Magnetic Filler Loading (wt%)	Additive Type	Additive Loading (wt%)
Base resin	-	-	-
10SF	10	-	-
10SF-0.5BYK	10	BYK-7410ET	0.5
10SF-2.0BYK	10	BYK-7410ET	2

Flow curves derived from rheological characterization experiments for the magnetic suspensions were used to interpret the suspension behavior. In material jetting processes, extrusion of the developed formulation is driven by applied pressure. Shear forces break the network structure developed in the resin material by the rheological additive. The flow of the liquid formulation that is ejected from the nozzle is primarily governed by equations for incompressible, laminar flow through a circular tube of constant cross-section [26]. The Hagen–Poiseuille equation defines the pressure drop Δp, as indicated in Equation (1) [27].

$$\Delta p = \left(\frac{8\eta QL}{\pi r^4}\right) \tag{1}$$

where Q is the volumetric flow rate, η is the formulation viscosity, r and L are the radius and length of a circular tube, respectively. The shear stress τ at any point inside the circular tube is given by Equation (2) [27].

$$\tau = -\frac{\Delta p}{2L}r \tag{2}$$

The wall shear rate, $\dot{\gamma}_w$, in terms of pressure drop, is given by Equation (3).

$$\dot{\gamma}_w = -\frac{\Delta p}{\eta L}\frac{r}{2} \tag{3}$$

The volumetric flow rate Q is expressed as [26]:

$$Q = \pi r^2 V = \pi\left(\frac{n}{3n+1}\right)\left(-\frac{\Delta p}{2mL}\right)^{\frac{1}{n}} r^{\frac{3n+1}{n}} \tag{4}$$

In Equation (4), n is the power-law index, and m is the consistency index, or viscosity, obtained from a mathematical analysis of rheological data through a power-law model. From the theoretical equations, we interpret that $-\Delta p \propto Q^n$, i.e., the pressure gradient is less sensitive for a shear-thinning fluid than for a Newtonian fluid [27]. The dependence of fluid viscosity on the shear rate $\dot{\gamma}$ is expressed utilizing a power law, as indicated in Equation (5) [28].

$$\eta = m\dot{\gamma}^{n-1} \tag{5}$$

The magnitude of the power-law index n indicates the degree of pseudoplasticity in the characterized material. The power-law model is a two-parameter model and is used extensively to enable a fundamental understanding of fluid behavior. An expression for the actual shear rate experienced by the fluid inside the cylinder, as expressed in the technical literature, is indicated by Equation (6) [26].

$$\dot{\gamma} = \frac{Vr^{(2+n)}}{\left(\frac{n}{3n+1}\right)r^{\frac{3n+1}{n}}} \tag{6}$$

It has been additionally understood that a formulation with low viscosity experiences higher shear rates during the dispensing process [26]. Overall, from the above theoretical equations, the importance of rheological modification of formulations is well established. The yield strength of the magnetic suspension that characterizes the behavior of the material at rest was determined using the Herschel–Bulkley equation that is well suited for Non-Newtonian fluids. The Hershel–Bulkley model equation is expressed as follows:

$$\tau = \tau_0 + C\dot{\gamma}^k \tag{7}$$

where τ_0 is the yield point or yield strength, C is the consistency index, and k is the Herschel–Bulkley index. The Herschel–Bulkley index primarily determines the behavior of the suspension as follows: $k < 1$ for shear thinning behavior, $k > 1$ for shear thickening behavior, and $k = 1$ for Bingham behavior [29,30].

3.2. Thixotropic Flow Behavior Analysis

Thixotropy refers to reversible changes in fluid behavior from a flowable liquid to a solid elastic gel. Liquids with a microstructure exhibit thixotropy, given the time needed to reversibly go from a given microstructural state to another. These materials exhibit structural decomposition at high shear rates and structural regeneration at low shear rates. Stresses experienced by the fluid play a dominant role in the breakdown of thixotropic network structures [31]. The influence of rheological additives on the thixotropic behavior of magnetic suspensions was characterized using a step test consisting of three intervals. Low shear rate conditions simulate the sample behavior under stationary conditions, and high shear rate conditions simulate sample behavior under the influence of external forces. Experiments were conducted using controlled shear rate conditions. In the first interval, viscosity was measured at $\dot{\gamma} = 1 \cdot s^{-1}$ for 20 s, followed by viscosity measurement at $\dot{\gamma} = 300 \cdot s^{-1}$ for 50 s, and finally, viscosity was measured again at $\dot{\gamma} = 1 \cdot s^{-1}$ for 40 s. This test was used to characterize the structural decomposition and regeneration behavior of the magnetic suspensions incorporated with rheological additives. The thixotropy index that characterizes time-dependent viscosity recovery was calculated using Equation (8).

$$\textit{Thixotropy index} = \frac{\eta_2 - \eta_1}{t_2 - t_1} = \frac{\Delta\eta}{\Delta t} \tag{8}$$

where η_2, η_1 are viscosities in the recovery phase at two different times t_1 and t_2.

3.3. Magnetic Particle-Reinforced Resin Behavior in Magnetic Field

To investigate the stability of magnetic particle loaded polymer resin droplets in a magnetic field, an in-house developed experimental system was utilized [21]. A computer control system programmed with a graphical user interface was used to control the mechanical and electrical components of the system. Conditions for the experiments were set using the graphical user interface. The experimental system allowed adjusting the separation between alignment magnets as well as the magnetization time. Droplets were deposited using an Ultimus V deposition system (Nordson EFD, East Providence, RI, USA). Real-time optical microscopy was used to capture the droplet behavior on the substrate.

3.4. Magnetic Particle Aggregation Control in Photopolymers

To evaluate the behavior of magnetic particles dispersed in the UV curable polymer, optical microscopy was utilized. The magnetic filler loading in the formulations was maintained at 0.5 wt% to enable light optical microscopy. Particle aggregation due to interparticle magnetic interactions, degree of particle alignment as a function of resin viscosity, and particle chain misorientations were captured and understood in this analysis. Magnetic particles were dispersed in a suspension aggregate due to the magnetic forces that are a function of particle size and magnetization of the particle [31]. Mathematically, the interaction between particles can be characterized using the following equation:

$$W_{\text{mag}} = \mu_0 M^2 a^3 \tag{9}$$

where W_{mag} is the interaction energy between two magnetic particles, M is the particle magnetization, a is the particle radius, and μ_0 is the permeability of a vacuum. To understand the role of additives in controlling particle aggregation, a droplet of the prepared suspension was dispensed within a nylon washer. The state of the particles within the dispensed droplet was captured after 15 min using an optical microscope.

3.5. Manufacturing Scenarios for Particle Structuring, and Influence of Resin Viscosity on Particle Alignment

During the process of magnetic particle structuring under an external magnetic field, the dispersed magnetic particles experience forces that are a function of several parameters, i.e., magnetic, gravity, and viscous drag forces. The particle motion is expressed mathematically in Equation (10) [32–35].

$$\left(\frac{4}{3}\pi a^3 \rho_p\right)\frac{dV_p}{dt} = \left[F_m + F_d + F_g \right] \tag{10}$$

where the individual force terms are expressed as indicated in the following equations:

$$Magnetic\ force: F_m = \mu_0\left(\frac{4}{3}\pi a^3\right)\chi\frac{1}{2}\nabla[H{\cdot}H] \tag{11}$$

$$Viscous\ drag\ force: F_d = 6\pi a\eta V_p \tag{12}$$

$$Gravitational\ force: F_g = \left(\frac{4}{3}\pi a^3\right)\left(\rho_p - \rho_l\right)g \tag{13}$$

In the above equations, μ_0 is the magnetic permeability of free space, H is the magnetic field, a is particle radius, χ is magnetic susceptibility, g is the acceleration due to gravity, V_p is the particle velocity, ρ_p and ρ_l are, respectively, the particle and fluid densities, and η is the fluid viscosity. The above equations enable understanding the mechanics of particle alignment, with the most important variables that can be manipulated being the magnetic field (H) and resin viscosity (η). In previous work, the influence of a magnetic field and novel methods of particle alignment have been already investigated [21]. Simple design changes in the material jetting equipment and alignment methodology have been illustrated, adopting two different manufacturing scenarios. Note that in the original design of the 3D printer, the deposition of the magnetic resin and particle alignment is coupled [23]. Moreover, UV curable resin would cure at the nozzle tip, hindering the deposition process. In the present work, the need for de-coupling printing processes through simple component design changes and manufacturing methodology is described. To understand the influence of resin viscosity on particle alignment, a single layer of the designed sample geometry (15 mm × 15 mm × 1 mm) was dispensed using the material jetting printer and, subsequently, cured using UV light. The height of the magnetic alignment jig from the substrate was maintained at around 10 mm, as further lowering the jig would interfere with the substrate. Optical microscopy was used as a tool to investigate the aforementioned aspects of the experiments. Directionality analysis using ImageJ software (National Institute of Mental Health, Bethesda, MD, USA) was used to quantify the degree of particle alignment in field-structured composites [36].

3.6. Additive Manufacturing of Magnetic Polymer Composites and Magnetic Characterization

To manufacture field-structured magnetic composites, a material jetting 3D printer was utilized to deposit the ferromagnetic resin on the substrate. An in-house developed material jetting 3D printer controlled using the LabVIEW programming environment (National Instruments, Austin, TX, USA) was employed to deposit the material to the designed geometry. A graphical user interface enabled controlling the different movements of the 3D printer. The sample geometry was designed in SolidWorks (Dassault Systems, Vélizy-Villacoublay, France), and the open-source software Sli3cr was used to generate the G-code for the nozzle deposition path [37]. The generated G-code was further modified for proper positioning of the magnetic alignment jig above the deposited material and, subsequently, curing every deposited layer.

The field-structured magnetic composite magnetic properties were characterized using a SQUID magnetometer (MPMS XL-7 Evercool, Quantum Design, San Diego, CA, USA) for its in-plane and out-of-plane magnetic characteristics, i.e., magnetic properties were measured along and perpendicular to the direction of particle structuring. A small piece from a 3D printed part was first weighed, placed in a gelatin capsule, and further inserted into a transparent diamagnetic plastic straw. Magnetization reversal loops were performed by applying a magnetic field with a strength of $\mu_0 H = \pm 7$ Tesla at a temperature of 300 K. Saturation magnetization was determined at an applied field strength of 7 Tesla. Remanence and coercivity were obtained, respectively, at zero applied field and as the applied field yielding zero magnetization from the hysteresis data. The magnetic properties were determined by averaging the values obtained through both magnetization and de-magnetization cycles.

4. Results and Discussion

4.1. Rheological Behavior Analysis of Ferromagnetic Polymers

Rheological properties of filler-modified resin formulations have great significance in extrusion-based additive manufacturing processes [38,39]. In the present rheological study, changes in material behavior as a result of two different additive loadings are investigated. Materials developed and utilized for additive manufacturing processes are subjected to various types of shear rates and deformations during storage and/or the manufacturing process. The study of viscosity properties at different shear rates provides useful information on the properties of the developed formulations. It is well-known that the rheological properties of the polymer formulations strongly depend on the characteristics of the fillers, volume fraction of fillers, dispersion quality, and network structure within the polymer [40,41]. As observed in Figure 2, apart from the base resin (pure PR-48) that exhibits Newtonian behavior (viscosity is independent of shear rate), all other formulations exhibit non-Newtonian behavior, which is confirmed by observing a decrease in viscosity with increasing shear rates. First, adding strontium ferrite to the base resin increases the viscosity of the suspension and makes the mixture exhibit non-Newtonian behavior. Additionally, an increase in low shear viscosity is observed in suspensions modified using the rheological additive. It is observed that the additive loading significantly influences the magnitude of an increase in low shear viscosity. Such observations where changes in mechanical behavior are imposed on the resin system by introducing additives are deemed important for extrusion-based additive manufacturing processes [42]. The BYK 7410ET additive is a polyuria-based thixotropic additive material system. Such material, when dispersed in a polymeric matrix, results in the formation of a network structure by hydrogen bonding [43]. The development of a structural network within the polymeric binder is supported by the enhancement in low shear (at $1 \cdot s^{-1}$) viscosity observed in Figure 2. With increasing shear rate, the viscosity decreases due to the disruption of this structural network. By fitting the viscosity versus shear rate data to a power-law model (using Origin software, Version 2020, OriginLab Corporation, Northampton, MA, USA), the formulation properties (power-law index and viscosity) are determined to understand the influence of additive loading (see Table 2).

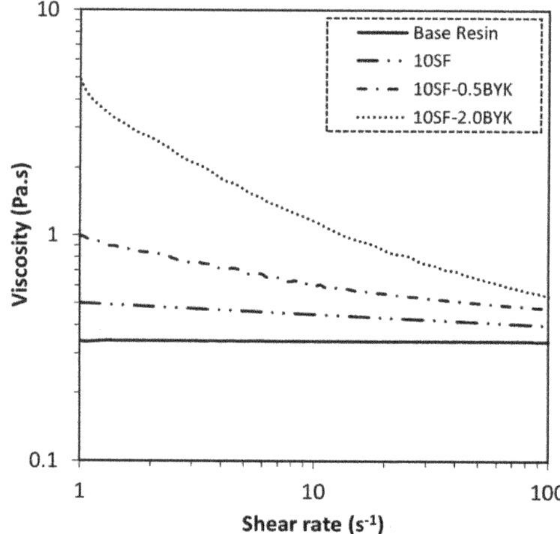

Figure 2. Viscosity as a function of shear rate for material formulation listed in Table 1.

Table 2. Rheological properties derived using Equation (5).

Material Code	n - Power Law Index	m - Viscosity (Pa·s)
Base resin	-	-
10SF	0.95	0.50
10SF-0.5BYK	0.87	0.88
10SF-2.0BYK	0.61	3.23

As observed in Table 2, curve fitting using the power-law model for all the formulations except for the base resin confirms shear thinning or pseudoplastic material behavior with power-law index $n < 1$. Additionally, it is observed that the additive loading significantly influences the degree of pseudoplasticity. The 10SF-2.0BYK formulation with the lowest power-law index exhibits the highest material viscosity, which is advantageous for extrusion-based additive manufacturing.

Yield point or yield stress, defined as the shear stress at zero shear rate, determined by curve fitting of shear stress versus shear rate data using Equation (7) are listed in Table 3. With Herschel–Bulkley indices being less than unity, it is additionally confirmed that all suspensions exhibit pseudoplastic behavior. As far as the yield strengths of the suspensions are concerned, they are observed to be dependent on additive loading. The additive is found to be efficient in terms of yield strength enhancement, as indicated by the results shown in Table 3. The enhancement in yield strength is caused by the thixotropic network structure as a result of additive incorporation within the magnetic particle reinforced formulations.

Table 3. Yield stress predictions using the Herschel–Bulkley model (Equation (7)).

Material Code	Herschel–Bulkley Model predictions		
	Yield Strength - τ_0 (Pa)	Consistency index - C (Pa·s)	Herschel-Bulkley index - k
Base resin	-	-	-
10SF	0.26	0.45	0.97
10SF-0.5BYK	0.65	0.59	0.93
10SF-2.0BYK	3.07	1.18	0.81

4.2. Thixotropic Flow Behavior Analysis

In material jetting additive manufacturing processes, the material experiences various forces at different processing stages, i.e., forces are imposed on the material during handling, pressure-induced material dispensing, and magnetic field exposure during particle structuring. Thixotropy analysis is further utilized to understand and determine the viscosity recovery in the magnetic suspensions. Results of the step test consisting of three intervals shown in Figure 3 indicate high initial suspension viscosity at $\dot{\gamma} = 1 \cdot \mathrm{s}^{-1}$, followed by instantaneous viscosity reduction when the shear rate is increased to $\dot{\gamma} = 300 \cdot \mathrm{s}^{-1}$. Finally, in the recovery phase, when the shear rate is again reduced to $\dot{\gamma} = 1\,\mathrm{s}^{-1}$, the magnetic suspensions exhibit time-dependent viscosity recovery behavior. The observed phenomena relate to the processes of structural decomposition at high shear rates and structural regeneration at low shear rates that are mainly controlled by the rheological additives. From this analysis, it is well understood that the additive imparts thixotropic properties to the magnetic suspension. Thixotropy indices calculated using Equation (8) are listed in Table 4.

The step test data indicates a strong influence of rheological additive content on the magnitude of viscosity recovery. This type of structural decomposition and regeneration is deemed to be one of the fundamental requirements for a material to be considered for material jetting-based additive manufacturing. Present results corroborate findings in the technical literature [26].

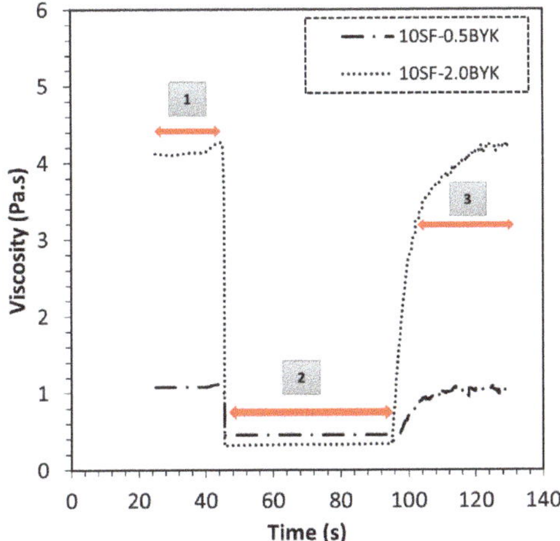

Figure 3. Three interval thixotropy tests for developed material formulations (1: Low shear phase $\dot{\gamma} = 1 \cdot \mathrm{s}^{-1}$; 2: High shear phase $\dot{\gamma} = 300 \cdot \mathrm{s}^{-1}$; 3: Low shear phase $\dot{\gamma} = 1 \cdot \mathrm{s}^{-1}$).

Table 4. Thixotropy index of developed magnetic suspensions.

Material Code	Thixotropy Index
10SF-0.5BYK	0.05
10SF-2.0BYK	0.32

4.3. Magnetic Particle-Reinforced Resin Behavior in Magnetic Field

In this study, the behavior of magnetically loaded polymer droplets is evaluated as a function of time at a magnet separation distance of 30 mm, where the magnetic flux density at the center between two magnets is 0.10 Tesla. From particle alignment experiments, it is evident that the degree of particle alignment is higher at a separation distance of 30 mm between the cube magnets [10]. However, it is observed that droplets dispensed on the substrate deform significantly under the influence of the magnetic field. Experiments are conducted using the base resin reinforced with just 10 wt% SrFeO particles to determine the time at which a droplet loses its stability on the substrate. Deformation of magnetic particle-reinforced polymer droplets obtained from experiments conducted, varying the time at a separation distance of 30 mm, is depicted in Figure 4.

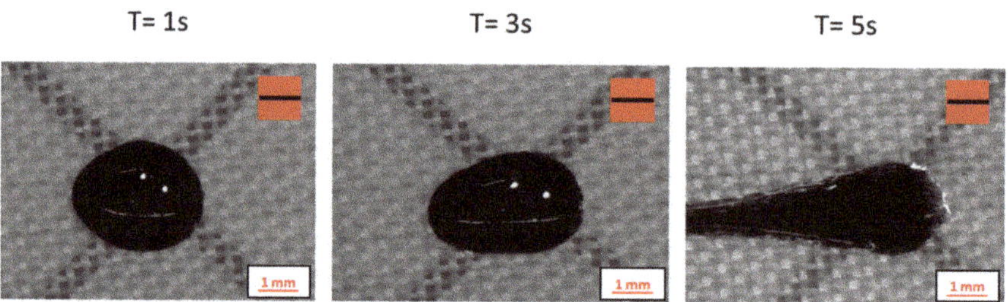

Figure 4. Optical microscopy images of droplet deformation for different magnetization times at magnet separation distance of 30 mm (Formulation: 10SF). Red squares with a black bar indicate the direction of the magnetic field.

The magnetic particle-reinforced resins engineered with 10 wt% magnetic particles loading and 0.5 wt% and 2 wt% of BYK 7410ET are tested for their stability on the substrate. As observed in Figure 5, the resin engineered with higher additive loading (2 wt%) is stable on the substrate in the presence of a magnetic field. This observed behavior corresponds well with the rheological analysis results, where the formulation engineered using 2 wt% of the additive exhibits enhanced low shear viscosity, yield strength, and thixotropic properties.

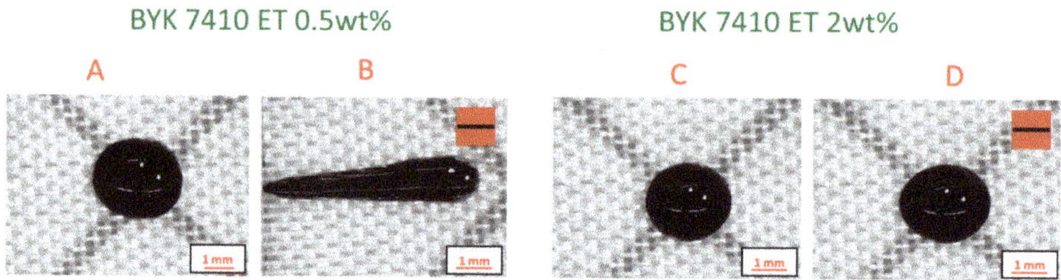

Figure 5. Droplet images, showing the influence of rheological additive loading: (**A**) and (**C**): Droplet images before magnetic field application; (**B**) and (**D**): Droplet images after field application. Red squares with a black bar indicate the presence and direction of the magnetic field.

4.4. Magnetic Particle Aggregation Control in Photopolymers

The present study also investigates the ability of magnetic particle-reinforced formulations to control particle aggregation due to interparticle interactions. To enable optical microscopy, the magnetic filler loading is kept low at 0.5 wt%. Table 5 lists the formulations prepared for evaluating the magnetic particle behavior in the prepared formulations.

Table 5. Suspensions prepared for optical microscopy analysis.

Sample Identifier	Magnetic Filler Loading (wt%)	BYK 7410ET Additive Loading (wt%)
A	0.5	0
B	0.5	0.5
C	0.5	1.0
D	0.5	2.0

First, the influence of additive loading toward mitigating particle aggregation is taken into consideration. According to Equation (9), magnetic particles attract each other due to the interaction energy and tend to form aggregates. The formation of aggregates is evident in the micrographs shown in Figure 6A,B. Aggregation is observed to reduce with increasing additive content in the base resin. Even though the micrograph in Figure 6B shows particle aggregation, the randomized chaining of magnetic particles is not as profound as the formulation without any additive (Figure 6A). Micrographs in Figure 6C,D exhibit uniform particle dispersion as the viscous drag due to the additive inhibits magnetic particle motion. This viscous drag is a result of thixotropic network formation within the formulation. The plastic fluidity, which is a result of additive incorporation, ensures good dispersion of the magnetic particles within the developed photopolymer formulation.

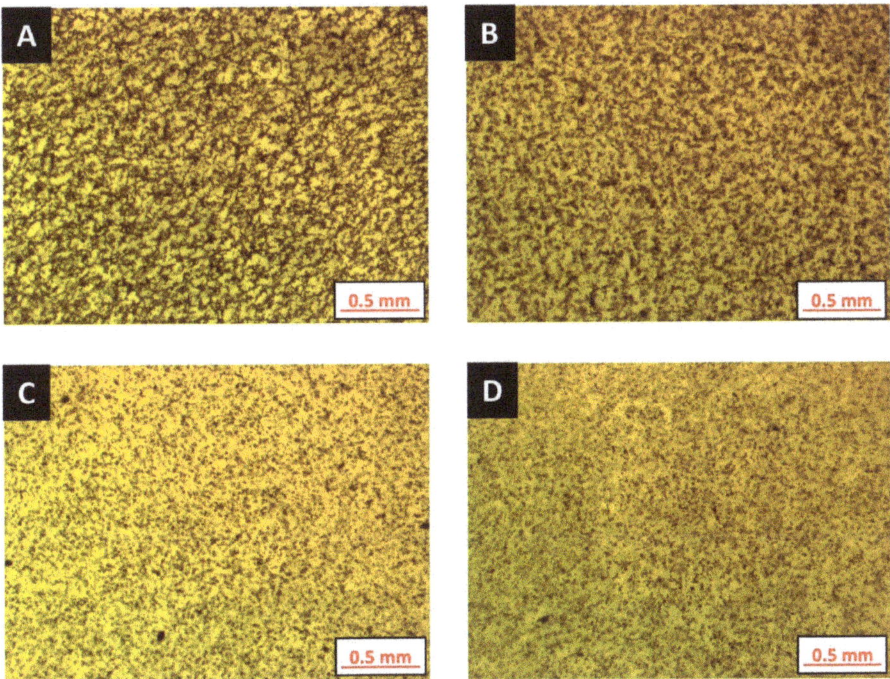

Figure 6. Particle aggregation decreasing from (**A–D**) in photosensitive polymer formulations, as listed in Table 5, (**A**), (**B**), (**C**) and (**D**) have different BYK 7410ET additive loading separately.

4.5. Manufacturing Scenarios and Influence of Resin Viscosity on Particle Alignment

To manufacture field-structured magnetic composites, two different manufacturing scenarios are tested. In scenario A, represented in Figure 7A, the permanent magnet alignment system is coupled with the dispensing system. In this manufacturing scenario, when the dispensing nozzle is in the presence of the magnetic field, some manufacturing issues, i.e., nozzle clogging, are encountered. These fabrication issues are overcome by adjusting the dispensing pressure. Once dispensed, the curing source moves to the dispensed material and solidifies the resin. In scenario B, depicted in Figure 7B, the alignment system is coupled with the curing system. Figure 8 shows the manufacturing system configuration of scenario B for the developed material jetting system. All machine movements are accomplished by programming appropriate G-codes with the respective wait times prior to the curing process. From the micrograph shown in Figure 9, it is understood that the alignment system coupled with the curing source enhances particle structuring (Figure 9B), whereas removing the field structuring setup during the curing process (Figure 9A) results in chain misalignment and a reconfigured microstructure.

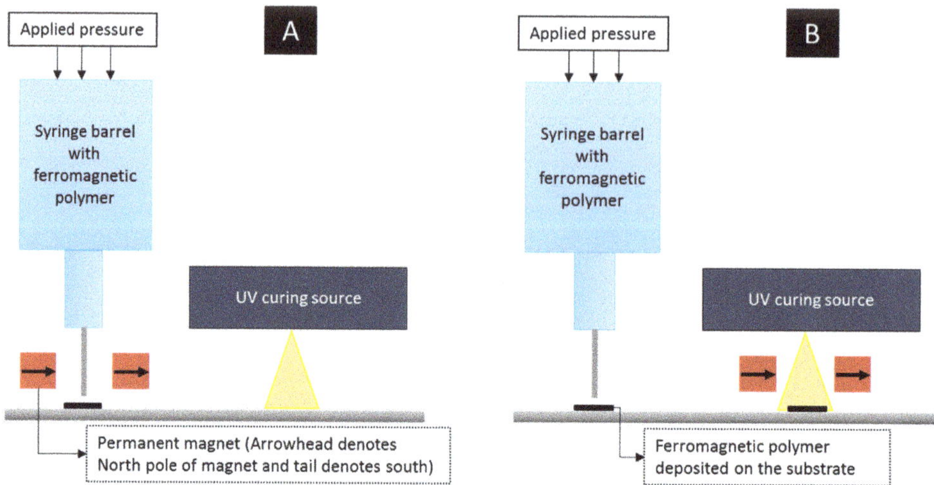

Figure 7. Graphical representation of adopted manufacturing scenarios: (**A**) field structuring setup decoupled from curing process, and (**B**) the alignment system coupled with curing source.

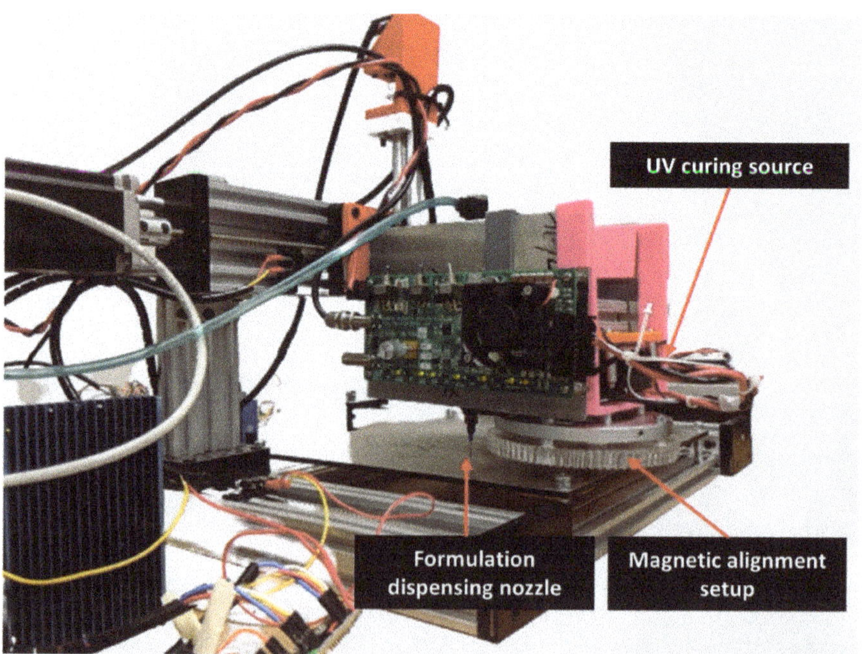

Figure 8. Material jetting equipment with the in-situ particle alignment system, representing manufacturing scenario (B) in Figure 7.

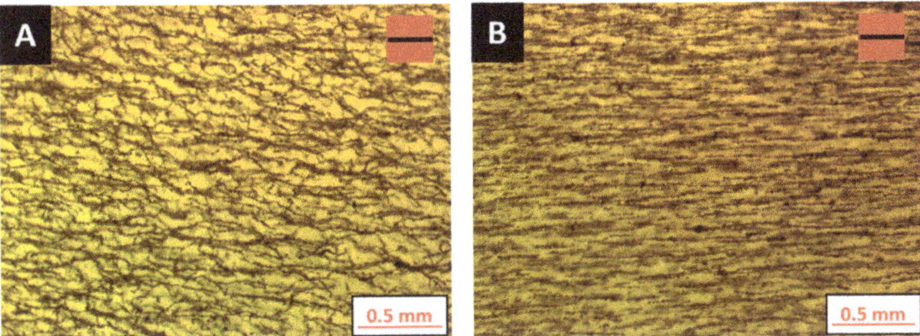

Figure 9. Micrographs obtained from the experiments adopting machine configuration scenarios (**A**) and (**B**), as indicated in Figure 7. Red squares with a black bar indicate the direction of the magnetic field.

The influence of additives on the degree of particle alignment is also studied, adopting scenario B as the machine setting and dispensing one single layer of the designed sample geometry. The height of the magnetic alignment jig is maintained at a target of 10 mm from the substrate. Maintaining this height requires increasing the time to allow particle chaining to occur prior to UV curing. The wait time prior to curing is maintained at 60 s. Optical micrographs obtained for samples manufactured using the formulations listed in Table 5 are shown in Figure 10. Results from the directionality analysis shown in Figure 11 indicate that increasing the additive loading decreases the particle chaining effect. The count of oriented structures, which enables quantification of filler directionality and the degree of filler orientation, is observed to decrease with increasing additive content. Furthermore, the observed behavior emphasizes that magnetic field strength, along with resin viscosity, are critical aspects of the manufacturing process, especially concerning the degree of particle alignment.

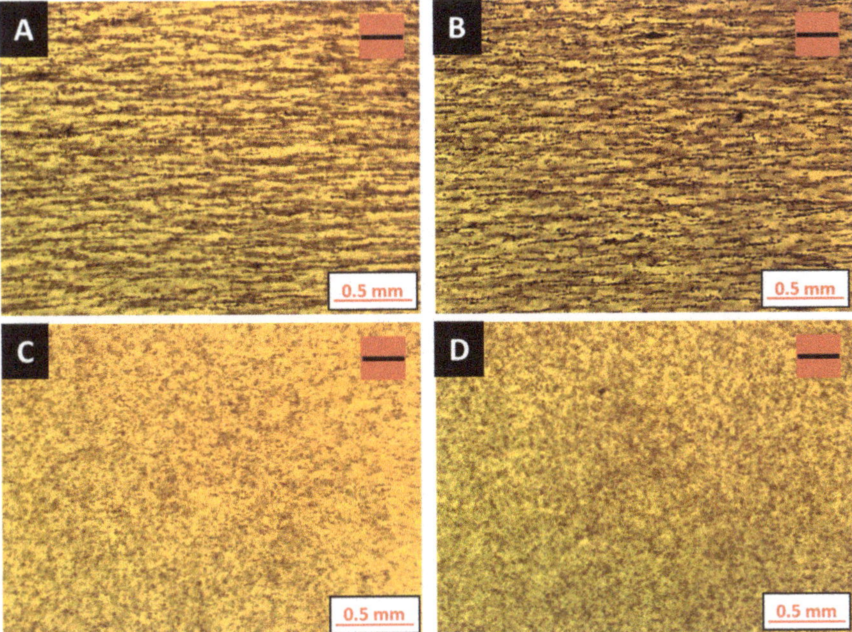

Figure 10. Influence of additive loading on particle alignment (**A–D** refer to suspensions listed in Table 5). Red squares with a black bar indicate the presence and direction of the magnetic field.

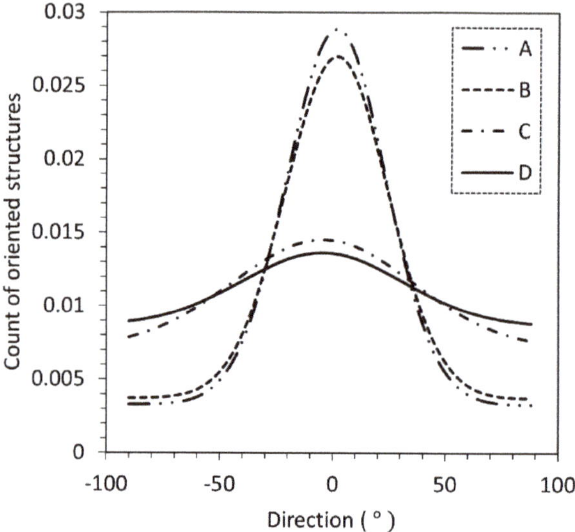

Figure 11. Particle alignment orientation obtained using image analysis (**A–D** refer to suspensions listed in Table 5).

4.6. Additive Manufacturing of Field-Structured Composites and Magnetic Characterization

Magnetic field-structured composites are printed using a formulation engineered with 10 wt% magnetic filler and 0.5 wt% BYK 7410ET additive. Printing magnetic composites is restricted to a simple geometry (flat plate, as indicated in Figure 12), as the primary focus of this study is to develop composites with field-structured microstructure. Additionally, experiments conducted at lower filler loadings enable establishing apt rheological additive loading and manufacturing scenarios as validated through optical microscopy. The extrusion pressure is set at 17 kPa (2.5 psi); layer thickness and deposition speed are 0.2 mm and 10 mm/s, respectively. Printer settings are modified to position the magnetic alignment jig and the curing source right above the deposited material. Such modifications are done to fabricate composite samples with three print layers of the deposited material. The times for magnetic field application and UV irradiation are maintained at 60 s and 20 s for every layer, respectively. As shown in Figure 12, small deformations caused by the magnetic field during the alignment process prior to curing are observed. These deformations occur due to a non-uniform magnetic flux density within the magnetic jig, which has been already reported in previous work [21].

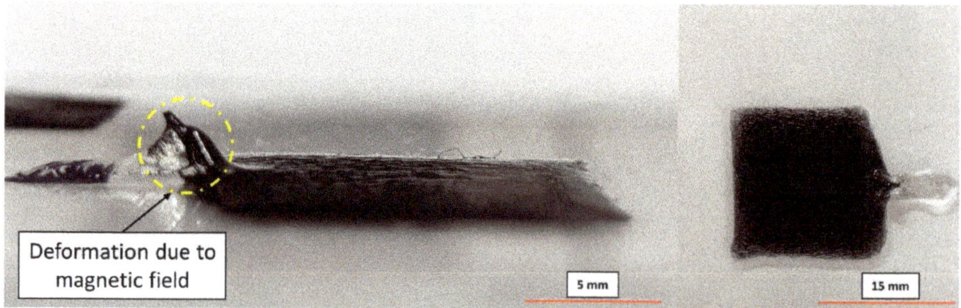

Figure 12. 3D printed magnetic composite with observed deformations caused by the magnetic field (deformation indicated by circles).

Magnetic composites are characterized by properties like saturation magnetization, magnetic remanence, and coercivity. These magnetic characteristics are derived from magnetization reversal loops. Figure 13A shows the magnetization reversal data for a sample evaluated along (in-plane) and perpendicular (out-of-plane) to the direction of particle structuring.

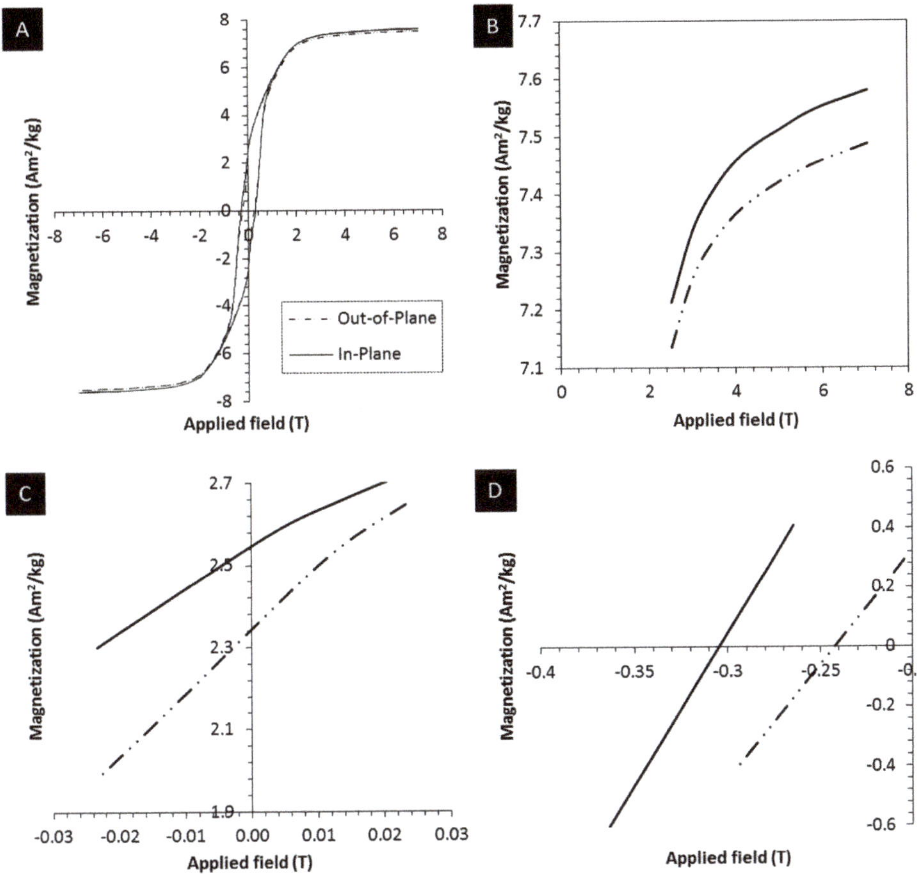

Figure 13. (**A**): Hysteresis data for magnetization versus applied magnetic field; (**B**), (**C**), and (**D**): Magnified views of hysteresis graph for saturation magnetization, magnetic remanence, coercivity, respectively.

The first observed characteristic in Figure 13A is the presence of hysteresis, attributed to the ferromagnetic SrFeO filler. The saturation magnetization of ~7.5 Am²/kg is consistent with the 10% of SrFeO with a saturation magnetization of ~70 Am²/kg, as previously reported [30]. Compared to the direction perpendicular to particle structuring, saturation magnetization along the direction of field structuring is observed to be higher by 0.10 Am²/kg, magnetic remanence by 0.13 Am²/kg, and coercivity by 0.06 Tesla, see Figure 13B–D, respectively. This behavior of remanence and coercivity is consistent with a magnetic easy axis along the direction of particle structuring and a magnetic hard axis perpendicular to it. This anisotropy is likely a combined effect of magnetocrystalline easy axis orientation, as well as shape anisotropy due to the chain-like configuration of the magnetic particles in the cured resin. The difference in saturation magnetization is most likely due to the samples not reaching full saturation even at 7 Tesla applied field, which is conceivable for magnetically highly anisotropic SrFeO particles. Nonetheless, the data indicates that magnetization is easier to achieve along the structuring direction. The area enclosed by the hysteresis curve for measurements along the direction of particle structuring is greater by approximately 5% compared to the hysteresis curve obtained for the direction perpendicular to the direction of field structuring. The present observations and findings are congruent with results in the technical literature for magnetic composites with aligned magnetic fillers [33,44]. This enhancement further confirms the alignment of the easy axis of magnetization of individual particles by orienting the fillers, resulting in the chain-like microstructures. Overall, magnetic composite structures printed by orienting ferromagnetic particles exhibit anisotropic magnetic properties due to particle assembly through the external magnetic field, which has the potential to unlock innovative approaches to build magnetic structures for a wide range of applications.

5. Conclusions

The present study expanded the understanding of material formulations for magnetic polymer composites in a multidisciplinary context. Rheology studies enabling microstructure control, field-structured composite manufacturing scenarios, and magnetic characterization of field-structured composites were successfully performed. It was observed that rheological additive materials enabled enhancing the low shear viscosity and yield strength through the formation of a thixotropic network within the prepared formulations. Mathematical analysis of rheological data enabled interpreting the formulation properties like flow index and yield strength. A flow index of less than unity for all formulations provided strong evidence for pseudoplastic material behavior. Additionally, a step test consisting of three intervals demonstrated the time-dependent viscosity recovery in the magnetic suspensions. It was observed that fluid properties like flow index, yield strength, and thixotropy index were dependent on the additive loading in the developed formulations. Experiments conducted using a formulation engineered with BYK 7410ET additive revealed that at the highest additive loading of 2 wt%, the deposited resin material was stable on the substrate in the presence of the magnetic field. However, optical microscopy at lower magnetic filler loading revealed that an increase in additive loading, while suppressing aggregation of magnetic particles, severely reduced the desired chaining effect in the presence of an applied magnetic field due to enhanced viscous drag in the magnetic formulations. Optical microscopy, coupled with image processing, enabled quantifying the degree of particle orientation in the polymer formulations. These fundamental studies were critical for providing an understanding of the role of material formulations in achieving a variety of material and manufacturing process-related goals for the additive manufacturing process in order to create magnetic particle-reinforced composites. Composites characterized using SQUID magnetometry revealed an enhancement in magnetic properties along the direction of particle structuring. Compared to the out-of-plane magnetic characteristics, the in-plane magnetic saturation, remanence, and coercive fields were observed to be enhanced. Ultimately, this research work provides the basis for devising robust material formulations and manufacturing processes to effectively form magnetic polymer composites with desired microstructure distribution during material jetting-based additive manufacturing.

Author Contributions: B.N., A.J.Q., and P.M. conceptualized the research framework. B.N. developed material formulations, conducted rheological measurements and analysis, developed the equipment framework, programmed g-codes, 3D printed samples, interpreted the results, and wrote the manuscript. M.A.W.S. conducted the magnetic characterization. S.T. provided SQUID facilities for magnetic characterization. P.M., A.J.Q., M.A.W.S., and S.T. reviewed and revised the manuscript. All authors have read and agreed to the published version of the manuscript.

Funding: This work is a part of the University of Alberta's Future Energy Systems program. This research and the APC were funded by the Canada First Research Excellence Fund with grant number Future Energy Systems T06-P03. The authors are thankful for the gracious funding to make this research possible.

Conflicts of Interest: The authors declare no conflict of interest.

References

1. Gibson, I.; Rosen, D.; Stucker, B. *Additive Manufacturing Technologies*; Springer: New York, NY, USA, 2015. [CrossRef]
2. Goswami, A.; Ankit, K.; Balashanmugam, N.; Umarji, A.M.; Madras, G. Optimization of rheological properties of Photopolymerizable alumina suspensions for ceramic Microstereolithography. *Ceram. Int.* **2014**, *40*, 3655–3665. [CrossRef]
3. Goh, G.L.; Saengchairat, N.; Agarwala, S.; Yeong, W.Y.; Tran, T. Sessile droplets containing carbon nanotubes: A study of evaporation dynamics and CNT alignment for printed electronics. *Nanoscale* **2019**, *11*, 10603–10614. [CrossRef]
4. Jabari, E.; Liravi, F.; Davoodi, E.; Lin, L.; Toyserkani, E. High speed 3D material-jetting additive manufacturing of viscous Graphene-based ink with high electrical conductivity. *Addit. Manuf.* **2020**, *35*, 101330. [CrossRef]
5. Oh, Y.; Bharambe, V.; Mummareddy, B.; Martin, J.; McKnight, J.; Abraham, M.A.; Walker, J.M.; Rogers, K.; Conner, B.; Cortes, P.; et al. Microwave dielectric properties of zirconia fabricated using nanoparticle Jetting^{TM}. *Addit. Manuf.* **2019**, *27*, 586–594. [CrossRef]
6. Saleh, E.; Zhang, F.; He, Y.; Vaithilingam, J.; Fernandez, J.L.; Wildman, R.; Ashcroft, I.; Hague, R.; Dickens, P.; Tuck, C. 3D inkjet printing of electronics using UV conversion. *Adv. Mater. Technol.* **2017**, *2*, 1700134. [CrossRef]
7. Derby, B. Additive manufacture of ceramics components by inkjet printing. *Engineering* **2015**, *1*, 113–123. [CrossRef]
8. Yang, H.; Lim, J.C.; Liu, Y.; Qi, X.; Yap, Y.L.; Dikshit, V.; Yeong, W.Y.; Wei, J. Performance evaluation of projet multi-material jetting 3D printer. *Virtual Phys. Prototyp.* **2017**, *12*, 95–103. [CrossRef]
9. Yap, Y.L.; Wang, C.; Sing, S.L.; Dikshit, V.; Yeong, W.Y.; Wei, J. Material jetting additive manufacturing: An experimental study using designed metrological benchmarks. *Precis. Eng.* **2017**, *50*, 275–285. [CrossRef]
10. Martin, J.; Venturini, E.; Odinek, J.; Anderson, R. Anisotropic magnetism in field-structured composites. *Phys. Rev. E* **2000**, *61*, 2818–2830. [CrossRef]
11. Erb, R.M.; Segmehl, J.; Charilaou, M.; Löffler, J.F.; Studart, A.R. Non-linear alignment dynamics in suspensions of platelets under rotating magnetic fields. *Soft Matter* **2012**, *8*, 7604–7609. [CrossRef]
12. Kokkinis, D.; Schaffner, M.; Studart, A.R. Multimaterial magnetically assisted 3D printing of composite materials. *Nat. Commun.* **2015**, *6*, 8643. [CrossRef] [PubMed]
13. Romero, J.J.; Cuadrado, R.; Pina, E.; de Hoyos, A.; Pigazo, F.; Palomares, F.J.; Hernando, A.; Sastre, R.; Gonzalez, J.M. Anisotropic polymer bonded hard-magnetic films for Microelectromechanical system applications. *J. Appl. Phys.* **2006**, *99*, 08N303. [CrossRef]
14. Xia, S.; Metwalli, E.; Opel, M.; Staniec, P.A.; Herzig, E.M.; Müller-Buschbaum, P. Printed thin magnetic films based on Diblock copolymer and magnetic nanoparticles. *ACS Appl. Mater. Interfaces* **2018**, *10*, 2982–2991. [CrossRef] [PubMed]
15. Speliotis, D. Magnetic recording beyond the first 100 years. *J. Magn. Magn. Mater.* **1999**, *193*, 29–35. [CrossRef]
16. Goc, K.; Gaska, K.; Klimczyk, K.; Wujek, A.; Prendota, W.; Jarosinski, L.; Rybak, A.; Kmita, G.; Kapusta, C. Influence of magnetic field-aided filler orientation on structure and transport properties of ferrite filled composites. *J. Magn. Magn. Mater.* **2016**, *419*, 345–353. [CrossRef]
17. Gao, M.; Kuang, M.; Li, L.; Liu, M.; Wang, L.; Song, Y. Printing 1D assembly array of single particle resolution for Magnetosensing. *Small* **2018**, *14*, 1800117. [CrossRef] [PubMed]
18. Ho, M.; Nagaraja, S.P.M.; Tok, R.U.; Rangchian, A.; Kavehpour, P.; Wang, Y.E.; Candler, R. Additive manufacturing with strontium Hexaferrite-photoresist composite. *IEEE Trans. Magn.* **2020**. [CrossRef]
19. Pullar, R.C. Hexagonal ferrites: A review of the synthesis, properties and applications of Hexaferrite ceramics. *Prog. Mater. Sci.* **2012**, *57*, 1191–1334. [CrossRef]
20. Lu, L.; Guo, P.; Pan, Y. Magnetic-field-assisted projection Stereolithography for three-dimensional printing of smart structures. *J. Manuf. Sci. Eng.* **2017**, *139*, 071008. [CrossRef]
21. Nagarajan, B.; Eufracio Aguilera, A.F.; Wiechmann, M.; Qureshi, A.J.; Mertiny, P. Characterization of magnetic particle alignment in photosensitive polymer resin: A preliminary study for additive manufacturing processes. *Addit. Manuf.* **2018**, *22*, 528–536. [CrossRef]

22. Nagarajan, B.; Aguilera, A.F.E.; Qureshi, A.; Mertiny, P. Additive manufacturing of magnetically loaded polymer composites: An experimental study for process development. In *ASME 2017 International Mechanical Engineering Congress and Exposition*; American Society of Mechanical Engineers: New York, NY, USA, 2017; p. V002T02A032.

23. Eufracio Aguilera, A.F.; Nagarajan, B.; Fleck, B.A.; Qureshi, A.J. Ferromagnetic particle structuring in material jetting—Manufacturing control system and software development. *Procedia Manuf.* **2019**, *34*, 545–551. [CrossRef]

24. Nagarajan, B.; Arshad, M.; Ullah, A.; Mertiny, P.; Qureshi, A.J. Additive manufacturing ferromagnetic polymers using Stereolithography—Materials and process development. *Manuf. Lett.* **2019**, *21*, 12–16. [CrossRef]

25. Berndlmaier, R. Rheology additives for solvent and Solventless coating. In *Handbook of Coating Additives*; CRC Press: Boca Raton, FL, USA, 2004; pp. 363–364. [CrossRef]

26. Bastola, A.K.; Paudel, M.; Li, L. Development of hybrid Magnetorheological elastomers by 3D printing. *Polymer* **2018**, *149*, 213–228. [CrossRef]

27. Chhabra, R.P.; Richardson, J.F. *Non-Newtonian Flow and Applied Rheology: Engineering Applications*; Butterworth-Heinemann: Oxford, UK, 2011.

28. Eberhard, U.; Seybold, H.J.; Floriancic, M.; Bertsch, P.; Jiménez-Martínez, J.; Andrade, J.S.; Holzner, M. Determination of the effective viscosity of non-Newtonian fluids flowing through porous media. *Front. Phys.* **2019**, *7*, 1–9. [CrossRef]

29. Mezger, T.G. The rheology handbook. *Pigment. Resin Technol.* **2009**, *38*. [CrossRef]

30. Nagarajan, B.; Kamkar, M.; Schoen, M.A.W.; Sundararaj, U.; Trudel, S.; Qureshi, A.J.; Mertiny, P. Development and characterization of stable polymer formulations for manufacturing magnetic composites. *J. Manuf. Mater. Process.* **2020**, *4*, 4. [CrossRef]

31. Barnes, H.A. Thixotropy—A Review. *J. Nonnewton. Fluid Mech.* **1997**, *70*, 1–33. [CrossRef]

32. López-López, M.T.; Kuzhir, P.; Bossis, G.; Mingalyov, P. Preparation of well-dispersed Magnetorheological fluids and effect of dispersion on their Magnetorheological properties. *Rheol. Acta* **2008**, *47*, 787–796. [CrossRef]

33. Al-Milaji, K.N.; Hadimani, R.L.; Gupta, S.; Pecharsky, V.K.; Zhao, H. Inkjet printing of magnetic particles toward anisotropic magnetic properties. *Sci. Rep.* **2019**, *9*, 16261. [CrossRef]

34. Banerjee, U.; Bit, P.; Ganguly, R.; Hardt, S. Aggregation dynamics of particles in a Microchannel due to an applied magnetic field. *Microfluid. Nanofluidics* **2012**, *13*, 565–577. [CrossRef]

35. Ganguly, R.; Puri, I.K. Field-assisted self-assembly of Superparamagnetic nanoparticles for biomedical, MEMS and BioMEMS applications. *Adv. Appl. Mech.* **2007**, *41*, 293–335.

36. Schindelin, J.; Arganda-Carreras, I.; Frise, E.; Kaynig, V.; Longair, M.; Pietzsch, T.; Preibisch, S.; Rueden, C.; Saalfeld, S.; Schmid, B.; et al. Fiji: An open-source platform for biological-image analysis. *Nat. Methods* **2012**, *9*, 676–682. [CrossRef] [PubMed]

37. Ranellucci, A. Slic3r. Available online: https://slic3r.org/ (accessed on 30 September 2019).

38. Ajinjeru, C.; Kishore, V.; Chen, X.; Hershey, C.; Lindahl, J.; Kunc, V.; Hassen, A.A.; Duty, C. Rheological survey of carbon fiber-reinforced high-temperature thermoplastics for big area additive manufacturing tooling applications. *J. Thermoplast. Compos. Mater.* **2019**. [CrossRef]

39. Ajinjeru, C.; Kishore, V.; Liu, P.; Hassen, A.A.; Lindahl, J.M.; Kunc, V.; Duty, C.E. *Rheological Evaluation of High. Temperature Polymers to Identify Successful Extrusion Parameters*; Oak Ridge National Lab (ORNL): Oak Ridge, TN, USA, 2017.

40. Kamkar, M.; Nourin Sultana, S.M.; Patangrao Pawar, S.; Eshraghian, A.; Erfanian, E.; Sundararaj, U. The key role of processing in tuning nonlinear viscoelastic properties and microwave absorption in CNT-based polymer Nanocomposites. *Mater. Today Commun.* **2020**, *24*, 101010. [CrossRef]

41. Hoseini, A.H.A.; Arjmand, M.; Sundararaj, U.; Trifkovic, M. Significance of interfacial interaction and agglomerates on electrical properties of polymer-carbon nanotube Nanocomposites. *Mater. Des.* **2017**, *125*, 126–134. [CrossRef]

42. Kunc, V.; Lee, A.; Mathews, M.; Lindahl, J. Low cost reactive polymers for large scale additive manufacturing. In Proceedings of the CAMX—The Composites and Advanced Materials Expo, Dallas, TX, USA, 15–18 October 2018.

43. Deka, A.; Dey, N. Rheological studies of two component high build epoxy and polyurethane based high performance coatings. *J. Coat. Technol. Res.* **2013**, *10*, 305–315. [CrossRef]
44. Song, H.; Spencer, J.; Jander, A.; Nielsen, J.; Stasiak, J.; Kasperchik, V.; Dhagat, P. Inkjet printing of magnetic materials with aligned anisotropy. *J. Appl. Phys.* **2014**, *115*, 17E308. [CrossRef]

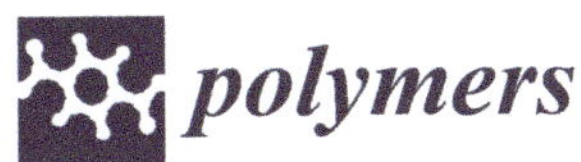 *polymers*

Article

Helicoidally Arranged Polyacrylonitrile Fiber-Reinforced Strong and Impact-Resistant Thin Polyvinyl Alcohol Film Enabled by Electrospinning-Based Additive Manufacturing

Rahul Sahay [1,†], Komal Agarwal [1,†], Anbazhagan Subramani [1], Nagarajan Raghavan [1], Arief S. Budiman [2,*] and Avinash Baji [3,*]

[1] Engineering Product Development, Singapore University of Technology and Design, Singapore 487372, Singapore; rahul@sutd.edu.sg (R.S.); komal_agarwal@mymail.sutd.edu.sg (K.A.); M130148@e.ntu.edu.sg (A.S.); nagarajan@sutd.edu.sg (N.R.)

[2] Industrial Engineering Department, BINUS Graduate Program—Master of Industrial Engineering, Bina Nusantara University, Jakarta 11480, Indonesia

[3] Department of Engineering, La Trobe University, Bundoora, VIC 3086, Australia

* Correspondence: suriadi@alumni.stanford.edu (A.S.B.); a.baji@latrobe.edu.au (A.B.)

† These authors contributed equally to this work.

Received: 13 August 2020; Accepted: 2 October 2020; Published: 15 October 2020

Abstract: In this study, we demonstrate the use of parallel plate far field electrospinning (pp-FFES) based manufacturing system for the fabrication of polyacrylonitrile (PAN) fiber reinforced polyvinyl alcohol (PVA) strong polymer thin films (PVA SPTF). Parallel plate far field electrospinning (also known as the gap electrospinning) is generally used to produce uniaxially aligned fibers between the two parallel collector plates. In the first step, a disc containing PVA/H_2O solution/bath (matrix material) was placed in between the two parallel plate collectors. Next, a layer of uniaxially aligned sub-micron PAN fibers (filler material) produced by pp-FFES was directly collected/embedded in the PVA/H_2O solution by bringing the fibers in contact with the matrix. Next, the disc containing the matrix solution was rotated at 45° angular offset and then the next layer of the uniaxial fibers was collected/stacked on top of the previous layer with now 45° rotation between the two layers. This process was continued progressively by stacking the layers of uniaxially aligned arrays of fibers at 45° angular offsets, until a periodic pattern was achieved. In total, 13 such layers were laid within the matrix solution to make a helicoidal geometry with three pitches. The results demonstrate that embedding the helicoidal PAN fibers within the PVA enables efficient load transfer during high rate loading such as impact. The fabricated PVA strong polymer thin films with helicoidally arranged PAN fiber reinforcement (PVA SPTF-HA) show specific tensile strength 5 MPa · cm^3· g^{-1} and can sustain specific impact energy (8 ± 0.9) mJ · cm^3· g^{-1}, which is superior to that of the pure PVA thin film (PVA TF) and PVA SPTF with randomly oriented PAN fiber reinforcement (PVA SPTF-RO). The novel fabrication methodology enables the further capability to produce even further smaller fibers (sub-micron down to even nanometer scales) and by the virtue of its layer-by-layer processing (in the manner of an additive manufacturing methodology) allowing further modulation of interfacial and inter-fiber adherence with the matrix materials. These parameters allow greater control and tunability of impact performances of the synthetic materials for various applications from army combat wear to sports and biomedical/wearable applications.

Keywords: impact resistance; bioinspired; helicoidal structure; electrospinning

---

1. Introduction

Research has been ongoing to study the mechanical properties of micron/sub-micron thin polymer films due to their widespread applications such as protective and functional coatings, non-fouling surfaces, microfluidics, sensors, lubrication and friction modification [1–4]. Typically, the successful long-term performance and reliability of such thin films are governed by their mechanical properties [5–8]. Common strategies used for the fabrication of strong polymer thin films (SPTFs) include the addition of inorganic material such as carbon tubes, carbon fibers, 2D materials such as graphene oxide and metallic/metal oxide particles in the thin polymeric films [5,7,8]. Blending with strong polymeric material is also used for the fabrication of SPTFs [9,10]. Nevertheless, these methodologies may result in an increase in the specific weight, incompatibility with inorganic material, anisotropic mechanical properties due to nonuniform distribution of the filler material, void formations and/or high cost of the resultant composite [11], reducing their strength. Therefore, nano/sub-micron fibers are being used as fillers due to their large surface area per unit volume to be completely enclosed by the matrix materials thus avoiding the formation of voids or stress concentration zones. Further, it is now known that structural hierarchical arrangement of nano/sub-micron fibers in thin films improves its tensile and impact properties [12–14].

Therefore, researchers are taking inspiration from the multiscale (nano/sub-micron) hierarchical architectures found in natural structural materials, such as in mantis shrimps' dactyl club, nacre, lobster claw and butterfly wing, to design next-generation of strong and/or tough materials [15–17]. For instance, natural structural material found in mantis shrimps' dactyl club consists of helicoidally arranged chitin fiber mineralized by hydroxyapatite [18]. This helicoidal architecture enables mantis shrimp to sustain high impact energy while smashing its prey. The helicoidal arrangement of the chitin fibers embedded in hydroxyapatite allows deflection of cracks and dissipation of the impact energy along distinct planes and directions to avoid catastrophic failure during impact [19]. The remarkable mechanical properties such as high strength (over 100 MPa) [20], impact resistance (33 J·m^{-1} for 10 ball drop impact) [21,22] and toughness of such natural structural materials strongly depend on the synergistic effects from architecture interfacial interactions as well as their multiscale hierarchical design [23–26]. Such geometrical inspiration from the natural fibrous structure can be employed to improve the strength/toughness of microscopic polymeric thin films. In this article, such helicoidal geometry has been taken as an inspiration to fabricate helicoidally aligned sub-micron sized fibers to reinforce the SPTFs.

Techniques such as conventional prepreg/ply stacking and 3D printing have been used in the past for mimicking the helicoidal structures [27–31]. However, conventional prepreg stacking usually produces macroscopic sample and 3D printing can only achieve microscopic fiber with the limitation of available constituent materials. In addition, the materials with smaller scale (sub-micron down to nanometer) structural features have received wide attention in the past two decades [32–34]. Nanowires [35], nanopillars [36] or bulk materials with nanoscale grain structure [37] or nanolayers [38–41] have shown excellent mechanical properties, including the combination of strength and toughness. The desirable properties of the molecular orientation and mechanical properties (strength, toughness and impact-resistance) occurring due to the smaller size effect have motivated fabrication of the structural geometries in the sub-micron/nano level [32,36,42]. Therefore, electrospinning has been used in this article for its capability of producing sub-micron to nanometer-scale fibers [43,44], for having the ability to use various types of constituent materials (polymers, metal particles, carbon nanotubes (CNTs), cellulose nanocrystals (CNCs), ceramics, metal powders and nanoparticles) [42] and for allowing tunability of the electrospinning parameters (such as voltage, solution viscosity, fiber size and nozzle to collector distance).

The objective of this article is to perform extended studies towards achieving nano-sized helicoidal architecture using a single-step fabrication method. In this study, a new parallel plate far field electrospinning (pp-FFES) based methodology was used for the first time to fabricate complex 3D helicoidal geometrical composites within a single fabrication step. During a typical electrospinning process, electrical forces are applied to a polymer solution dispensing out of a nozzle to produce randomly oriented 3D fibrous membrane. The extruding polymer solution experiences bending and whipping instabilities due to the applied electrical forces and lays down fibers in random fashion on top of the collector, whereas parallel plate electrospinning setup is a common technique used to generate uniaxially aligned fibrous layers, by reducing the bending and whipping instabilities by placing sharp edged parallel plates within the electrical field [45,46] (see Figure A6). However, to the best of our knowledge, pp-FFES has not been used for generating complex hierarchical geometrical network of nano fibers so far, let alone the helicoidal architecture. In this study, a matrix solution/bath (PVA/H_2O) was placed in between the two parallel plate collectors, such that, as soon as the fibers formed between the two-plates, the fibers were brought in contact with the matrix solution. Initially, a single layer of uniaxial arrays of PAN fiber was obtained between the two parallel plates, which was then directly collected into the matrix solution by bringing fibers in contact with the PVA/H_2O solution in the disc. The extra fibers hanging outside the periphery of the disc were trimmed off neatly. Then, the disc containing the matrix solution was rotated at 45° angular offset, and, again, the next layer of the uniaxial arrays of PAN fibers was collected/stacked on top of the previous layer with now 45° angular offset between the two layers. This process was continued with progressive 45° angular offsets between the layers until a periodic pattern was achieved. In total, 13 such layers were laid within the matrix solution. This layer-by-layer process is indeed vital for completely embedding PAN fibers in the PVA/H_2O solution and providing better surface interaction between the fibers and the matrix material, as compared to the conventional method of post impregnation of fibers in the matrix materials. This methodology is thus superior to the state of the art where pre-prepared helicoidal fiber structures are embedded in the matrix material, which may not allow complete surrounding of filler fibers by matrix [47]. Moreover, the current state of the art used for fabrication of helicoidal geometry in the composites is limited by constituent materials, pre-prepared fillers, prepregs and the fiber size. This study eliminates these limitations for generation of the sub-micron to nano sized helicoidal geometrical architecture. The authors propose that the smaller fibers (sub-micron to nanometer scale) in the helicoidal composites could enable further enhancement in the impact performance (measured by its ability to protect the glass coverslip lying underneath it from fracturing) as the more interfacial surfaces (between fibers as well as layers) per unit volume of the materials would allow very efficient dissipation of energy, thus greatly enhancing the impact resistance of the materials. However, electrospinning methodologies are relatively a newer platform for fabrication of hierarchical complex structural geometries in nano range. Hence, furthermore advancements in electrospinning techniques are desired to obtain even more dense uniaxial fibers, even better alignment and uniformity for fabricating superior helicoidal composites in the future.

The relationship between architecture, interfacial interactions and the mechanical properties of samples were systematically analyzed in this study. The morphology, structure and composition of the samples were investigated through scanning electron microscopy (SEM) and Fourier-transform infrared spectroscopy (FTIR). Thermogravimetric analysis (TGA) and differential scanning calorimetry (DSC) were used to analyze the thermal properties of the samples. The mechanical properties of the samples were studied through uniaxial tensile tests and free-falling steel ball impact test [12,48].

2. Material and Methods

2.1. Materials

Polyvinyl alcohol (PVA, Mw = 130,000 Da), polyacrylonitrile (PAN, Mw = 154,000 Da) and dimethyl formaldehyde (DMF) were purchased from Merck & Co. (Kenilworth, NJ, USA) and used without any further purification. The syringe and needles used during electrospinning were purchased from Terumo, Tokyo, Japan.

Polyvinyl alcohol was used as a matrix material in this study due to its ability to dissolve in water (allowing to control the viscosity of the matrix solution) and the ability to form thin films.

PVA also serves a wide variety of applications including wound dressing, biomedical and food packaging [49–52]. Nevertheless, PVA thin films suffer from low tensile strength (~10 MPa) [53]. Various strategies are being employed to improve the mechanical properties of PVA thin films such as the addition of starch/critic acid [54], nanofibrillated cellulose [55] and halloysite nanotubes [52]. However, these filler materials were randomly distributed in the thin films. The size of the thin film is also governed by the size of the fibers or particles added to improve its mechanical properties. Therefore, helicoidally arranged polyacrylonitrile sub-microfibers were used for reinforcing PVA to improve its mechanical properties [48,56]. Further, polyacrylonitrile filler sub-microfibers were used for reinforcing PVA due to their high Young's modulus (~3.1 GPa) [57] and compatibility with PVA. PAN and PVA interact through hydrogen bonding [58].

2.2. Methods

Typically, fabricated PVA strong polymer thin film (PVA SPTF) consisted of PAN fibers embedded in PVA and two types of samples were made: (i) helicoidally arranged PAN fibers (diameter = (0.93 ± 0.1) µm) embedded in PVA thin film (PVA SPTF-HA) (see Figures 2 and A4); and (ii) randomly oriented PAN fibers (diameter = (3.31 ± 1) µm) embedded in PVA thin film (PVA SPTF-RO) (see Figures A2 and A5 in Appendix A, respectively). Initially, 10 wt% PAN was dissolved in DMF using a hot-plate magnetic stirrer (Thermo Fisher Scientific, Waltham, MA, USA) at 60 °C. This solution was fed into the syringe for electrospinning (Nanospinner-24, Inovenso, Turkey). A custom-built parallel plate fiber collection setup was used for the arrangement of uniaxial arrays of PAN fibers in a hierarchical helicoidal geometry with 45° angular offsets directly within the matrix solution (see Figure 1). A digital syringe pump (Premier Solution, Singapore) was used to extrude the PAN solution at a flow rate of 3 mL·h^{-1}. The electric field used during electrospinning was 1 kV·cm^{-1} between the spinneret and the parallel plate collectors (see Figure A1, Appendix A). The distance between the spinneret to the parallel plate was 15 cm and the distance between the parallel plates was 6 cm. A disc of 50 mm diameter and 3 mm depth filled with PVA/H$_2$O matrix solution was placed in between the two plates such that the uniaxial fibers get directly deposited/embedded in the matrix solution by bringing it in contact with the solution in the disc. Once an array of uniaxial PAN fibers were collected in the matrix bath, the disc was manually rotated progressively at 45° angular rotations to align the uniaxial arrays of the fibers in a helicoidal stack. A periodic pattern of continuous 45° offsets (i.e., 0°, 45°, 90°, 135°, 180°, 225°, 270°, 315°, 360/0°, until the next 180°) from the previously deposited layers were followed until three pitches of helicoidal geometry were obtained (one pitch means half a turn of the helix, i.e., 0° to 180°, then 225° to 360/0° and, lastly, 45° to 180°). Thus, 13 layers of uniaxial arrays of PAN fibers were stacked one above the other with 45° angular offsets (see Figure 2). After electrospinning each layer of uniaxial fibers inside the disc with the matrix solution, the excess fibers hanging between the disc and the edge of the parallel plate were carefully trimmed off using a pair of scissors. This was done to detach the fibers from the parallel plates while avoiding any stretching, twisting and misalignment when rotating the disc at 45° offset to collect the next set of uniaxial layers.

Various weight% (1 to 5 wt%) of PVA/H$_2$O solutions were prepared to embed aligned electrospun fibers. Nevertheless, only 5 wt% PVA/H$_2$O solution was able to form uniform thin film with PAN fibrous layers embedded in it. Hence, 5 wt% PVA/H$_2$O solution was chosen in the fabrication of all the samples for all the experiments in this article. The authors believe that layer by layer insertion of uniaxially aligned PAN fibers into the PVA bath/solution would provide better surface interaction between the impregnating solution and the PAN fibers as compared to post impregnation of fully formed helicoidally arranged fibrous structure. Later, the disc containing the PVA/H$_2$O soaked PAN films was kept undisturbed at the room temperature for drying (by the simple evaporation process) for 3 days to remove water and result in the formation of PVA strong polymer thin films with helicoidally arranged PAN fiber reinforcement (PVA SPTF-HA). The randomly oriented PAN fiber reinforced PVA strong polymer thin films (PVA SPTF-RO) were fabricated using typical random fibrous film prepared by conventional electrospinning setup, while following the above electrospinning parameters and without using the parallel plates. The time (for electrospinning) required for the fabrication of random PAN fibrous film for PVA SPTF-RO was same as for the fabrication of 13 uniaxially aligned PAN fibrous layers for PVA SPTF-HA so as to keep the amount of PAN same for both PVA SPTF-RO and PVA SPTF-HA. The fiber weight percentage in both the type of composites was kept ∼1 wt%.

A pure/neat PVA thin film (PVA TF) was also fabricated by pouring the 5 wt% PVA/H_2O solution on a flat circular disk of dimension 50 mm diameter and 3 mm depth and allowing it to dry for 3 days. Once the PVA film dried due to the evaporation process, it was peeled from the disc and used for experiments for comparison with PVA SPTFs. Next, the pure/neat PAN fibers were collected on an aluminum foil in random orientation, using the same PAN solution and electrospinning parameters as above, but without the parallel plates or the disc containing the matrix material. In this case, the PAN fibers were electrospun for 30 mins to obtain a film thick enough to easily peel it away from the foil. These PAN fibers were then dried in an oven at 60 °C for a day to obtain a dried electrospun neat PAN thin film (PAN TF). This PAN TF sample was later used for determining its chemical and thermal properties for comparison with PVA SPTFs.

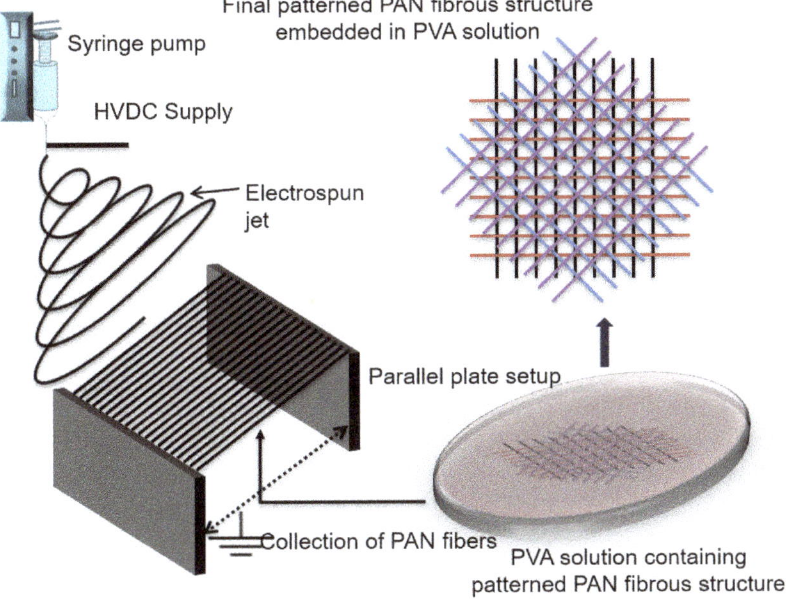

Figure 1. Fabrication of polyvinyl alcohol (PVA) strong polymer thin films with helicoidally arranged polyacrylonitrile (PAN) fiber reinforcement (i.e., PVA SPTF-HA). Uniaxial arrays of PAN fibers generated using parallel plate far field electrospinning (pp-FFES) were embedded directly into a circular disc containing PVA/H_2O solution/bath by bringing them in contact with solution in the disc. The disc was then rotated progressively at 45° angular offsets to stack three pitches of helix to form PVA SPTF-HA.

2.3. Characterization

A field-emission scanning electron microscope (FESEM)(JEOL (JSM—6700F), Tokyo, Japan) was used to observe the morphology and arrangement of the PAN fibers in PVA SPTF samples. The thickness of the samples was measured from their cross-sectional images obtained through FESEM. The density of the samples was calculated with and without the fibrous reinforcement from the weight and the volume measurements of the samples. The density of samples was calculated by dividing the total mass of the fabricated sample by the volume of the sample. Fourier transform infrared spectroscopy (FTIR, Bruker Alpha spectrometer, Billerica, MA, USA) were recorded at a resolution of 4 cm^{-1} at room temperature. Thermogravimetric analysis (TGA, TA instruments (Q50), New Castle, DE, USA) was used to document the thermal stability of the samples at a heating rate of 10 °C·min^{-1}. Differential scanning calorimetry (DSC, TA instruments (Q20), New Castle, DE, USA) tests were performed from −60 °C to 250 °C with a ramp rate of 3°C·min^{-1} in a nitrogen atmosphere (flow rate 40 mL·min^{-1}).

The mechanical properties of the samples were investigated using a tensile testing machine (MTS Criterion Model 43, MTS Systems Corporation, Eden Prairie, MN, USA). The typical sample dimensions used for testing were thickness ∼50 μm, width 5 mm and length 25 mm. The tensile tests were performed at a constant strain rate of 1 mm·min^{-1} and the clamp pressure was maintained at 25*psi*. Neat PVA thin film (PVA TF), helicoidally arranged PAN fiber reinforced PVA SPTF (PVA SPTF-HA) and randomly oriented PAN fiber reinforced PVA SPTF (PVA SPTF-RO) were tested for their mechanical properties. The weight percentage of the PAN fibers in the composites was kept around ∼1 wt%. The tests were performed for at least 5 specimens to check the repeatability of the results. A customized setup was used to determine the impact-resistance of the samples. The impact test was carried out using a free-falling steel ball to impact the sample with a given potential energy [12,48], and the performance of the samples was determined by the ability of the samples to protect a glass coverslip placed underneath the samples. The glass coverslip (14 mm diameter, 0.16 mm thickness) was made of borosilicate glass (*D263*, deckglaser, Trade 21, Singapore). The samples were adhered to the glass coverslip at the diagonal edge points so as to eliminate the effects of glue on the on the measurements and avoid the sample from moving from its place upon impact. The potential energy of the steel ball can be varied either by changing the mass or impact height of the steel balls. The mass of the steel ball used for the measurement was 0.51 g and its diameter was 4 mm, whereas the height from which steel ball is dropped was varied. The height from which the steel ball is dropped was initially increased with an increment of 5 cm. The initial experiments enabled to obtain initial tentative height required for the fracture of the glass coverslip with/without the samples. Later, the height in the vicinity of already obtained initial tentative fracture height was varied in the increments of 1 cm to obtain the exact value of height required to fracture the glass coverslip with/without the samples. Later, the potential energy of the steel ball required to fracture the glass coverslip with/without the samples were documented. All the samples used for impact tests were ∼50 μm thick and had a cross-sectional area of 1.1 × 1.1 cm^2. Dynamic mechanical analysis (DMA, DMA Q800, TA Instruments, New Castle, DE, USA) was used to understand the specific viscoelastic properties of the samples. The scanning temperature range was selected from 0 °C to 150 °C at a strain ramp rate of 1%· min^{-1} and 1 Hz frequency in the multi strain mode (see Appendix A and Figure A3).

3. Results and Discussion

3.1. Physical Characterization of PVA SPTF

First, the parallel-plate far field electrospinning was used for producing uniaxial arrays of PAN fibers. Then, these uniaxial PAN fibers were arranged helicoidally to obtain PVA SPTF-HA samples. Figure 2a,b shows the morphologies of the helicoidally arranged uniform and smooth PAN fibers (diameter = (0.93 ± 0.1) μm, see Figure A4, Appendix A) that were embedded within the PVA thin films (sample PVA SPTF-HA). The arrangement of layers consisting of aligned fibers one above the other at 45° angle within the PVA matrix is evident from these figures. Similar observations are also reported in the literature [12,48]. Due to the small size of the fibers, densely aligned fibers were not observed in the SEM images—similar to other reports in the literature, especially on a smaller range of fibers size [12]. It is significant that this novel technique has allowed the fabrication of helicoidally arranged fibers architectures with diameters of (0.93 ± 0.1) μm, while many other previous reports have conducted their studies with much larger fibers [59–61]. This technique would thus allow us to further investigate the small size effects (sub-micron and down to nanometer scales) of the fibers on the mechanical properties of the composites. Figure A2 (in Appendix A) shows the SEM image of the randomly oriented PAN fibers reinforced in PVA thin films (sample PVA SPTF-RO). This sample shows that the pp-FFES methodology gives better control in fiber orientation than compared to the electrospinning done without using parallel plate setup.

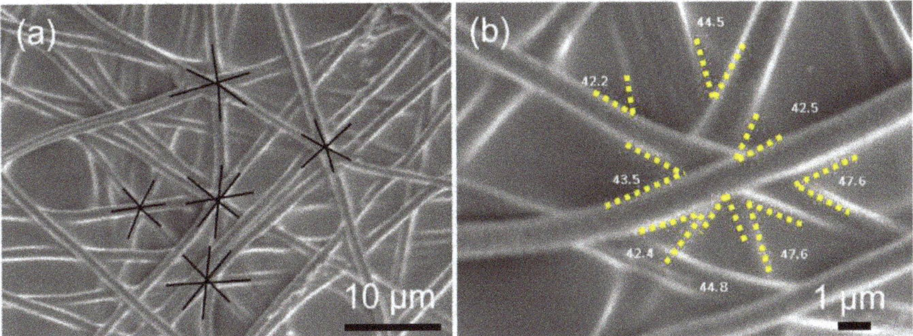

Figure 2. (**a**) SEM image depicting the morphology of fabricated PVA SPTF-HA; and (**b**) magnified image of PVA SPTF-HA with helicoidally arranged PAN fibers. Here, 13 helicoidally arranged PAN fibrous layers with 45° ± 3° offset angle are embedded layer by layer in 5 wt% PVA/H_2O solution and later dried to fabricate PVA SPTFs.

Next, the fabricated PVA SPTF samples are chemically, and thermally characterized and compared with the neat PVA thin film (PVA TF). These characterization techniques help in identifying the components in the samples as well as confirmation of the interfacial interaction between PAN and PVA. The good interfacial adhesion between the matrix and the fibers is expected to improve the mechanical deformation behavior of the PVA SPTF. The mechanical properties of PVA SPTF-HA samples are then characterized using uniaxial tensile and impact resistance experiments and compared with the PVA SPTF-RO and PVA TF, as discussed in the subsequent sections. The comparison of the mechanical properties helps in understanding the toughening and impact-resistant mechanisms of the hierarchical helicoidal arrangement of the fibers in the matrix, over the neat matrix and randomly oriented fiber reinforced thin films.

3.2. Chemical and Thermal Properties of PVA SPTF

3.2.1. Fourier-Transform Infrared Spectroscopy (FTIR)

The FTIR scan of neat PVA shows the characteristic peaks of PVA associated with -OH- stretching and –OH– bending peaks that are recorded at 3265 and 1410 cm^{-1} (see Figure 3c). The asymmetric stretching of methylene group –CH_2- is recorded at 2940 cm^{-1}. The peaks at 1090 and 1660 cm^{-1} are associated with –C–O– and –C–C– bond stretching, respectively [62]. FTIR scan recorded for the PAN fibers demonstrate the characteristic peaks associated with PAN, which include stretching vibration peaks of –CH_2- at 2930 cm^{-1}. The peak at 2244 cm^{-1} corresponds to the stretching vibrations of –C≡N–. In the case of PVA SPTF, the characteristic peaks associated with –OH– stretching decreased with the addition of PAN in PVA SPTF, which is associated with hydrogen bonding [63]. Typically, stretch peaks redshift to lower wavenumber, whereas bending peaks blue shift to a higher wavenumber due to the presence of hydrogen bonding initiating groups. The –OH– stretching peak for PVA shifts to a lower wavenumber (3255 cm^{-1}) and –OH– bending peaks shifted to a higher wavenumber (1420 cm^{-1}) when PAN is incorporated into the composite. This is attributed to the formation of hydrogen bonds between the hydroxyls of PVA and the nitriles of PAN.

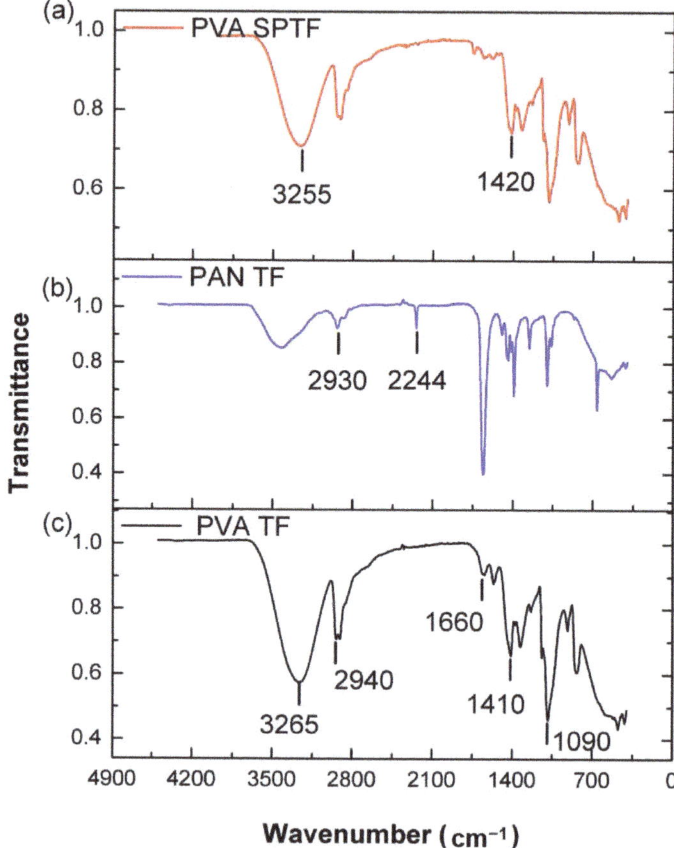

Figure 3. FTIR plots of: (**a**) PVA SPTF; (**b**) neat PAN thin film (PAN TF); and (**c**) PVA TF.

3.2.2. Thermogravimetric Analysis (TGA)

Figure 4 shows the TGA curves of neat PAN, PVA and PVA SPTF. An initial weight loss of 11% is seen for the PAN sample between 25 and 183 °C, which is attributed to the loss of moisture as well as the solvent present in the sample. Later, the PAN sample lost 3% weight in the temperature range of 183–279 °C, which can be attributed to dehydrogenation reactions and release of volatile compounds [64]. Then, the PAN macromolecule started degrading with an increase in the applied temperature from 279 to 700 °C with the final weight loss of around 60% [65]. At 700 °C, it is observed that the PAN fibers get partially carbonized. Neat PVA depicts a weight loss of 13% from room temperature to 210 °C. This weight loss can be further divided into two temperature ranges: one from room temperature to 163 °C corresponding to the loss of moisture and the other for 163–210 °C corresponding to the condensation of the hydroxyl groups in the PVA TF [66]. Later, the degradation of PVA macromolecule commences at 210 °C and finishes at 450 °C with a reduction to 10% of its initial weight [62]. The final weight loss of the PVA thin film is 93 wt% at 595 °C [67].

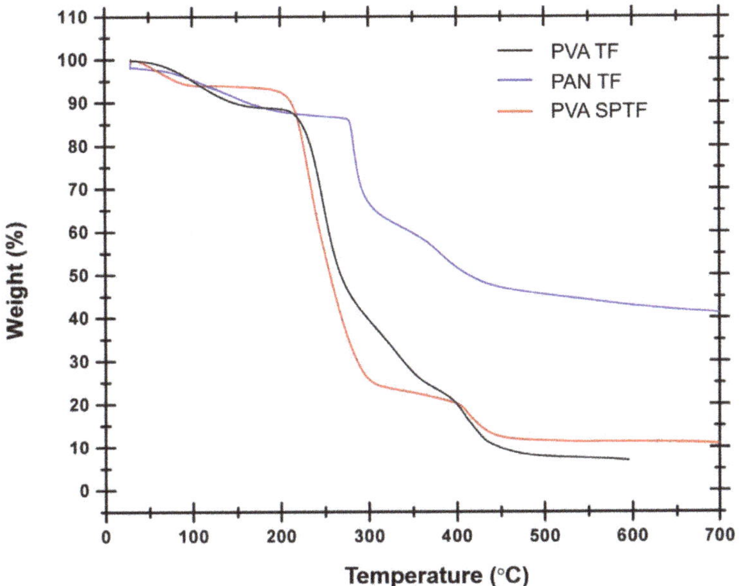

Figure 4. TGA thermograms of PVA SPTF, PAN TF and PVA TF.

Similarly, the PVA SPTF lost 9% weight during 25–210 °C. This weight loss can be further divided into two temperature ranges. The first temperature ranges from 25 to 108 °C corresponds to the loss of moisture, and the second from 108 to 210 °C corresponds to the condensation of the hydroxyl groups in the PVA SPTF. Further, the PVA SPTF show a higher degradation temperature 700 °C (9% residual weight) in comparison to the neat PVA film 595 °C (6% residual weight) due to the presence of PAN fibers.

3.2.3. Differential Scanning Calorimetry (DSC)

DSC scans are then recorded to determine the glass transition temperature (T_g) and melting point (T_m) of the samples (see Figure 5). Broadly, DSC thermograms show two endothermic peaks. The first peak at around 90 °C depicts the melting of the crystallites of the cross-linked system (PVA/H_2O) [66]. The second endothermic peak occurring at a relatively higher temperature is attributed to the melting of the PVA [68]. The lower melting temperature pertaining to the cross-linking of PVA/H_2O decreased from 91 °C for neat PVA to 75 °C for the PVA SPTF. The reduction in the melting temperature pertaining to the cross-linking is attributed to the presence of PAN fibers which hinders the cross-linking in the PVA/H_2O system. The second endothermic peak in the DSC scan of the PVA TF shows 227 °C as its melting temperature. The endothermic peak at 227 °C recorded for the neat PVA thin film shifts to a slightly higher temperature ~229 °C in the case of the PVA SPTF. This may indicate a slight increase in the crystallinity of PVA with the addition of PAN in the PVA SPTF. The dynamic mechanical analysis shows higher storage modulus for PVA SPTF as compared to PVA TF (see the Appendix A).

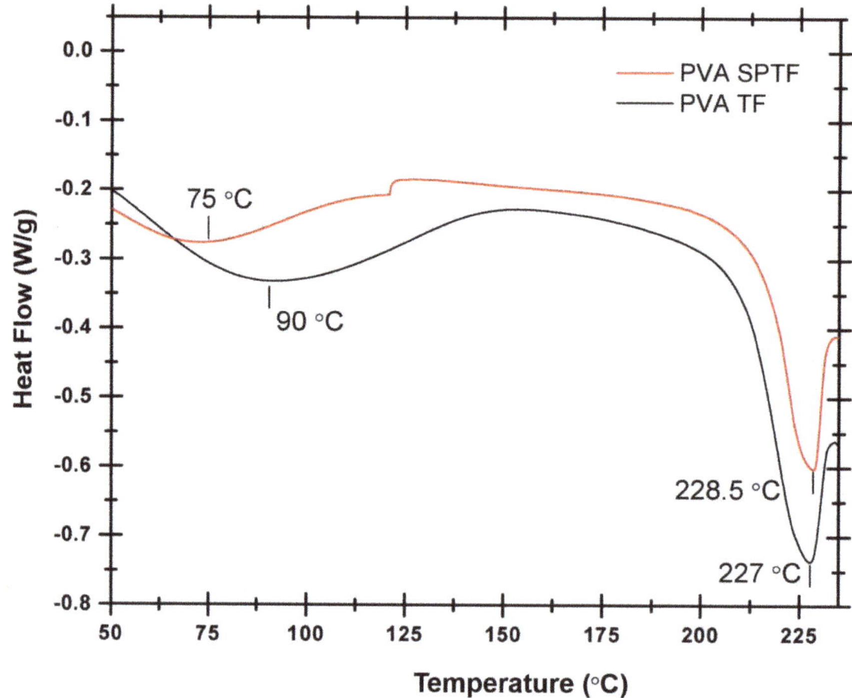

Figure 5. DSC plots of PVA TF and PVA SPTF.

3.3. Mechanical Properties of PVA SPTF

3.3.1. Tensile Properties

The mechanical properties of the reinforced thin films are influenced by the filler type and filler orientation within the matrix. Typically, polymeric thin films reinforced with unidirectionally aligned fibers demonstrate anisotropic mechanical properties, which restricts their applications, while the PVA SPTFs reinforced with a helicoidally arranged PAN fibrous structure show predominately isotropic mechanical properties, making them suitable for varying load conditions. The mechanical deformation behavior of the PVA SPTF reinforced with the helicoidally arranged PAN fibrous structure (PVA SPTF-HA) was evaluated and compared with that of the PVA thin films (PVA TF) and randomly oriented PVA SPTF (PVA SPTF-RO). Figure 6 shows the specific stress versus strain curves for the samples. Specific stress is determined by dividing the measured stress with the density of the sample to account for variation in density between composite as well as neat samples. This methodology is typically followed for fiber reinforced composites to compare them with unreinforced neat samples [48,69].

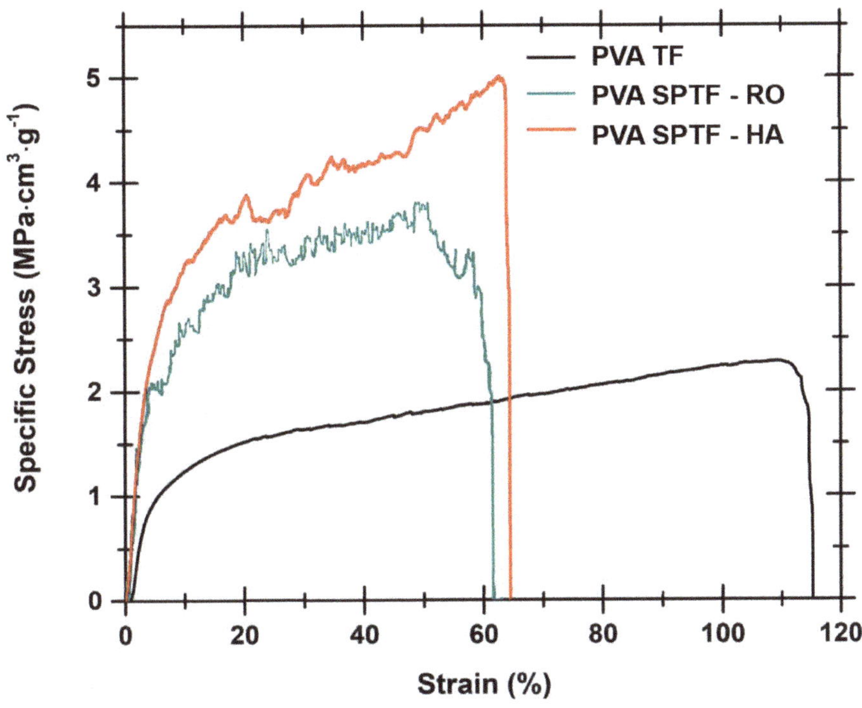

Figure 6. Specific engineering stress vs strain curves for PVA TF, PVA SPTF-HA and PVA SPTF-RO.

PVA thin film displays large elongation (Figure 6), which is typical of neat thermoplastic polymers [70,71], whereas PVA SPTF-HA exhibits higher specific tensile strength in comparison to the neat PVA thin film and PVA SPTF-RO. The high specific tensile strength of PVA SPTF-HA is attributed to the helicoidally arranged PAN fibers. In PVA SPTF-HA, hydrogen bonding between PAN and PVA can contribute towards transferring load from the stiff PAN fibers to the soft PVA membrane during tensile as well as impact loading. During uniaxial tensile loading in the sample PVA SPTF-HA, the load transfer from PAN to PVA could allow the alignment of the PAN fibers along the tensile stress direction, which can contribute to the increased specific strength of the samples (see Figure 7). Further, the helicoidal arrangement of the PAN fibers, as well as the presence of alternate soft and stiff layer along the thickness of PVA SPTF-HA, generates disparity in modulus along the plane as well as across the planes of the PVA SPTF-HA. This disparity in modulus can hinder the propagation of crack formed during loading, thereby can increase the stress required to propagate crack and cause the failure of PVA SPTF-HA. The ultimate specific tensile strength of the PVA SPTF-HA is determined to be (5 ± 0.3) MPa·cm³·g⁻¹. This is 52% and 127% higher than the specific tensile strength of PVA SPTF-RO $((3.3 \pm 0.4)$ MPa·cm³·g⁻¹$)$ and PVA TF $((2.2 \pm 0.2)$ MPa·cm³·g⁻¹$)$, respectively. Here, ultimate specific tensile strength is equal to the maximum specific tensile stress obtained from the specific stress–strain curve (see Figure 6). In addition, the specific toughness values determined by measuring the area under the specific stress and strain curve is found to be much higher for PVA SPTF-HA $((2.5 \pm 0.4)$ J/g$)$ as compared to PVA SPTF-RO $((1.5 \pm 0.4)$ J/g and PVA TF $((2 \pm 0.3)$ J/g$)$, respectively (see the Table A1 in Appendix A). The low specific toughness for PVA SPTF-RO can be attributed to the randomness of the fibers in the PVA matrix. These random fibers may not align along the tensile stress direction as in the case of PVA SPTF-HA and thus could hinder the elongation of PAN fibers along the tensile stress direction. Further, random fibers give rise to fibers entanglement and/or fiber bundles (see Figure A2 (Appendix A)). These fibers entanglement could give rise to stress concentration zones, which can initiate micro-cracks formation under tensile loading and thereby can result in early failure of PVA SPTF-RO also confined by its low specific toughness values in comparison to both PVA SPTF-HA and PVA TFs. It is interesting to note that the PVA SPTF-HA samples are filled with only ∼1 wt% PAN fibers. Nevertheless, they displayed substantial enhancement of specific tensile strength in comparison to the PVA TF and PVA SPTF-RO samples.

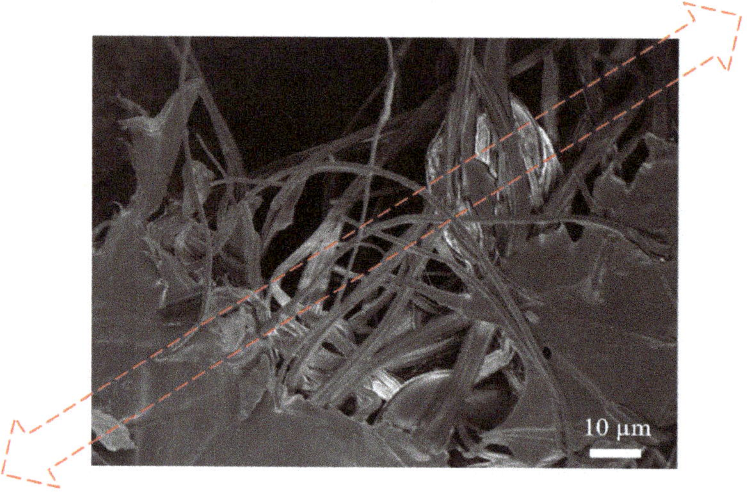

Figure 7. Representative image of the PVA SPTF-HA after the uniaxial tensile test. The figure depicts stretching as well as alignment of the PAN fibers along the tensile loading direction (shown by the red-dashed arrow in the image).

The calculated value of specific tensile strength of the PVA SPTF is found to be comparable (same order of magnitude) to the data available in the literature for different PVA composites [48,56,72,73] and other types of polymer composites [11,72,74,75]. For instance, Ding et al. [72] prepared PVA hydrogels with 2 wt% of PVA, 0.4 wt% borax (PB) and cellulose nanofibers-polypyrrole (CNFs-PPy/PB) and compared it with hydrogel synthesized in the absence of polypyrrole (CNF/PB). They found that the specific compressive stress of CNF-PPy/PB was 18.5 MPa·cm^3·g^{-1}, 1850 times higher than that of CNF/PB (0.01 MPa·cm^3·g^{-1}). Further, silica micrfiber reinforced polybenzoxazine achieved specific tensile strength 90 MPa·cm^3·g^{-1} for the silica microfiber loading as high as 40% by volume [11]. Wu et al. [74] used thermally expandable microspheres and silicon-containing arylacetylene (SA) resin to fabricate poly(silicon-arylacetylene) syntactic foams. Furthermore, the PSA foam was reinforced with attapulgite (ATT) nanoparticles. The specific compressive strength of the PSA syntactic foam was as high as 25.6 MPa·cm^3·g^{-1}. Zhang et al. [76] obtained low-density and high-performance EMI shielding carbon foam through direct carbonization of the phthalonitrile (PN)-based polymer foam with specific compressive strength of 6.0 MPa·cm^3·g^{-1}. Pilla et al. [75] fabricated and documented the effects of hyperbranched polyesters (HBPs) and nanoclay on the mechanical properties of both solid and microcellular polylactide (PLA). The addition of hyperbranched polyesters (HBPs) and nanoclay increased the specific toughness but resulted in the reduction of specific strength compared with neat PLA samples. The maximum specific strength obtained for PLA samples was 50 MPa·cm^3·g^{-1}.

3.3.2. Impact Resistance

Impact resistance of the PVA SPTF is determined using free-falling spherical steel ball impact tests [12,48]. Although for smaller samples, we lack the appropriate testing methodology (standards such as ASTM are only documented for bulk or large scale samples) to compare impact performance of the novel materials, we used free-falling ball impact testing [48] methodology to compare at least qualitatively (with other parameters kept constant) and understand the effects of the modifications we have made in terms of the synthesis of the novel materials and thus their microstructural features (such as interfacial adherence between layers as well as between fibers) on the resulting impact performances. In this case, the potential energy of the impacting steel balls is used to test the impact resistance of the samples. The impact energy can be varied either by changing the mass or height of the impacting steel balls [12]. In the first step, the samples are adhered on top of a glass coverslip. Following this, the height and mass of the steel ball required to initiate a macroscopic crack in the glass coverslip that is protected by the samples are measured. For simplicity, the mass (m) of the spherical steel ball used is kept constant at 0.51 g (diameter 4 mm). A bare glass coverslip is used as a control sample. Here, the data are summarized from the impact test of five individual samples. This control sample is seen to break when the steel ball falls from an average height (h) of (20 ± 2) cm. In the case of PVA TF, the glass coverslip is seen to fracture when the steel ball is dropped from an (80 ± 5) cm height (see Figure 8) and for PVA SPTF-RO the glass coverslip fractures at (84 ± 4) cm (see the Table A2 in Appendix A). In contrast, glass coverslip covered with the PVA SPTF-HA fractures when the steel ball is dropped from a (121 ± 7) cm height (see Figure 8). The high impact energy required for the PVA SPTF-HA $((8 \pm 0.9) \text{ mJ} \cdot \text{cm}^3 \cdot \text{g}^{-1})$ in comparison to the PVA SPTF-RO $((4.2 \pm 0.5) \text{ mJ} \cdot \text{cm}^3 \cdot \text{g}^{-1})$ and PVA TF $((3.4 \pm 0.3) \text{ mJ} \cdot \text{cm}^3 \cdot \text{g}^{-1})$ to break the glass coverslip is attributed to both its efficacy and efficiency in dissipating energy through alternatively arranged layers of the soft PVA and relatively stiff PAN fibers in a plane as well as across the plane. Trends from both of these data sets (height when glass coverslip's fracture is observed and impact energy required) are consistent with our earlier observation with the specific toughness (significantly higher for PVA SPTF-HA compared to other samples), which is also a measure of capacity of the samples to absorb damage/energy during deformation.

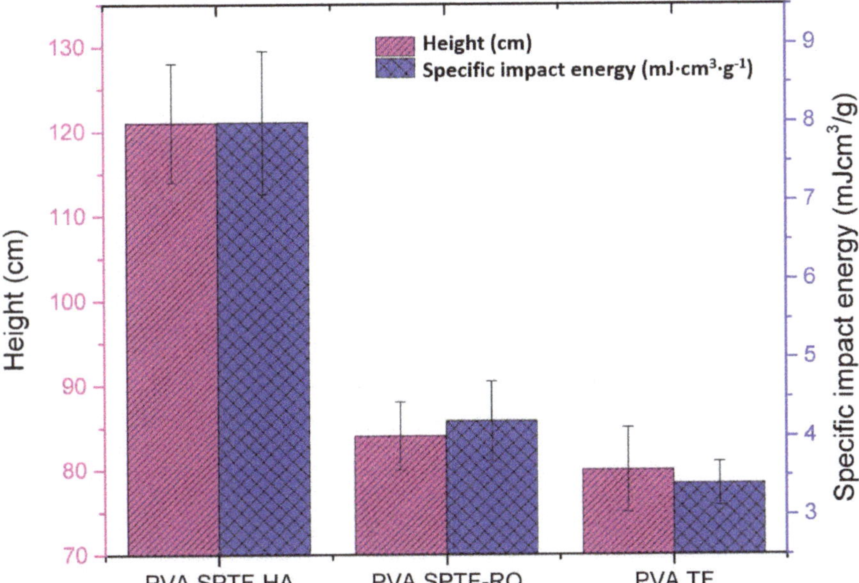

Figure 8. The plot show heights of the free-falling ball, and its corresponding impact energy required to break the glass coverslip protected by the samples. Each bar in the plot depicts the data summarized from the impact test of five individual samples. The error bars represent the range of the data presented in the bar chart.

This phenomenon can be further explained by assuming that the potential energy of the impacting steel ball was sufficient to deform the first stiff PAN fibrous layer along the thickness of the PVA

SPTF-HA. After deforming the PAN fibrous layer, a portion of available potential energy typically dissipates to the adjacent soft PVA matrix material. Therefore, the remaining potential energy may not be enough for deforming the next stiff PAN fibrous layer. Further, these helicoidally arranged PAN fibers intensify the disparity in modulus along the thickness of the PVA SPTF-HA, making it difficult for the deformation induced crack to propagate through it, thus providing high impact resistance in comparison to the PVA thin films as well as PVA SPTF-RO. In the case of PVA SPTF-RO, random PAN fibers could not provide sufficient resistance to the applied potential energy as was the case for PVA SPTF-HA. The applied potential energy is predominately used for the deformation of soft PVA matrix material, which later transfers the energy to the glass slip (protected by PVA SPTF-RO) and resulting in the initiation of cracking in the glass slip at a relatively low applied potential energy.

By virtue of the layer-by-layer approaches of additive manufacturing, the novel fabrication methodology as reported in this present manuscript has shown its broad scope to allow control or modulation in the interfacial adherence between fibers and between layers. Such a capability is important for the design and development of novel materials with highly tunable impact resistances.

4. Conclusions

High strength and lightweight PVA SPTF were prepared by embedding helicoidally arranged uniaxially aligned electrospun PAN fibers in PVA through pp-FFES. Tensile and impact tests were performed to verify the specific strength and impact resistance of the PVA SPTF-HAs. The result shows that the PVA SPTF-HAs possess better specific tensile strength and higher resistance to impact compared to the corresponding neat PVA thin films and randomly fiber oriented PVA SPTF. The excellent mechanical properties such as high specific strength and impact resistance of the PVA SPTF-HAs are attributed to its efficient dissipation of energy as well as load transfer during impact and/or uniaxial loading by virtue of its helicoidally arranged PAN fibers network in the PVA thin films (enabled by electrospinning-based processing as an additive manufacturing methodology). Such high specific strength and impact resistant helicoidally arranged fiber reinforced PVA SPTF may have a wide variety of applications ranging from army combat wear and sports gear to novel polycarbonate-based solar photovoltaics (PV) modules.

Author Contributions: Conceptualization, A.B. and A.S.B.; experiments, K.A. and R.S. (equal contribution); writing and review, K.A. and R.S. (equal contribution); editing, A.B. and A.S.B.; experimental assistance, A.S.; and resources and project administration, N.R. All authors have read and agreed to the published version of the manuscript.

Funding: This work is supported by La Trobe University Leadership RFA Grant, La Trobe University Start-up Grant and the Collaboration and Research Engagement (CaRE) Grant offered by the School of Engineering and Mathematical Sciences (SEMS), La Trobe University. The authors would like to acknowledge the funding from the Ministry of Education (MOE) Academic Research Funds Tier 2 titled "Materials with Tunable Impact Resistance via Integrated Additive Manufacturing" (MOE2017-T2-2-175). The authors also acknowledge receipt of funding support from TEMASEK LAB@SUTD Singapore, through its SEED grant program for the project IGDS S15 01011 and SMART (Singapore-MIT Alliance for Research and Technology) through its Ignition grant program for the project SMART ING-000067 ENG IGN.

Acknowledgments: The authors gratefully acknowledge the critical support and infrastructure provided by the Singapore University of Technology and Design (SUTD). A.S.B. gratefully acknowledges PT Impack Pratama Indonesia for providing materials, exploratory supports and technical discussions related to possible implementations of the materials development for the applications in building-integrated PV (photovoltaics) technology/design.

Conflicts of Interest: The authors declare no conflict of interest.

Abbreviations

The following abbreviations are used in this manuscript:

pp-FFES	Parallel plate far field electrospinning
TF	Thin films
SPTF	Strong polymer thin films
PAN	Polyacrylonitrile
PVA	polyvinyl alcohol
PVA SPTF-RO	PVA strong polymer thin films with randomly oriented PAN fibers
PVA SPTF-HA	PVA strong polymer thin films with helicoidally arranged PAN fibers

Appendix A

The calculations were based on cross-sectional area = $t \times w$, where t is the thickness of the sample, w is the width of the sample, tensile stress = applied load/cross-sectional area, density of the sample = mass/volume and specific tensile stress = tensile stress/density. Ultimate specific tensile strength used for the calculations of specific toughness is equal to maximum specific tensile stress obtained from the specific stress -strain curve and the stain is measured at complete failure.

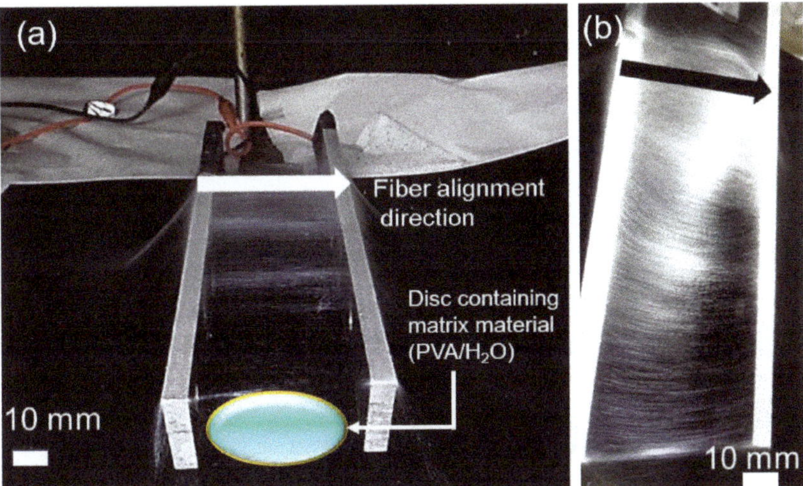

Figure A1. (a) Parallel plate far field electrospinning methodology used for the fabrication of uniaxially aligned PAN fibers. (b) These uniaxially aligned fibers were then collected in the PVA solution. The process is repeated 13 times to generate PVA SPTF-HA. The arrow in the figure shows the fiber alignment direction.

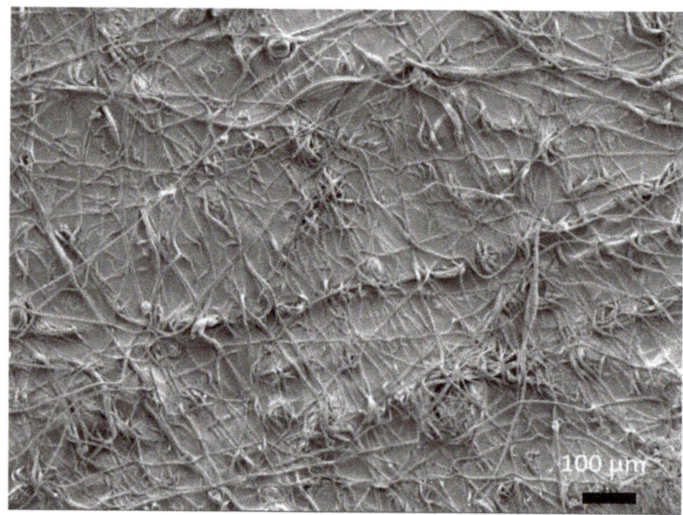

Figure A2. PVA SPTF-RO prepared using conventional electrospinning methodology.

Table A1. Specific tensile properties of the samples PVA TF, PVA SPTF-RO and PVA SPTF-HA.

Sample	Ultimate Specific Tensile Strength (MPa·cm^3·g^{-1})	Strain (%)	Specific Toughness (J·g^{-1})	Density (g·cm^{-3})
PVA TF	2.2 ± 0.2	114 ± 3	2.2 ± 0.3	1.1
PVA SPTF-RO	3.3 ± 0.4	61.4 ± 6	1.5 ± 0.4	1
PVA SPTF-HA	5 ± 0.3	64.4 ± 14	2.5 ± 0.4	0.7

Table A2. Specific tensile properties of the samples PVA TF, PVA SPTF-RO and PVA SPTF-HA.

Sample	Impact Height at Which Glass Beneath Sample Breaks (cm)	Specific Impact Energy (mJ·cm^3·g^{-1})
Coverslip	20 ± 2	-
PVA TF	80 ± 5	3.4 ± 0.3
PVA SPTF-RO	84 ± 4	4.2 ± 0.5
PVA SPTF-HA	120 ± 7	8 ± 0.9

The impact energy of the glass sample is 1 mJ, for PVA TF is 4 mJ, for PVA SPTF-RO is 4 mJ and for PVA SPTF-HA is 6 mJ.

Dynamic mechanical analysis (DMA): The viscoelastic properties of the PVA SPTF and PVA TF are shown in Figure A3. The storage modulus is an indicator of the energy stored during the deformation. The specific viscoelastic properties of the material are determined by dividing the modulus by the density of the samples. The specific storage modulus of the PVA SPTF is determined to be 2650 MPa·cm^3·g^{-1}, while for PVA TF the value is determined to be 1700 MPa·cm^3·g^{-1}. Over the entire temperature range, the specific storage modulus and the loss modulus values of PVA SPTF are higher than PVA TF. For PVA TF, the first tan delta peak is observed at 32 °C corresponding to the α phase relaxation, and there is a second one at 112 °C, corresponding to the β relaxation. The tan delta peaks of the PVA SPTF are seen at 29 °C and 99 °C. The T_g of the PVA SPTF reduces in comparison to the PVA TF. These results agree with our DSC results indicating that the presence of PAN fibers hinders in the crosslinking of PVA/H$_2$O system.

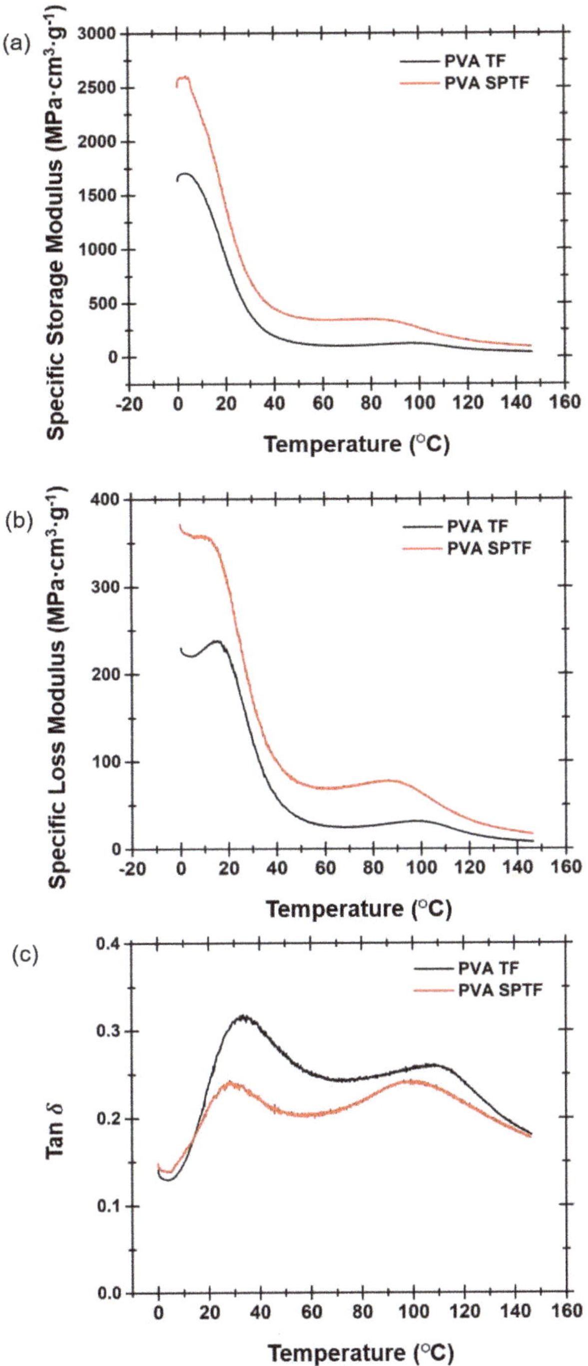

Figure A3. Specific viscoelastic properties of the PVA SPTF and PVA TF- (**a**) storage modulus; (**b**) loss modulus; (**c**) tan delta.

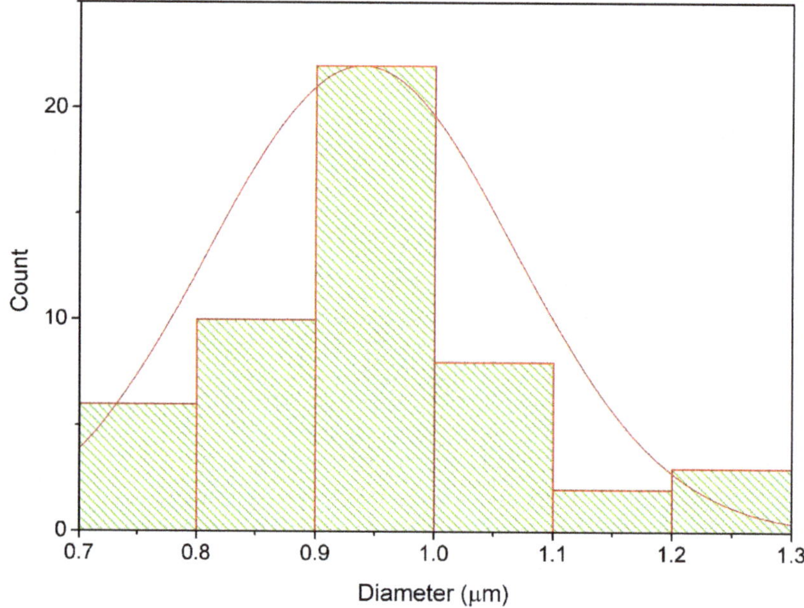

Figure A4. The diameter distribution of PVA SPTF-HA prepared through parallel plate far field electrospinning setup.

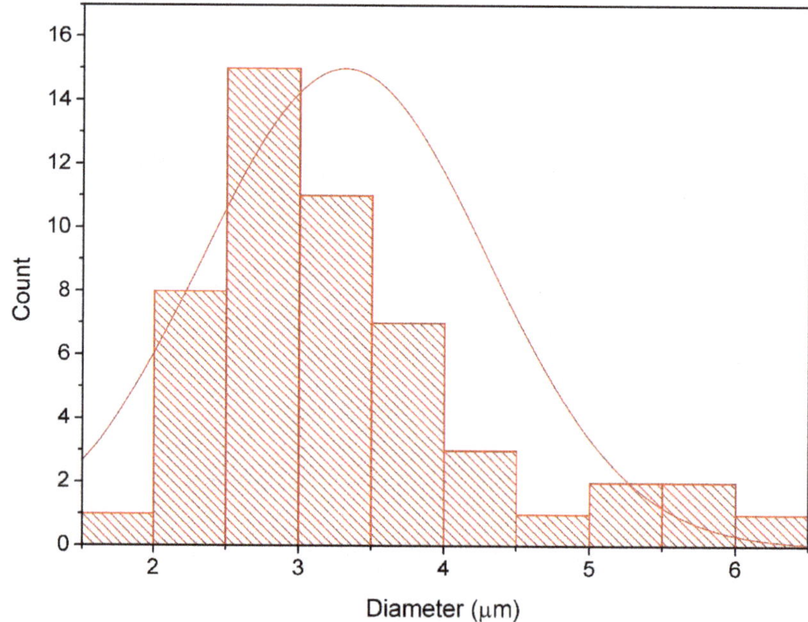

Figure A5. The diameter distribution of PVA SPTF-RO prepared using conventional electrospinning methodology.

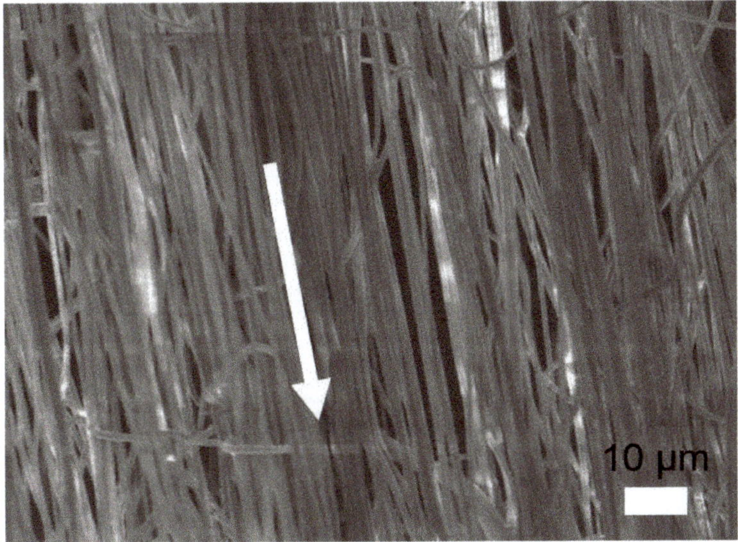

Figure A6. The SEM image of thick aligned PAN fibrous membrane obtained after electrospinning performed for an extended period of time using the parallel-plate setup shown in Figure A1. The white arrow shows the direction of fiber alignment.

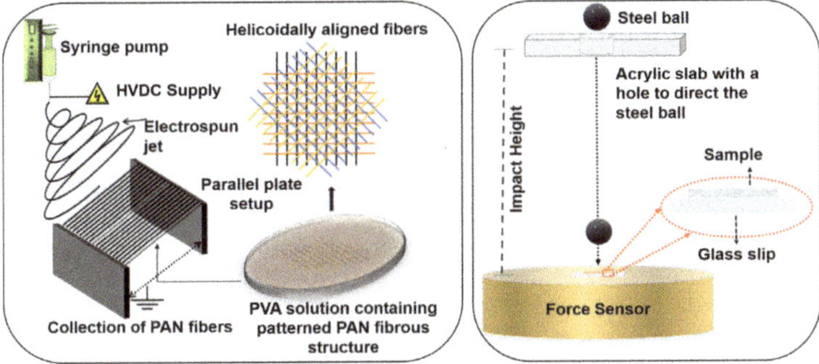

Figure A7. Graphical Abstract.

References

1. Klauk, H. *Organic Electronics: Materials, Manufacturing, and Applications*; John Wiley & Sons: Hoboken, NJ, USA, 2006.
2. Ouyang, J.; Chu, C.W.; Szmanda, C.R.; Ma, L.; Yang, Y. Programmable Polymer Thin Film and Non-Volatile Memory Device. *Nat. Mater.* **2004**, *3*, 918–922. [CrossRef] [PubMed]
3. Wu, J.; Li, P.; Dong, C.; Jiang, H.; Xue, B.; Gao, X.; Qin, M.; Wang, W.; Chen, B.; Cao, Y. Rationally Designed Synthetic Protein Hydrogels with Predictable Mechanical Properties. *Nat. Commun.* **2018**, *9*, 1–11. [CrossRef]
4. Baji, A.; Agarwal, K.; Oopath, S.V. Emerging Developments in the Use of Electrospun Fibers and Membranes for Protective Clothing Applications. *Polymers* **2020**, *12*, 492. [CrossRef]
5. Vadukumpully, S.; Paul, J.; Mahanta, N.; Valiyaveettil, S. Flexible Conductive Graphene/Poly (vinyl chloride) Composite Thin Films With High Mechanical Strength and Thermal Stability. *Carbon* **2011**, *49*, 198–205. [CrossRef]
6. Xu, X.; Thwe, M.M.; Shearwood, C.; Liao, K. Mechanical Properties and Interfacial Characteristics of Carbon-Nanotube-Reinforced Epoxy Thin Films. *Appl. Phys. Lett.* **2002**, *81*, 2833–2835. [CrossRef]

7.   Sakurai, H.; Moriguchi, K.; Katayama, Y. Inorganic Filler-Incorporated Ethylene Polymer Film. U.S. Patent 4,219,453, 26 August 1980.

8.   Campoy-Quiles, M.; Ferenczi, T.; Agostinelli, T.; Etchegoin, P.G.; Kim, Y.; Anthopoulos, T.D.; Stavrinou, P.N.; Bradley, D.D.; Nelson, J. Morphology Evolution via Self-Organization and Lateral and Vertical Diffusion in Polymer: Fullerene Solar Cell Blends. *Nat. Mater.* **2008**, *7*, 158–164. [CrossRef] [PubMed]

9.   Cramm, J.A. Increased Strength Polymer-Blended Membranes. U.S. Patent 4,609,468, 2 September 1986.

10. Lovinger, A.J.; Williams, M. Tensile Properties and Morphology of Blends of Polyethylene and Polypropylene. *J. Appl. Polym. Sci.* **1980**, *25*, 1703–1713. [CrossRef]

11. Santhosh Kumar, K.; Reghunadhan Nair, C.; Ninan, K. Silica Fiber-Polybenzoxazine-Syntactic Foams; Processing and Properties. *J. Appl. Polym. Sci.* **2008**, *107*, 1091–1099. [CrossRef]

12. Chen, R.; Liu, J.; Yang, C.; Weitz, D.A.; He, H.; Li, D.; Chen, D.; Liu, K.; Bai, H. Transparent Impact-Resistant Composite Films with Bioinspired Hierarchical Structure. *ACS Appl. Mater. Interfaces* **2019**, *11*, 23616–23622. [CrossRef]

13. Bosia, F.; Abdalrahman, T.; Pugno, N.M. Investigating the Role of Hierarchy on the Strength of Composite Materials: Evidence of a Crucial Synergy Between Hierarchy and Material Mixing. *Nanoscale* **2012**, *4*, 1200–1207. [CrossRef]

14. Ji, B.; Gao, H. Mechanical Properties of Nanostructure of Biological Materials. *J. Mech. Phys. Solids* **2004**, *52*, 1963–1990. [CrossRef]

15. Sakhavand, N.; Shahsavari, R. Universal Composition-Structure-Property Maps for Natural and Biomimetic Platelet-Matrix Composites and Stacked Heterostructures. *Nat. Commun.* **2015**, *6*, 1–13.

16. Patek, S.; Caldwell, R. Extreme Impact and Cavitation Forces of a Biological Hammer: Strike Forces of the Peacock Mantis Shrimp Odontodactylus Scyllarus. *J. Exp. Biol.* **2005**, *208*, 3655–3664. [CrossRef] [PubMed]

17. Agarwal, K.; Sahay, R.; Baji, A.; Budiman, A.S. Biomimetic Tough Helicoidally Structured Material Through Novel Electrospinning Based Additive Manufacturing. *MRS Adv.* **2019**, *4*, 2345–2354. [CrossRef]

18. Patek, S.N.; Korff, W.; Caldwell, R.L. Deadly Strike Mechanism of a Mantis Shrimp. *Nature* **2004**, *428*, 819–820. [CrossRef] [PubMed]

19. Wilts, B.D.; Whitney, H.M.; Glover, B.J.; Steiner, U.; Vignolini, S. Natural Helicoidal Structures: Morphology, Self-Assembly and Optical Properties. *Mater. Today Proc.* **2014**, *1*, 177–185. [CrossRef]

20. Weaver, J.C.; Milliron, G.W.; Miserez, A.; Evans-Lutterodt, K.; Herrera, S.; Gallana, I.; Mershon, W.J.; Swanson, B.; Zavattieri, P.; DiMasi, E.; et al. The Stomatopod Dactyl Club: A Formidable Damage-Tolerant Biological Hammer. *Science* **2012**, *336*, 1275–1280. [CrossRef]

21. Shun, S.J. Study on the Structural Strength of the Mantis Shrimp. Ph.D. Thesis, National University of Singapore, Singapore, 2014.

22. Taylor, J.; Patek, S. Ritualized Fighting and Biological Armor: The Impact Mechanics of the Mantis Shrimp's Telson. *J. Exp. Biol.* **2010**, *213*, 3496–3504. [CrossRef]

23. Launey, M.E.; Buehler, M.J.; Ritchie, R.O. On the Mechanistic Origins of Toughness in Bone. *Annu. Rev. Mater. Res.* **2010**, *40*, 25–53. [CrossRef]

24. Fratzl, P. Biomimetic Materials Research: What Can We Really Learn from Nature's Structural Materials? *J. R. Soc. Interface* **2007**, *4*, 637–642. [CrossRef]

25. Aizenberg, J.; Fratzl, P. Biological and Biomimetic Materials. *Adv. Mater.* **2009**, *21*, 387–388. [CrossRef]

26. Bar-Cohen, Y. Biomimetics—Using Nature to Inspire Human Innovation. *Bioinspiration Biomimetics* **2006**, *1*, P1. [CrossRef] [PubMed]

27. Cheng, L.; Thomas, A.; Glancey, J.L.; Karlsson, A.M. Mechanical Behavior of Bio-Inspired Laminated Composites. *Compos. Part A Appl. Sci. Manuf.* **2011**, *42*, 211–220. [CrossRef]

28. Grunenfelder, L.; Suksangpanya, N.; Salinas, C.; Milliron, G.; Yaraghi, N.; Herrera, S.; Evans-Lutterodt, K.; Nutt, S.; Zavattieri, P.; Kisailus, D. Bio-Inspired Impact-Resistant Composites. *Acta Biomater.* **2014**, *10*, 3997–4008. [CrossRef]

29. Ribbans, B.; Li, Y.; Tan, T. A Bioinspired Study on the Interlaminar Shear Resistance of Helicoidal Fiber Structures. *J. Mech. Behav. Biomed. Mater.* **2016**, *56*, 57–67. [CrossRef]

30. Tan, T.; Ribbans, B. A Bioinspired Study on the Compressive Resistance of Helicoidal Fibre Structures. *Proc. R. Soc. A Math. Phys. Eng. Sci.* **2017**, *473*, 20170538. [CrossRef] [PubMed]

31. Yaraghi, N.A.; Guarín-Zapata, N.; Grunenfelder, L.K.; Hintsala, E.; Bhowmick, S.; Hiller, J.M.; Betts, M.; Principe, E.L.; Jung, J.Y.; Sheppard, L.; et al. A Sinusoidally Architected Helicoidal Biocomposite. *Adv. Mater.* **2016**, *28*, 6835–6844. [CrossRef] [PubMed]

32. Nix, W.D.; Gao, H. Indentation Size Effects in Crystalline Materials: A Law For Strain Gradient plasticity. *J. Mech. Phys. Solids* **1998**, *46*, 411–425. [CrossRef]

33. Greer, J.R.; Oliver, W.C.; Nix, W.D. Size Dependence of Mechanical Properties of Gold at the Micron Scale in the Absence of Strain Gradients. *Acta Mater.* **2005**, *53*, 1821–1830. [CrossRef]

34. Greer, J.R.; De Hosson, J.T.M. Plasticity in Small-Sized Metallic Systems: Intrinsic Versus Extrinsic Size Effect. *Prog. Mater. Sci.* **2011**, *56*, 654–724. [CrossRef]

35. Budiman, A.; Nix, W.; Tamura, N.; Valek, B.; Gadre, K.; Maiz, J.; Spolenak, R.; Patel, J. Crystal Plasticity in Cu Damascene Interconnect Lines Undergoing Electromigration as Revealed by Synchrotron X-Ray Microdiffraction. *Appl. Phys. Lett.* **2006**, *88*, 233515. [CrossRef]

36. Budiman, A.; Han, S.; Greer, J.; Tamura, N.; Patel, J.; Nix, W. A Search for Evidence of Strain Gradient Hardening in Au Submicron Pillars Under Uniaxial Compression Using Synchrotron X-Ray Microdiffraction. *Acta Mater.* **2008**, *56*, 602–608. [CrossRef]

37. Lu, K.; Lu, L.; Suresh, S. Strengthening Materials by Engineering Coherent Internal Boundaries at the Nanoscale. *Science* **2009**, *324*, 349–352. [CrossRef] [PubMed]

38. Radchenko, I.; Anwarali, H.; Tippabhotla, S.; Budiman, A. Effects of Interface Shear Strength During Failure of Semicoherent Metal-Metal Nanolaminates: An Example of Accumulative Roll-Bonded Cu/Nb. *Acta Mater.* **2018**, *156*, 125–135. [CrossRef]

39. Ali, H.P.A.; Radchenko, I.; Li, N.; Budiman, A. The Roles of Interfaces and Other Microstructural Features in Cu/Nb Nanolayers as Revealed by In Situ Beam Bending Experiments Inside an Scanning Electron Microscope (SEM). *Mater. Sci. Eng. A* **2018**, *738*, 253–263.

40. Ali, H.P.A.; Budiman, A. Advances in In situ Microfracture Experimentation Techniques: A Case of Nanoscale Metal-Metal Multilayered Materials. *J. Mater. Res.* **2019**, *34*, 1449–1468.

41. Tippabhotla, S.K.; Radchenko, I.; Rengarajan, K.N.; Illya, G.; Handara, V.; Kunz, M.; Tamura, N.; Budiman, A.S. Synchrotron X-ray Micro-Diffraction-Probing Stress State in Encapsulated Thin Silicon Solar Cells. *Procedia Eng.* **2016**, *139*, 123–133. [CrossRef]

42. Jiang, S.; Chen, Y.; Duan, G.; Mei, C.; Greiner, A.; Agarwal, S. Electrospun Nanofiber Reinforced Composites: A Review. *Polym. Chem.* **2018**, *9*, 2685–2720. [CrossRef]

43. Sahay, R.; Teo, C.; Chew, Y. New Correlation Formulae for the Straight Section of the Electrospun Jet from a Polymer Drop. *J. Fluid Mech.* **2013**, *735*, 150. [CrossRef]

44. Sahay, R.; Reddy, V.J.; Ramakrishna, S. Synthesis and Applications of Multifunctional Composite Nanomaterials. *Int. J. Mech. Mater. Eng.* **2014**, *9*, 25. [CrossRef]

45. Li, D.; Xia, Y. Electrospinning of Nanofibers: Reinventing the Wheel? *Adv. Mater.* **2004**, *16*, 1151–1170. [CrossRef]

46. Yan, H.; Liu, L.; Zhang, Z. Alignment of Electrospun Nanofibers Using Dielectric Materials. *Appl. Phys. Lett.* **2009**, *95*, 143114. [CrossRef]

47. Ray, B. Temperature Effect During Humid Ageing on Interfaces of Glass and Carbon Fibers Reinforced Epoxy Composites. *J. Colloid Interface Sci.* **2006**, *298*, 111–117. [CrossRef]

48. Agarwal, K.; Sahay, R.; Baji, A.; Budiman, A.S. Impact Resistant and Tough Helicoidally Aligned Ribbon Reinforced Composite with Tunable Mechanical Properties via Integrated Additive Manufacturing Methodologies. *ACS Appl. Polym. Mater.* **2020**, *2*, 3491–3504 [CrossRef]

49. Pal, K.; Banthia, A.K.; Majumdar, D.K. Preparation and Characterization of Polyvinyl Alcohol-Gelatin Hydrogel Membranes for Biomedical Applications. *AAPS PharmSciTech* **2007**, *8*, E142–E146. [CrossRef] [PubMed]

50. Zhang, X.; Liu, T.; Sreekumar, T.; Kumar, S.; Moore, V.C.; Hauge, R.H.; Smalley, R.E. Poly (vinyl alcohol)/SWNT Composite Film. *Nano Lett.* **2003**, *3*, 1285–1288. [CrossRef]

51. Kamoun, E.A.; Kenawy, E.R.S.; Chen, X. A Review on Polymeric Hydrogel Membranes for Wound Dressing Applications: PVA-Based Hydrogel Dressings. *J. Adv. Res.* **2017**, *8*, 217–233. [CrossRef]

52. Gaaz, T.S.; Sulong, A.B.; Akhtar, M.N.; Kadhum, A.A.H.; Mohamad, A.B.; Al-Amiery, A.A. Properties and Applications of Polyvinyl alcohol, Halloysite Nanotubes and Their Nanocomposites. *Molecules* **2015**, *20*, 22833–22847. [CrossRef]

53. Ismail, H.; Zaaba, N.F. Effect of Polyvinyl Alcohol on Tensile Properties and Morphology of Sago Starch Plastic Films. In Proceedings of the 2011 National Postgraduate Conference, Kuala Lumpur, Malaysia, 19–20 September 2011; pp. 1–3.

54. Wu, Z.; Wu, J.; Peng, T.; Li, Y.; Lin, D.; Xing, B.; Li, C.; Yang, Y.; Yang, L.; Zhang, L.; et al. Preparation and Application of Starch/Polyvinyl alcohol/Citric Acid Ternary Blend Antimicrobial Functional Food Packaging Films. *Polymers* **2017**, *9*, 102. [CrossRef]

55. Noshirvani, N.; Ghanbarzadeh, B.; Fasihi, H.; Almasi, H. Starch-PVA Nanocomposite Film Incorporated with Cellulose Nanocrystals and MMT: A Comparative Study. *Int. J. Food Eng.* **2016**, *12*, 37–48. [CrossRef]

56. Agarwal, K.; Sahay, R.; Baji, A. Tensile Properties of Composite Reinforced with Three-Dimensional Printed Fibers. *Polymers* **2020**, *12*, 1089. [CrossRef] [PubMed]

57. Brandrup, J.; Immergut, E.H.; Grulke, E.A. *Handbook, Polymer*; John Wiley & Sons: Hoboken, NJ, USA, 1999.

58. Zhu, G.; Wang, F.; Xu, K.; Gao, Q.; Liu, Y. Study on Properties of Poly (vinyl alcohol)/Polyacrylonitrile Blend Film. *Polímeros* **2013**, *23*, 146–151. [CrossRef]

59. Raney, J.R.; Compton, B.G.; Mueller, J.; Ober, T.J.; Shea, K.; Lewis, J.A. Rotational 3D Printing of Damage-Tolerant Composites with Programmable Mechanics. *Proc. Natl. Acad. Sci. USA* **2018**, *115*, 1198–1203. [CrossRef] [PubMed]

60. Feilden, E.; Ferraro, C.; Zhang, Q.; García-Tu nón, E.; D'Elia, E.; Giuliani, F.; Vandeperre, L.; Saiz, E. 3D Printing Bioinspired Ceramic Composites. *Sci. Rep.* **2017**, *7*, 1–9. [CrossRef] [PubMed]

61. Studart, A.R. Additive Manufacturing of Biologically-Inspired Materials. *Chem. Soc. Rev.* **2016**, *45*, 359–376. [CrossRef] [PubMed]

62. Mishra, S.K.; Kannan, S. Development, Mechanical Evaluation and Surface Characteristics of Chitosan/Polyvinyl alcohol Based Polymer Composite Coatings on Titanium Metal. *J. Mech. Behav. Biomed. Mater.* **2014**, *40*, 314–324. [CrossRef]

63. Zhou, W.; Yan, X.; Jiang, M.; Liu, P.; Xu, J. Study of Flame-Resistant Acrylic Fibers Reinforced by Poly (vinyl alcohol). *J. Appl. Polym. Sci.* **2016**, *133*. [CrossRef]

64. Ribeiro, R.F.; Pardini, L.C.; Alves, N.P.; Júnior, B.; Rios, C.A. Thermal Stabilization Study of Polyacrylonitrile Fiber Obtained by Extrusion. *Polimeros* **2015**, *25*, 523–530. [CrossRef]

65. Zhang, W.X.; Wang, Y.Z.; Sun, C.F. Characterization on Oxidative Stabilization of Polyacrylonitrile Nanofibers Prepared by Electrospinning. *J. Polym. Res.* **2007**, *14*, 467–474. [CrossRef]

66. Guo, Z.; Zhang, D.; Wei, S.; Wang, Z.; Karki, A.B.; Li, Y.; Bernazzani, P.; Young, D.P.; Gomes, J.; Cocke, D.L.; et al. Effects of Iron Oxide Nanoparticles on Polyvinyl alcohol: Interfacial Layer and Bulk Nanocomposites Thin Film. *J. Nanopart. Res.* **2010**, *12*, 2415–2426. [CrossRef]

67. Yang, C.C. Synthesis and Characterization of the Cross-Linked PVA/TiO_2 Composite Polymer Membrane for Alkaline DMFC. *J. Membr. Sci.* **2007**, *288*, 51–60. [CrossRef]

68. Agrawal, l.L.; Awadhia, A. DSC and Conductivity Studies on PVA Based Proton Conducting Gel Electrolytes. *Bull. Mater. Sci.* **2004**, *27*, 523–527. [CrossRef]

69. Qiu, Y.; Xu, W.; Wang, Y.; Zikry, M.A.; Mohamed, M.H. Fabrication and Characterization of Three-Dimensional Cellular-Matrix Composites Reinforced with Woven Carbon Fabric. *Compos. Sci. Technol.* **2001**, *61*, 2425–2435. [CrossRef]

70. Billmeyer, F.W.; Billmeyer, F.W. *Textbook of Polymer Science*; Wiley: New York, NY, USA, 1984; Volume 19842.

71. Arends, C. *Polymer Toughening*; CRC Press: Boca Raton, FL, USA, 1996; Volume 30.

72. Ding, Q.; Xu, X.; Yue, Y.; Mei, C.; Huang, C.; Jiang, S.; Wu, Q.; Han, J. Nanocellulose-Mediated Electroconductive Self-Healing Hydrogels With High Strength, Plasticity, Viscoelasticity, Stretchability, and Biocompatibility Toward Multifunctional Applications. *ACS Appl. Mater. Interfaces* **2018**, *10*, 27987–28002. [CrossRef] [PubMed]

73. Blond, D.; Walshe, W.; Young, K.; Blighe, F.M.; Khan, U.; Almecija, D.; Carpenter, L.; McCauley, J.; Blau, W.J.; Coleman, J.N. Strong, Tough, Electrospun Polymer-Nanotube Composite Membranes with Extremely Low Density. *Adv. Funct. Mater.* **2008**, *18*, 2618–2624. [CrossRef]

74. Wu, J.; Dong, S.; Yuan, Q. *Preparation and Properties of Poly (Silicon-Containing Arylacetylene)/Attapulgite Foams*; IOP Conference Series: Materials Science and Engineering; IOP Publishing: Bristol, UK, 2020; Volume 735, p. 012055.

75. Pilla, S.; Kramschuster, A.; Lee, J.; Clemons, C.; Gong, S.; Turng, L.S. Microcellular Processing of Polylactide–Hyperbranched Polyester–Nanoclay Composites. *J. Mater. Sci.* **2010**, *45*, 2732–2746. [CrossRef]
76. Zhang, L.; Liu, M.; Roy, S.; Chu, E.K.; See, K.Y.; Hu, X. Phthalonitrile-Based Carbon Foam With High Specific Mechanical Strength and Superior Electromagnetic Interference Shielding Performance. *ACS Appl. Mater. Interfaces* **2016**, *8*, 7422–7430. [CrossRef]

Publisher's Note: MDPI stays neutral with regard to jurisdictional claims in published maps and institutional affiliations.

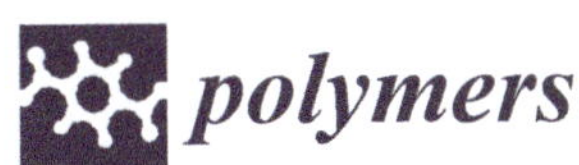 *polymers*

Article

Deformation Process of 3D Printed Structures Made from Flexible Material with Different Values of Relative Density

Paweł Płatek [1,*], Kamil Rajkowski [1], Kamil Cieplak [1], Marcin Sarzyński [1], Jerzy Małachowski [2], Ryszard Woźniak [1] and Jacek Janiszewski [1]

[1] Faculty of Mechatronics and Aerospace, Military University of Technology, 2 Gen. S. Kaliskiego Street, 00-908 Warsaw, Poland; kamil.rajkowski@wat.edu.pl (K.R.); kamil.cieplak@wat.edu.pl (K.C.); marcin.sarzynski@wat.edu.pl (M.S.); ryszard.wozniak@wat.edu.pl (R.W.); jacek.janiszewski@wat.edu.pl (J.J.)

[2] Faculty of Mechanical Engineering, Military University of Technology, 2 Gen. S. Kaliskiego Street, 00-908 Warsaw, Poland; jerzy.malachowski@wat.edu.pl

* Correspondence: pawel.platek@wat.edu.pl; Tel.: +48-261-839-657

Received: 29 August 2020; Accepted: 14 September 2020; Published: 17 September 2020

Abstract: The main aim of this article is the analysis of the deformation process of regular cell structures under quasi-static load conditions. The methodology used in the presented investigations included a manufacturability study, strength tests of the base material as well as experimental and numerical compression tests of developed regular cellular structures. A regular honeycomb and four variants with gradually changing topologies of different relative density values have been successfully designed and produced in the TPU-Polyflex flexible thermoplastic polyurethane material using the Fused Filament Fabrication (FFF) 3D printing technique. Based on the results of performed technological studies, the most productive and accurate 3D printing parameters for the thermoplastic polyurethane filament were defined. It has been found that the 3D printed Polyflex material is characterised by a very high flexibility (elongation up to 380%) and a non-linear stress-strain relationship. A detailed analysis of the compression process of the structure specimens revealed that buckling and bending were the main mechanisms responsible for the deformation of developed structures. The Finite Element (FE) method and Ls Dyna software were used to conduct computer simulations reflecting the mechanical response of the structural specimens subjected to a quasi-static compression load. The hyperelastic properties of the TPU material were described with the Simplified Rubber Material (SRM) constitutive model. The proposed FE models, as well as assumed initial boundary conditions, were successfully validated. The results obtained from computer simulations agreed well with the data from the experimental compression tests. A linear relationship was found between the relative density and the maximum strain energy value.

Keywords: 3D printing; Fused Filament Fabrication; thermoplastic polyurethane; energy absorption; dynamic compression; crashworthiness; Simplified Rubber Material; Ls Dyna

---

1. Introduction

Over the last two decades, a growing development of additive manufacturing (AM) techniques has been observed [1–4]. The diversity of techniques available and a wide range of materials used with different physical properties make this type of manufacturing methods very attractive to many industrial sectors [5]. In addition, the design freedom offered by AM allows for manufacturing of objects with complex shapes that are otherwise not possible with conventional subtractive production methods [6,7]. Among the most popular AM techniques offered in the market are Fused Deposition Modelling (FDM) and Fused Filament Fabrication (FFF), which came into being after the Stratasys

Corp. (Eden Prairie, MN, USA) patents expired. The two techniques are commonly referred to as 3D printing [8–10]. This name comes from the technical solution used during the object building process, where the thermoplastic material (filament) is melted and then extruded through a nozzle to build up the layers that gradually make up a complete part. The FDM technology marketed by Stratasys Corp is generally intended for professional applications and uses a limited number of materials. However, their quality and mechanical properties are certified by the producer. Fused Filament Fabrication 3D printers offered in the market by many manufacturers have generally become very popular due to low prices and wide range of materials available. Acrylonitrile-butadiene-styrene-ABS, polylactic acid-PLA, and polyamide are among the most popular and most frequently used [11–15]. In general, this is caused by optimised parameters of the technological process and widely discussed methods of reducing the negative effects of thermal shrinkage. In recent years, an increasing number of special materials (flexible, conductive, magnetic, ferromagnetic, with shape-memory) dedicated to the FFF 3D printing technique have been observed [16–19]. They come as a response to new technological needs. Some of them have multifunctional properties that offer the possibility of using them in cutting-edge products in the aerospace industry [11], the automotive industry, robotics [14], electronics, and biotechnology [20,21]. Reinforced with fibres, they exhibit a high mechanical strength, significantly higher than typical ABS or PLA materials [22–24]. Some of the representatives of the mentioned group of special filaments are flexible and high plasticity materials [17,25]. They demonstrate a high range of deformation under various loading conditions, where the damage mechanism is usually caused by buckling, bending, and shearing [26]. These properties are particularly important for energy absorption and crashworthiness applications, where a wide range of material deformations without a damage mechanism such as cracking suddenly occurring are required [27–32].

Based on the literature study conducted, it was found that thermoplastic materials are widely used in the process of manufacturing regular cell structures, demonstrating high efficiency in terms of energy absorption [33–35]. The advantages of the additive structure manufacturing make it possible to tailor the deformation process of the structural specimen according to the applied material and elementary singular cell topology [36–39]. Many available papers present results obtained for 2D structural specimens with different topologies made from ABS or PLA [40,41]. Hedayati et al. [34] presented a mechanical response of 3D printed honeycomb specimens made from PLA under quasi-static in-plane compression tests. They analysed the influence of the specimen wall thickness value on the deformation process. Kucewicz et al. [42] studied the behaviour of various topologies of ABS structures under quasi-static and impact loading conditions. Furthermore, they developed FE models taking into account the material damage mechanism, which resulted in a good agreement with the experimental studies [43]. Yang et al. [44] carried out tensile and compression tests on specimens, taking auxetic topologies into account. In addition, they analysed the mechanical response of a structure with gradually changing elementary singular cell topology to the deformation process. Chang et al. [45] focused on the in-plane crushing response of tetra-chiral (TC) honeycombs exhibiting auxetic properties. They carried out experimental and numerical investigations taking into account the impact load and quasi-static conditions. The main conclusion of their study is that the relative density has a strong influence on the in-plane crush strength value of the sample. Tabacu and Ducu in [46] presented the results of their research where they evaluated multicellular ABS inserts subjected to compression tests. Using an experimental and numerical approach, they discovered that ABS, multicellular inserts with an outer aluminium tube, designed by them, gave a progressive profile of crushing force during out-plane compression. The application of the proposed solution allows to significantly increase the level of protection of thin-walled structures used in transport systems. Bodaghi et al. [47] have conducted interesting studies on soft PLA material manufactured additively with the use of FFF. They revealed that the singular cell shape, direction, type, and magnitude of mechanical load affects the metamaterial's anisotropic response and its instability characteristics.

With reference to the work mentioned, it can be observed that additive manufacturing attracts the attention of many scientists in the field of crashworthiness and energy absorption studies. However, only a few of them were made with the use of flexible 3D printing materials. This is generally due to the fact that the 3D printing process with flexible material is more demanding and cannot be realized with all FFF 3D printers available in the market. Furthermore, the determination of suitable technological parameters requires additional, time-consuming optimisation studies.

The main aim of this paper is to present results regarding the mechanical response of 3D printed regular cellular structures made from thermoplastic polyurethane material (TPU 95 Polyflex, sold by Polymaker Corp. Shanghai, China), subjected to quasi-static in-plane compression tests. The main feature of the applied material is high flexibility that results in a high deformation range. This property, combined with the correct topology of the regular cell structure, can lead to high efficiency in energy absorption and crashworthiness applications. For this reason, the authors decided to conduct studies in which the honeycomb topology and its modification were used to analyse the deformation plot of the structure under quasi-static load conditions. The proposed research consists of two parts. The first is related to experimental tests, in which few variants of honeycomb topologies (regular and gradually changing) with various relative density values were manufactured using the 3D printing technique and then subjected to in-plane compression tests. The other part of the research involved computer simulations performed using finite element analysis. On the basis of the collected results, conclusions were formulated on the influence of the proposed topologies on the deformation process of the structure. The research carried out by the authors as a result could make it possible to find an effective energy absorption solution that could be applied in military protection systems, as well as in automotive, aerospace, electronics, and biotechnology applications. The main steps of the applied research methodology used to conduct the studies are shown in Figure 1.

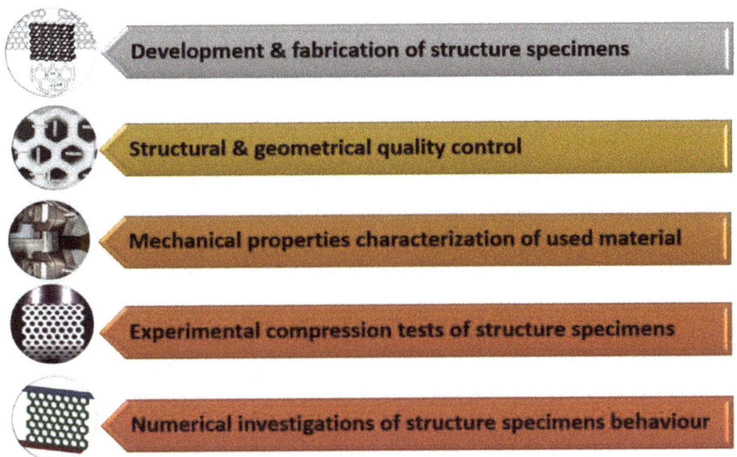

Figure 1. The main steps of applied research methodology.

The first step was related to the development and manufacture of structural specimens with different relative density values. The next task was the geometric quality control of fabricated structure specimens. Then, the characterisation of the mechanical properties of the applied TPU 95 Polyflex filament was performed. The last two stages were related to investigations of the deformation process of the structural specimens. The former stage was based on experimental compression tests, while the latter stage was based on the application of FE analysis and Ls-Dyna software. Based on the results obtained, the influence of the structure topology on its deformation history plot was discussed.

2. Development and Fabrication of Structural Specimens

The main idea considered during the structural specimen design process involved the application of typical honeycomb topology (HC) as the reference model. In Figure 2, other topology variants were defined as gradient models developed based on the proposed geometric features of honeycomb specimen. The main geometric assumptions made during this phase are the following: The total size of the specimen must be less than 40 mm, with a minimum of seven cells in a single layer of the structure. The total specimen's dimension value of less than 40 mm is the result of the diameter of the Split Hopkinson Pressure bar laboratory stand that is planned to be used for further dynamic compression tests. Taking into account guidelines regarding the investigation of regular cell structures, it is recommended to use a minimum of seven cells in a single structure layer to properly evaluate the influence of topology on the specimen deformation history plot. For this reason, the specimens were defined with the same overall dimensions (height-31.4 mm, width-34.4 mm, depth-20 mm), but with different dimensions of the elementary singular cells. As a result, it was possible to obtain different values of the structure relative density ρ_{rel}.

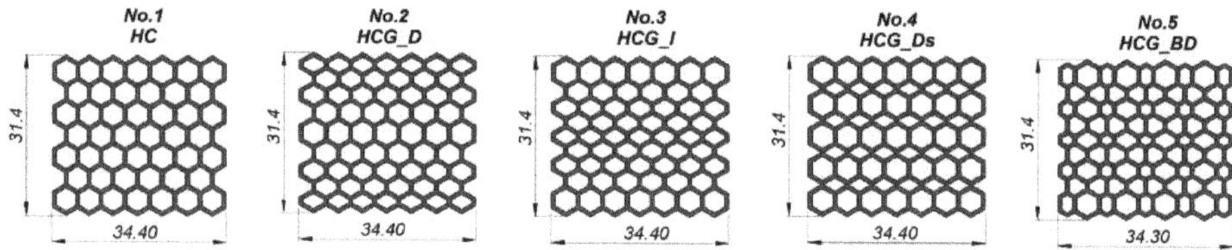

Figure 2. The main view of developed structure specimens with different topologies.

Detailed information on the dimensions of the developed structural specimens has been presented in Figure 3 and Table 1. Specimens No.2 and No.3 are unidirectional, linear gradients of the honeycomb topology (HCG). The main difference between them is the direction of decrease (HCG_D) or increase (HCG_D) of the value of the elementary singular cell size. Specimen No.4 consists of two elementary singular cells that enable the definition of the topology with discrete gradients (HCG_Ds). The size of the particular singular cells was chosen so that the global dimensions of the specimen would be equal to the reference HC model. Specimen No.5 was characterised by a singular cell size that gradually changed in two directions (HCG_BD). The nominal wall thickness dimension of the specimen (1.0 mm) was established based on the results of preliminary technological studies in which a flexible material was used.

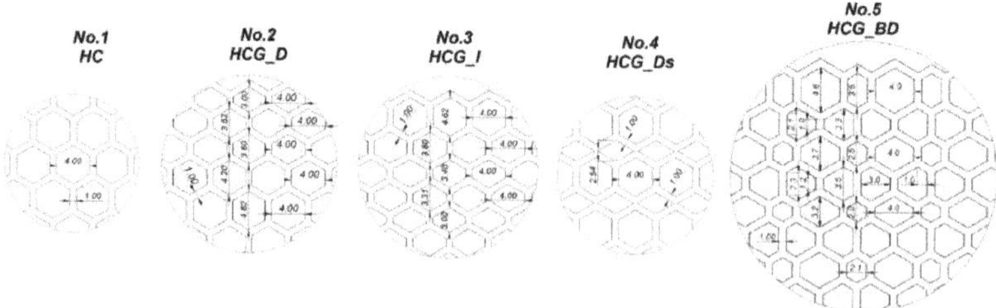

Figure 3. The detailed view of developed structure specimens with different topologies.

The next stage of the studies concerned the process of manufacturing the structural specimens. Based on preliminary test results, it was found that the flexible material (e.g., thermoplastic polyurethane) indicates prospective mechanical properties that could be used in energy absorption applications. Studies conducted using typical materials used in the FFF 3D printing techniques (ABS, PLA) showed a tendency for rapid cracking due to a low range of plastic deformation. Taking this feature into account, the authors decided to conduct a preliminary investigation with flexible materials, which are more demanding in terms of defining the appropriate parameters of the technological process but give a greater range of deformation. Considering the flexible materials available on the market, it was decided to use the thermoplastic polyurethane TPU 95-Polyflex distributed by Polymaker Corp. This material has a Shore hardness of 95 A and can elongate more than three times its original length. The mechanical properties of the material used provided by a producer are presented in Table 2.

The high flexibility of the proposed materials affects their manufacturability and limits the number of FFF 3D printers that can be used to conduct the manufacturing process. In general, devices with a direct filament feed mechanism are dedicated to carrying out this process. For 3D printers with a Bowden feed mechanism, where the distance between the feed mechanism and the end of the extruder is relatively greater, the use of a highly flexible filament is not recommended. The FFF 3D printer used to fabricate the developed structure specimens and samples used for the characterisation of materials is presented in Figure 4. It was a Prime 3D, a low-cost device manufactured by the Polish Monkeyfab Corp (Warsaw, Poland). However, the authors made some modifications to the extruder construction to improve the heat flux distribution. The hot-end block was made from an aluminium alloy and additional cooling fins were added to increase the cooling efficiency of the extruder tip. In addition, an extra fan was installed on the back of the extruder system with a directional airflow housing system. Due to this modification, the presented 3D printer allowed for manufacturing process with the use of a wide range of thermoplastic and flexible materials.

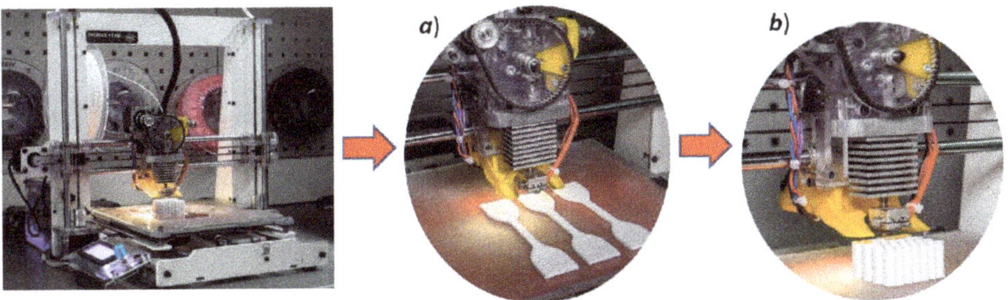

Figure 4. The general view of used 3D printer-Prime 3D (MonkeyFab) during the fabrication process of: (**a**) Dog bone samples, (**b**) structure specimens.

Table 1. Geometrical specifications of developed lattice structures.

No.	Wall Thickness [mm]	Dimensions of Specimen [mm]	Theoretical Relative Density ρ_{rel} [−]
Specimen No.1 (HC)	1.0	31.4 × 34.4 × 20.0	0.37
Specimen No.2 (HCG_D)	1.0	31.4 × 34.4 × 20.0	0.39
Specimen No.3 (HCG_I)	1.0	31.4 × 34.4 × 20.0	0.39
Specimen No.4 (HCG_Ds)	1.0	31.4 × 34.4 × 20.0	0.38
Specimen No.5 (HCG_BD)	1.0	31.4 × 34.4 × 20.0	0.42

Table 2. Mechanical properties of TPU 95 Polyflex material available on the producer website.

Mechanical properties	Density (ASTM D792)	Melt Index (210 °C, 1.2 kg)	Elastic Modulus (X-Y) ASTM D638	Tensile Strength (X-Y) ASTM D638	Elongation at Break (X-Y) ASTM D638	Shore Hardness ASTM D2240
TPU 95-Polyflex	1.20–1.24	3–6 (g/10 min)	9.4 ± 0.3 (MPa)	29.0 ± 2.8 (MPa)	330.1 ± 14.9 (%)	95 A

Rubber-like filaments are flexible and require time-consuming studies to find the most efficient 3D printing parameters that ensure high structural and geometric quality of manufactured objects.

For this reason, the specimens manufacturing process was preceded by additional technological studies. They focused on optimising the 3D printing process when using the TPU 95 Polyflex filament. Additional benchmark model presented in Figure 5a has been designed in a SolidWorks CAD system to perform these tests. It consists of a series of single walls, square tubes, and hexagonal singular cells. All elements were designed with different wall thicknesses. By changing the values of selected 3D printing parameters, such as material flow, nozzle and table temperatures, layer height, and a fill pattern, the quality of manufactured objects was evaluated. The main criteria that were considered during these studies were: A maximum ratio of filling of the model by the material, absence of voids and imperfections in the material, regular shape and height of a single layer, and minimal deviations in wall thickness. The geometrical measurements (the results are presented in Figure 5b,c) were carried out with the use of VI-2510 Venture (Baty Corp. Albany Court, UK) Optical Coordinate Measuring Machine (CMM).

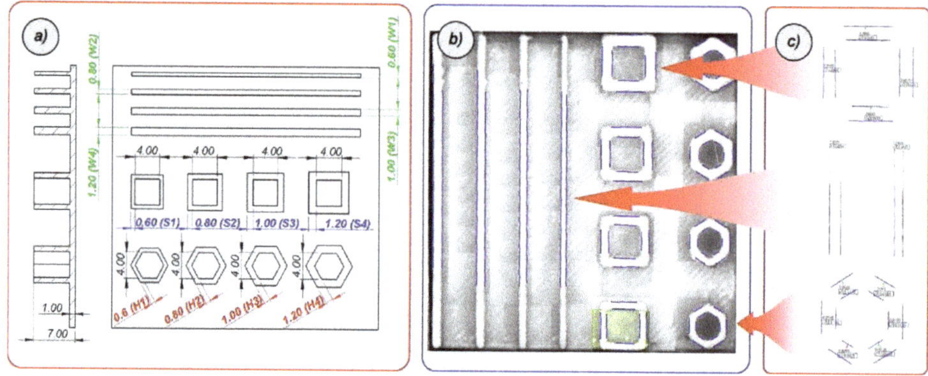

Figure 5. The process of optimisation the 3D printing parameters: (**a**) Developed benchmark model, (**b**) conducted measurements, (**c**) defined report with dimensional deviations.

The influence of the 3D printing parameters on the quality of the manufactured model can be discussed using the data presented in Tables 3 and 4. Table 3 contains sample sets of 3D printing parameter groups that were changed during tests. Table 4 presents the measured values of the characteristic dimensions (wall thickness) defined in Figure 4. From these data it can be seen that a very important parameter is the material flow ratio that affects the filling of a model. However, excessive material flow causes a significant dimensional deviation in the measured wall thickness. Based on the results of many tests carried out, it was found that in the case of the used 3D printer as well as the TPU 95 material, the most effective group of 3D printing parameters is the set number 3 presented in Table 3. This group of parameters was used in further fabrication processes of samples used in material characterisation, as well as structural specimens used during the experimental tests.

Table 3. Technological tests of 3D printing parameters (red dotted line—selected group of 3D printing parameters).

Parameters Group	Nozzle Temperature (°C)	Bed Temperature (°C)	Wall Printing Speed (mm/s)	Infill Printing Speed (mm/s)	Layer Height (mm)	Line Width (mm)	Flow Ratio (%)
Parameters set No.1	215.0 ± 0.2	60.0 ± 0.1	6.00	12.00	0.2	0.4	145.0
Parameters set No.2	215.0 ± 0.2	60.0 ± 0.1	6.00	12.00	0.2	0.4	125.0
Parameters set No.3	215.0 ± 0.2	60.0 ± 0.1	12.00	12.00	0.2	0.4	100.0

Table 4. Results of wall thickness measurements depending on applied 3D printing parameters set (green line—nominal dimensions, red lines – estimated values of wall thickness).

Set Groups	W1 [mm]	W2 [mm]	W3 [mm]	W4 [mm]	S1 [mm]	S2 [mm]	S3 [mm]	S4 [mm]	H1 [mm]	H2 [mm]	H3 [mm]	H4 [mm]
Nominal Values	0.60	0.80	1.00	1.2	0.60	0.80	1.00	1.2	0.60	0.80	1.00	1.2
Set No.1	0.89 ± 0.02	1.06 ± 0.02	1.13 ± 0.03	1.32 ± 0.03	0.94 ± 0.05	1.12 ± 0.06	1.27 ± 0.04	1.48 ± 0.09	0.94 ± 0.04	1.11 ± 0.03	1.28 ± 0.03	1.50 ± 0.05
Set No.2	0.80 ± 0.02	1.01 ± 0.02	1.13 ± 0.03	1.33 ± 0.03	0.75 ± 0.04	0.94 ± 0.04	1.21 ± 0.05	1.39 ± 0.06	0.73 ± 0.05	0.93 ± 0.06	1.11 ± 0.09	1.32 ± 0.09
Set No.3	0.72 ± 0.02	0.84 ± 0.02	1.28 ± 0.03	1.36 ± 0.03	0.66 ± 0.02	0.84 ± 0.03	1.13 ± 0.03	1.31 ± 0.04	0.67 ± 0.02	0.84 ± 0.03	1.11 ± 0.03	1.38 ± 0.04

3. Structural and Geometrical Quality Control

The next stage of the study carried out was related to structural and geometric quality control. The Keyence VHX-600 digital microscope (Keyence, Osaka, Japan) was used to evaluate the structural quality of fabricated specimens. Based on the photos taken, it was determined that the parameters of the 3D printing process were properly adopted. By looking at the upper surface of the specimen, a uniform way of filling the specimen structure with material can be seen (Figure 6a). The wall thickness was determined by three paths of the working nozzle. There are no visible voids or other imperfection of material filling (Figure 6b,c). In addition, Figure 7 shows the side view of the specimen. All filament layers are set correctly, they show a similar height, and no significant irregularities were observed after the manufacturing process.

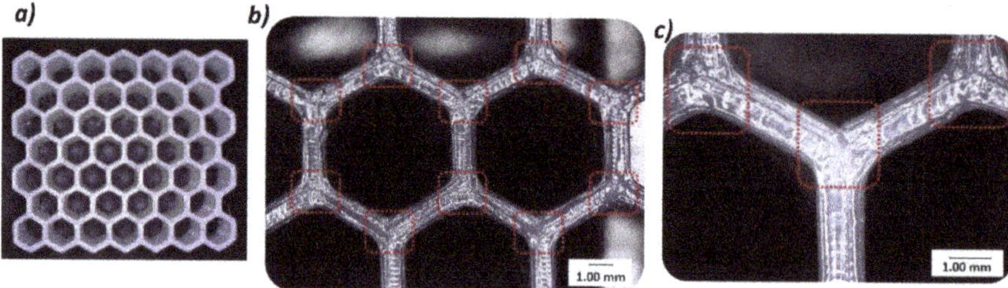

Figure 6. The view of the top surface of 3D printed honeycomb specimen with the use of thermoplastic polyurethane (TPU) material: (**a**) General view, (**b**) with magnitude ×10, (**c**) with magnitude ×20.

The other important issue evaluated during the quality control process was the ratio of wall filling by material, and verification of internal imperfections that could affect the mechanical response of the structure during compression tests. This evaluation was made based on the digital microscopic observation of the sectional views presented in Figures 8 and 9. These observations revealed the presence of small voids in the mid-section of the wall thickness of the sample. They have a regular character and were caused by the method of building up the wall thickness (three paths without additional filling). However, the low single-layer height and the high elasticity of the TPU filament material mean that the influence of the observed imperfection on the mechanical response of the structure under compression tests is rather negligible when compared to other materials, like ABS or PLA.

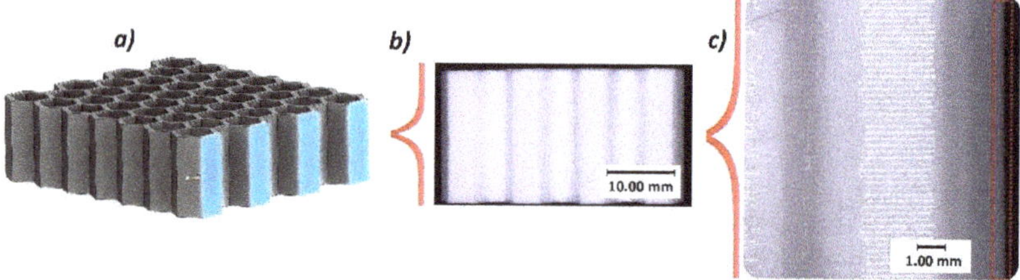

Figure 7. The view of the side surface of 3D printed honeycomb specimen: (**a**) 3D CAD model, (**b**) general view of specimen, (**c**) view with magnitude ×20.

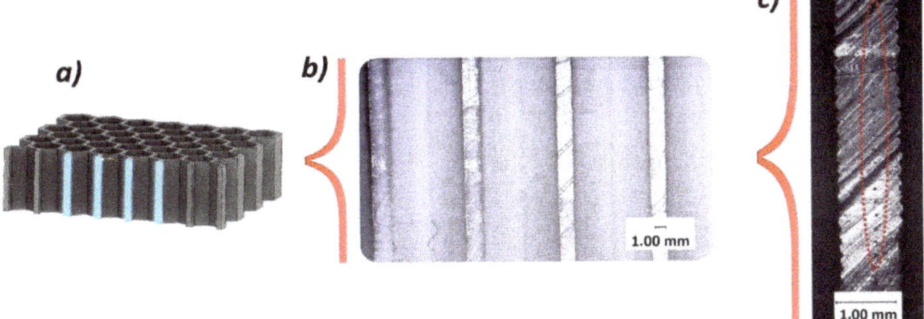

Figure 8. The normal sectional plane view of 3D printed honeycomb specimen: (**a**) 3D CAD model with sectional plane, (**b**) general view of specimen, (**c**) view with magnitude ×20.

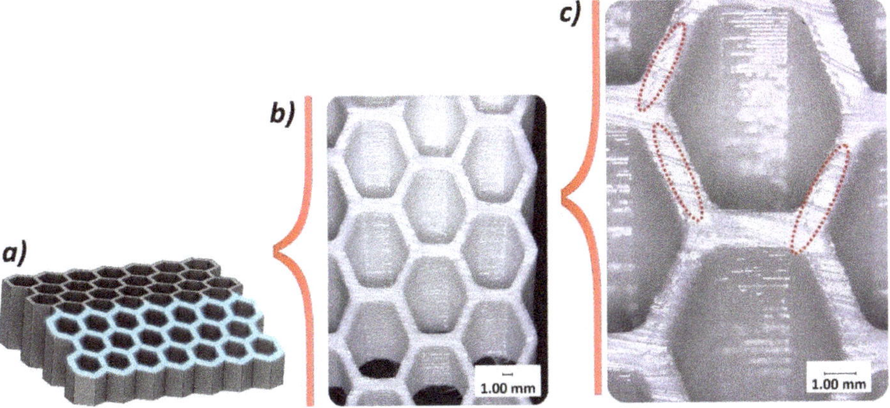

Figure 9. The angle sectional plane view of 3D printed honeycomb specimen: (**a**) 3D CAD model with sectional plane, (**b**) general view of specimen, (**c**) view with magnitude ×20.

Another step in the quality control process involved the determination of the dimensional deviations of the specimens. This problem is particularly critical in terms of defining accurate FE models of structures required to conduct computer simulations. Wall thickness measurements of the structure were performed using the Keyence VHX-600 digital microscope. Images were taken with a magnitude of ×20. Figure 10 shows the results of the measurements obtained for the honeycomb as a representative topology.

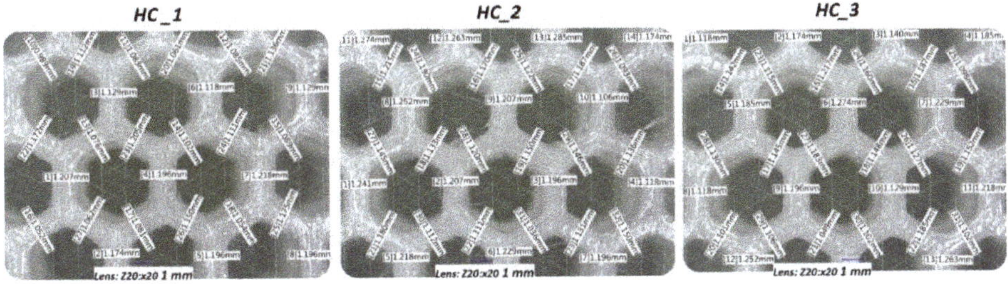

Figure 10. The results of honeycomb topology wall thickness measurements.

Despite the preliminary manufacturability tests performed to determine the correct parameters of the 3D printing process, a high geometric deviation of the fabricated wall thickness of the cellular samples can be observed. Based on carried out measurements with the use of digital microscopy, the estimated average value of structure wall thickness was 1.1 ± 0.065 mm. Measurements were performed based on three various specimens of regular honeycomb topology. This situation can be explained by a high material flow during the extrusion of filaments (controlled by the temperature value of the nozzle and the speed of the material feeding mechanism) which had to be used to guarantee a high structural quality. Observed discrepancies between nominal and real values of specimen wall thickness were included in the developed FE models that were used in the numerical simulations.

4. Determination of Mechanical Properties of the TPU Material

The further step of the investigation was related to the determination of the mechanical properties of the applied TPU 95 Polyflex material. Both the uniaxial compression and tensile tests were carried out in accordance to the PN-EN ISO 604:2006 and PN-EN ISO 37:2011 standards under normal temperature conditions. The shape and the dimensions of the samples used during the tests are shown in Figure 11. The tests were carried out using an MTS Criterion C45 universal strength machine. The loading velocity applied in both tests was 1.0 mm/s.

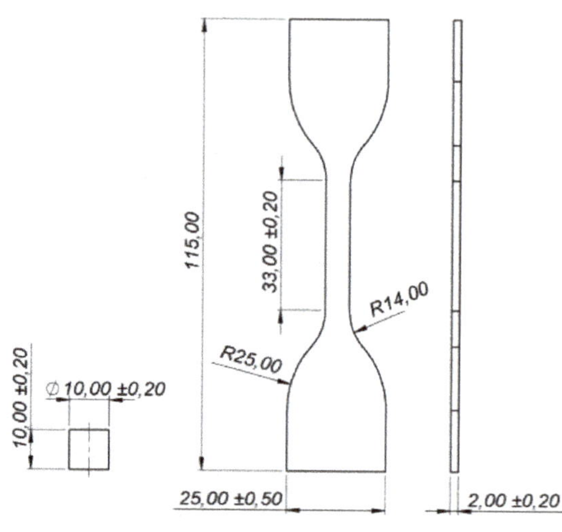

Figure 11. The view of the samples used during the determination of the mechanical properties of the TPU material under uniaxial compression and tensile tests.

The results obtained from the tensile tests are presented in Figure 12. Taking into consideration the repeatability of registered test data, carried out at room temperature, it has been decided that the number of attempts could be limited to three for compression and three for tensile tests. When analysing the history plot, the non-linear, hyper-elastic mechanical properties of TPU 95 Polyflex filament material could be observed. In addition, it showed a high deformation range during tensile tests. The tensile elongation of specimens was approximately 380 ± 0.36% and the maximum stress value was approximately 38.3 ± 0.21 MPa. These values differ from the data for the TPU 95 Polyflex filament. This discrepancy can be explained by the technological process parameters adopted during the manufacturing process, such as: The height of the layer and the type of specimen filling pattern applied. The obtained material characterisation results were used in subsequent research related to computer simulations.

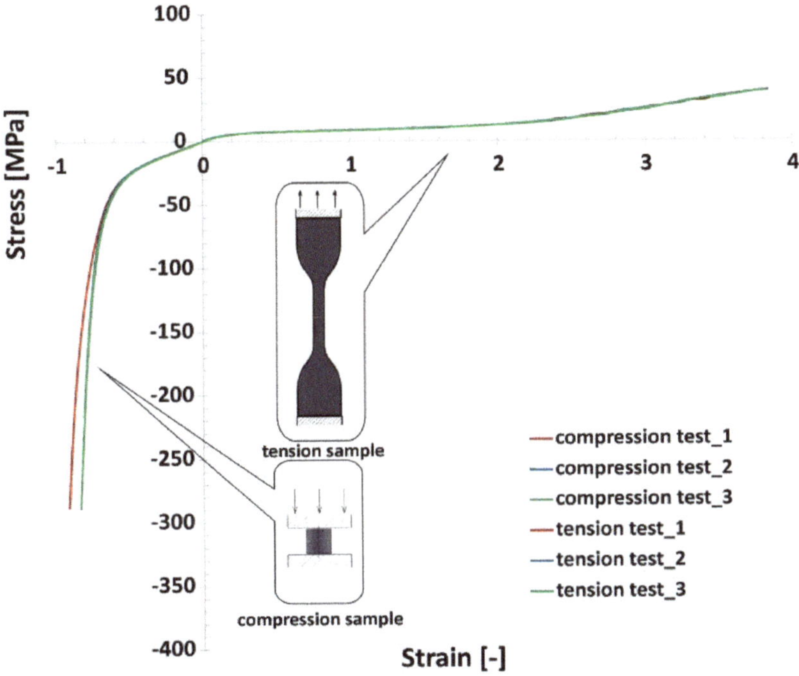

Figure 12. The uniaxial compression and tensile tests of the 3D printed TPU material under normal temperature conditions.

5. Experimental Investigations of Mechanical Properties of the Structural Specimens

The next stage of the investigations carried out focused on defining of the mechanical response of the proposed structure topologies under quasi-static load conditions. Compression tests were performed using the MTS Criterion C45 universal strength machine. The use of a digital camera made it possible to register the deformation process step by step. The structure orientation in relation to the direction of the load applied was the same in all tests. As a result, deformation history plots presented in Figures 13–17 were defined. Furthermore, thanks to the recorded videos it was possible to analyse in detail the mechanical behaviour of the structures during the compression tests. Taking into account the high repeatability of registered data, it was decided that the number of specimens used in these tests could be limited to three for particular structure topologies. By referring to the images and plots that recorded the mechanical response of the reference honeycomb (HC) topology (Figure 13), three main stages of deformation can be highlighted. The first stage (marked between points from No.1 to No.2) refers to a linear deformation resulting from the mechanical properties of the applied material and geometric stiffness of the singular cells. The next stage of deformation (highlighted between points from No.2 to No.3) is caused by the loss of stability and the arrival of buckling and bending mechanisms in one of the structural layers. A further deformation stage is reflected in the plot as a long plateau (marked between points No.3 and No.4). The final stage, represented by the rapid growth in the value of the deformation force, is linked to the collapse of all arrays inside the specimen and the final densification of the structure (marked between points from No. 4 to No.5).

Subsequent variations of the specimens subjected to compression tests were the gradual honeycomb topologies in which singular cell size changed linearly (decrease-HCG_D and increase-HCG_I). The two topologies show a similar relative density value (ρ_{rel}-0.39), which is greater than the representative honeycomb topology's value (ρ_{rel}-0.37). When analysing the stages of their deformation process (Figures 14 and 15), three main stages can be distinguished. However, the range of the plateau is relatively small compared to the honeycomb. The first stage, similar to the previous case, is linear and results from the mechanical properties of the applied TPU 95 material and geometric stiffness of the specimen (marked on the curve by points from No.1 to No.2). However, a larger number of singular cells that are in the direction of the acting load leads to a higher value of the registered deformation force. The next stage of deformation is related to the loss of structural stability caused by the buckling and bending mechanisms and the decrease in the deformation force value (highlighted on the curve by points from No.2 to No.3). Taking into account the large difference in stiffness between the largest singular cells and the rest, the array with the largest cells collapsed first, and this way the process of progressive structure deformation has been initiated (marked by points from No.3 to No.4). This process is caused by buckling and bending mechanisms, that arrived progressively in particular layers of the structure. The last stage marked on the curve by points from No.4 to No.5 is related to the deformation of the remaining arrays followed by the final densification. When comparing the deformation process of the HCG_D and HCG_I specimens, it can be determined that they show a similar history plot. However, due to the different locations of the arrays consisting of the larger singular cells, they collapsed differently—the structure HCG_D collapsed starting from the inside, and the HCG_I–starting from the outside.

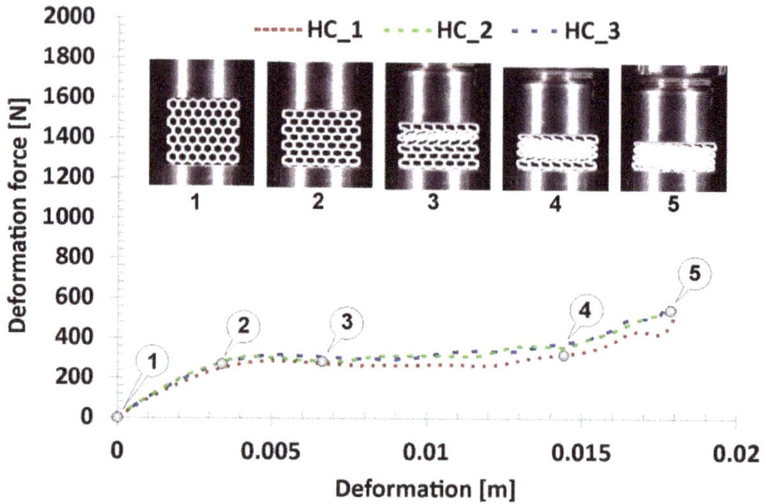

Figure 13. The results of the compression tests of the structural specimen with regular honeycomb topology.

The following results in Figure 16 refer to the honeycomb topology with discrete gradients (HCG_Ds). It consists of two types of singular cells that differ in size. When comparing the obtained results with the plots registered for previously discussed specimens, it can be noticed that the deformation force history plot is similar to the regular honeycomb topology. The range of the first, linear stage of the deformation process (marked on the curve by points from No.1 to No.2) results in a higher value of the deformation force, which is generally due to a higher geometric stiffness of the topology of a singular cell. The next stage was related to the loss of structural stability caused initially because of buckling and then bending mechanisms. This period of deformation begun at point No.2 and lasted until point No.3. However, the plateau range is shorter due to a limited number of larger cells located in the direction of the loading action (marked on the curve by points from No.3 to No.4). Furthermore, the buckling effect of the whole structure was observed in the final stage of the plateau before densification which begun in point No.4 and lasted until point No.5.

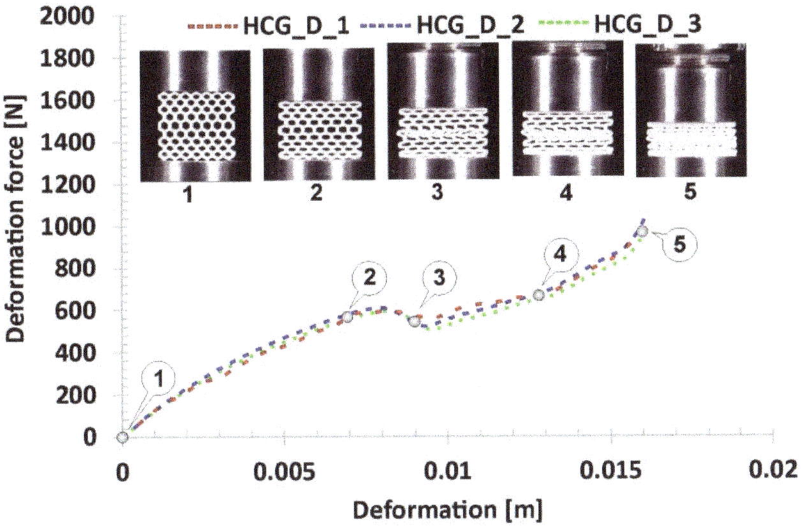

Figure 14. The results of the compression tests of the structural specimen with gradually decreasing honeycomb topology.

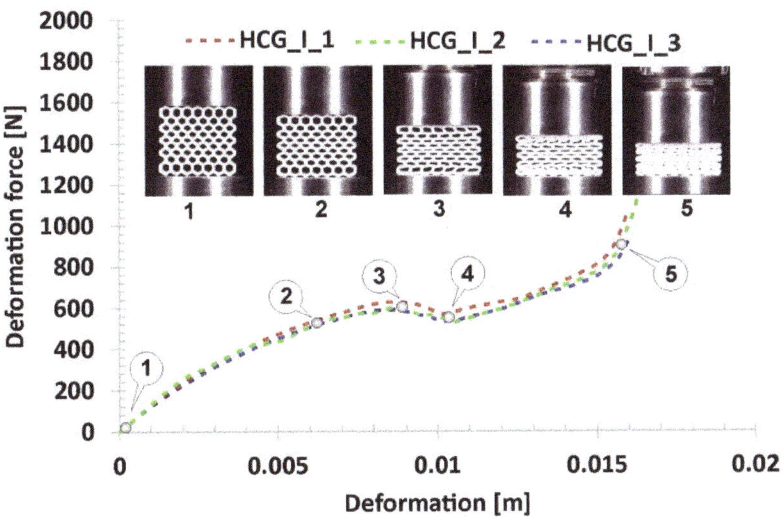

Figure 15. The results of the compression tests of the structural specimen with gradually increasing honeycomb topology.

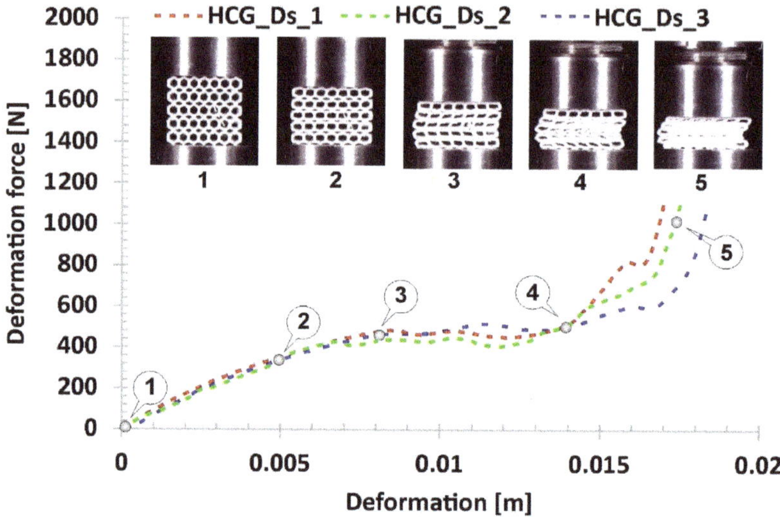

Figure 16. The results of the compression tests of the structural specimen with gradual discrete honeycomb topology.

The last of the results obtained during the compression tests refer to the bidirectional gradual honeycomb topology (HCG_BD) presented in Figure 17. Due to a higher value of the relative density (ρ_{rel}-0.42), the stiffness of the structure was relatively high compared to the representative regular honeycomb. The value of the deformation force registered in the first linear stage of the deformation process (marked on the curve by points from No.1. to No.2) was almost two times higher. After the loss of the structural stability, caused by buckling and bending mechanisms, the gradual decrease of deformation force was registered (highlighted by points from No.2 to No.3). This stage lasted until the collapse of the layers defined by cells with the largest size. Afterward, the progressive deformation of the whole structure could be observed (from point No.3 to point No.4). The final densification happened faster compared to other structural topologies due to a higher relative density value (marked on the curve from point No.4 to point No.5).

Based on the results of performed compression tests, it can be stated that the main mechanisms responsible for the deformation of the particular structures were buckling and bending. Taking into account the high flexibility of the applied base material, there was no presence of shearing or cracking mechanisms.

Results of experimental compression tests made it possible to compare the honeycomb-based topologies proposed by the authors in terms of energy absorption efficiency. The estimated values of the deformation energy are shown in Figure 18. The influence of the proposed topology on the deformation energy value as well as on the deformation range can be observed. In fact, it is interesting that the use of a gradually changing topology with a similar relative density value with respect to the reference honeycomb structure made it possible to obtain a higher value of the deformation energy. The same conclusion can be formulated from the data presented in Figure 19, where the value of the absorbed energy was related to the structural volume (SEA_v), defined by the Formula (1):

$$SEA_v = \frac{EA}{V_s} \tag{1}$$

where EA is the energy absorption and vs. is the volume of the structure specimen

Furthermore, the application of the TPU 95 Polyflex material, which exhibits hyper-elastic mechanical properties, made it possible to reduce the negative effects of the material cracking during the deformation process. After unloading, their dimensions were similar to those from before the tests.

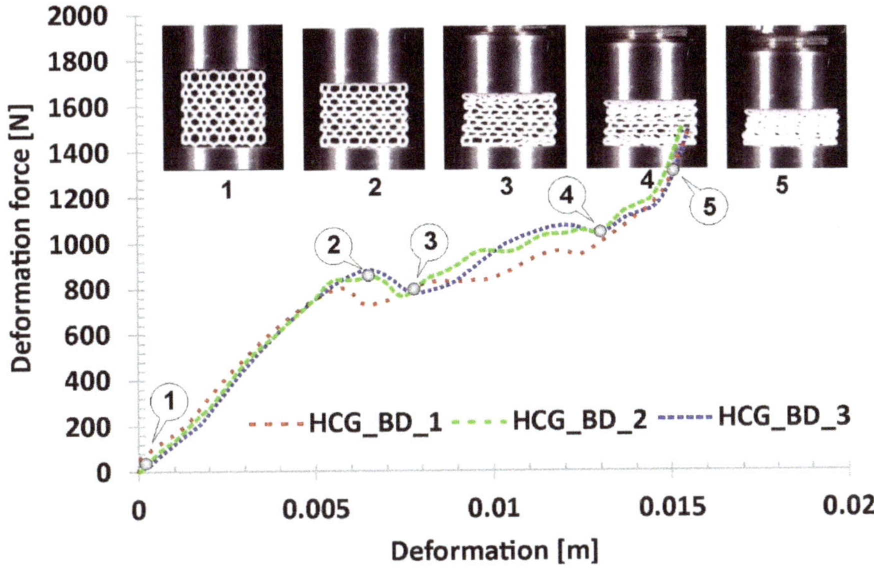

Figure 17. The results of the compression tests of the structural specimen with bidirectional gradual honeycomb topology.

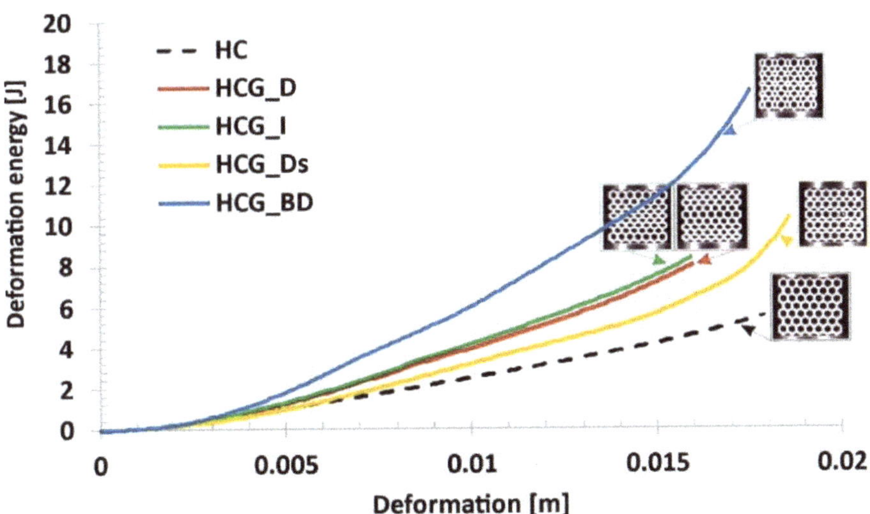

Figure 18. The comparison of the estimated values of the deformation energy versus the range of deformation.

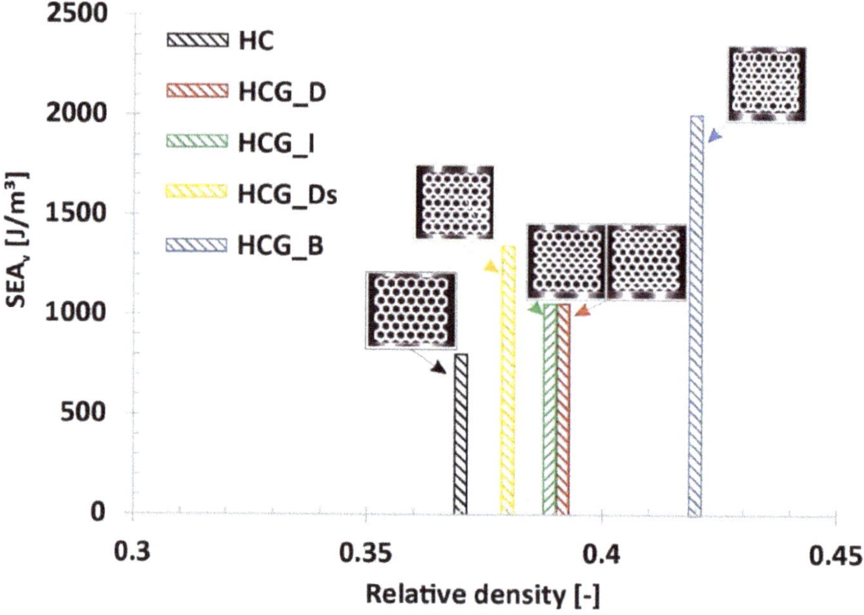

Figure 19. The comparison of the estimated values of the deformation energy in relation to the relative density.

6. Numerical Investigations of Mechanical Properties of Structural Specimens

The experimental compression tests conducted have allowed to draw a conclusion about the possibility of using gradient honeycomb topologies and the TPU 95 Polyflex materials as an interesting solution in terms of energy absorption. However, the experimental approach in further studies to optimise the structure topology is costly and time-consuming. For this reason, it was decided to use an additional numerical modelling approach based on the Finite Element Analysis (FEA). Computer simulations were conducted with the commercial Ls-Dyna hydrocode with Multi Parallel Processing (MPP) (Livermore Software Technology, Livermore, CA, USA) [48]. The main assumptions made during computer simulations were similar to those presented by the authors in [42,49,50]. The FE models required to run the simulation of the particular structural behaviour under the compression tests have been developed from the CAD models (Figure 20). Eight-node hexagonal elements with single integration point were used to create FE models. They were defined as a 2 mm sections in all structures. Based on the results of previous studies [43,49], this assumption allowed a significant reduction in simulation time and caused no impact on the final result's accuracy. Taking into account the geometric deviations of the manufactured specimens observed during the geometrical quality control, the wall thickness of the specimen was defined as 1.10 mm and consisted of four elements.

Detailed information on the type of computer simulation is shown in Figure 21. The applied boundary conditions corresponded to the quasi-static tests. The structural specimen was placed between two rigid planes. The bottom plane was fixed with the top plane moving in the 0Y direction. The plane motion $v(t)$ was prescribed by the Formula (2) used in previous papers [43,51].

$$v(t) = \frac{\pi}{\pi - 2} \frac{S_{max}}{T} \left[1 - \cos\left(\frac{\pi}{2T}\right) \right] \tag{2}$$

where T is the termination time of the simulation and S_{max} is the final displacement of the rigid surface.

A Single-Surface and Surface-to-Surface contact settings based on the penalty method were used to simulate the interactions between the planes and the structure, and the self-interaction of the structure cells. A Coulomb formulation was used to describe the tangential interaction between bodies [48] with the friction coefficient μ-0.65 for the TPU 95 Polyflex material.

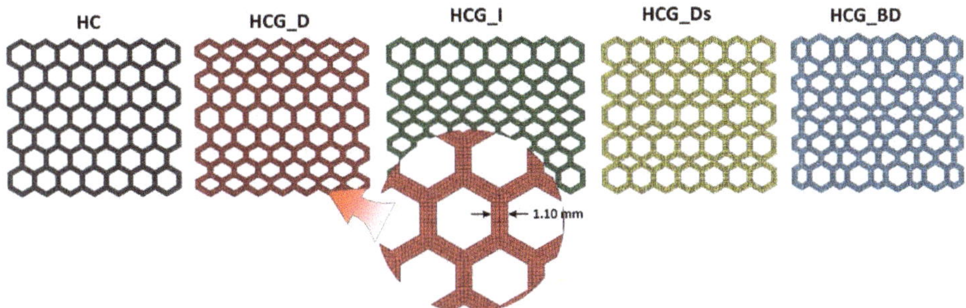

Figure 20. The main view of Finite Element (FE) models of developed structure specimens.

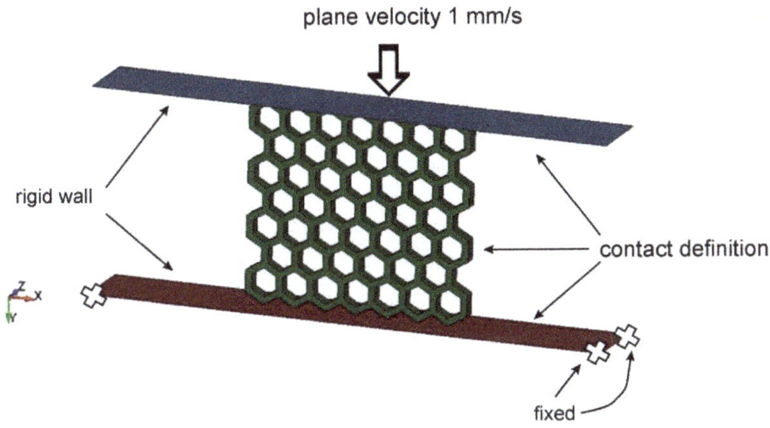

Figure 21. The main boundary conditions adopted for the numerical simulations to reproduce the experimental tests.

Ls-Dyna software allows for use of a few constitutive material models based on the Ogden formulation, which are suitable for describing hyperelastic materials. Nevertheless, based on the conclusions presented in the following articles [52–55], the authors decided to use the Simplified Rubber Material (SRM) model. It is defined on the basis of the experimentally determined tabulated stress-strain curve (Table 5). Furthermore, the SRM material model allows the effects of strain rate to be taken into account based on the series of stress–strain curves that have been experimentally defined under different load conditions. This feature is intended to be used in further studies related to dynamic compression tests. In addition, Table 6 presents the material parameters used to define the SRM.

Table 5. Adopted TPU 95 Polyflex stress–strain characteristics.

Strain [-]	−0.83	−0.78	−0.71	−0.69	−0.64	−0.43	−0.33	−0.23	−0.14	−0.09	−0.04	−0.03
Stress [MPa]	−270.48	−156.06	−83.28	−69.07	−50.23	−20.11	−14.95	−10.02	−6.02	−3.91	−1.96	−1.05
Strain [-]	0.0	0.04	0.1	0.23	0.71	1.45	2.34	2.67	2.96	3.24	3.55	3.81
Stress [MPa]	0.0	2.0	4.01	5.99	7.99	10.01	14.99	20.02	25.0	29.97	34.99	38.0

Table 6. Software parameters applied in Ls Dyna to describe the Simplified Rubber Material (SRM) model.

RO Density	KM Linear Bulk Modulus	SGL Specimen Gauge Length	SW Specimen Width	ST Specimen Thickness
1100.00 kg/m^3	1.650×10^9 Pa	1.00	1.00	1.00

The results obtained during the computer simulations were compared with the data recorded during the experimental tests to verify the accuracy of the initial boundary conditions adopted as well as the applied constitutive model describing the mechanical properties of the TPU 95 Polyflex. In Figure 22, the history of the deformation forces of the honeycomb topology is shown. The numerical results obtained, which are highlighted by a black dotted line, corresponds to the data recorded during the experimental tests. In addition, the same mechanisms, such as buckling and bending, which are responsible for the deformation process of the specimen, were observed. Applied material model without definition of failure criteria resulted in lack of shearing and cracking effects.

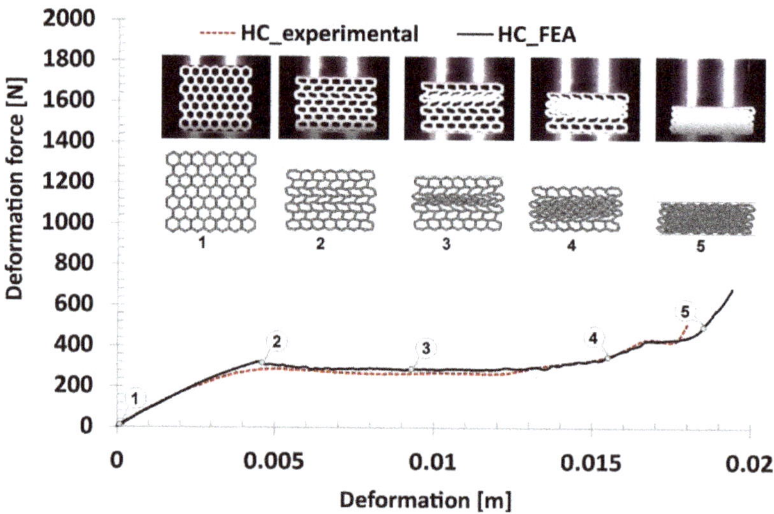

Figure 22. The results of the honeycomb compression tests obtained based on the FE analysis.

The following results refer to the honeycomb gradual topologies. In Figures 23 and 24, the results that reproduce the deformations of HCG_D and HCG_I are presented. Both agree well with the experimental data. The estimated maximum values of deformation forces at each compression stage are similar to each other. The main difference between the used topologies is the location of the initialisation of the specimen collapse process. In the case of the HCG_D topology, this process started in the inner layer of the specimen, while in the case of the HCG_I—in the outer layers. In the analysis of the main mechanisms responsible for the specimens deformation process, similar to regular honeycomb topology, the buckling and bending mechanisms were observed. Similar to experimental tests, there are no shearing or cracking effects.

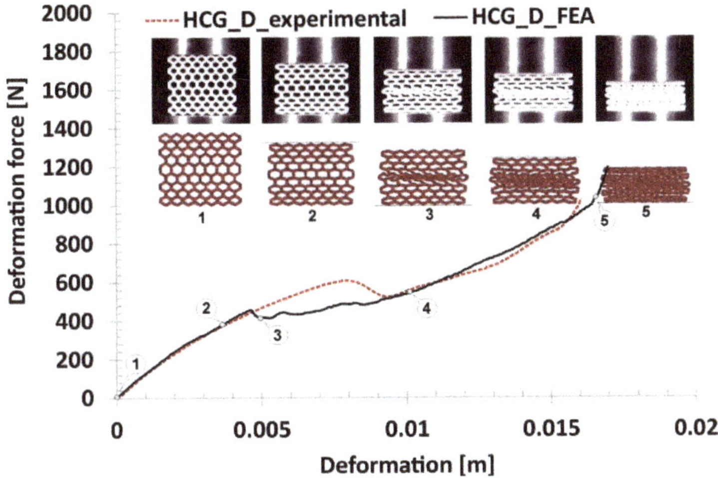

Figure 23. The results of the gradually decreasing honeycomb compression tests obtained based on the FE analysis.

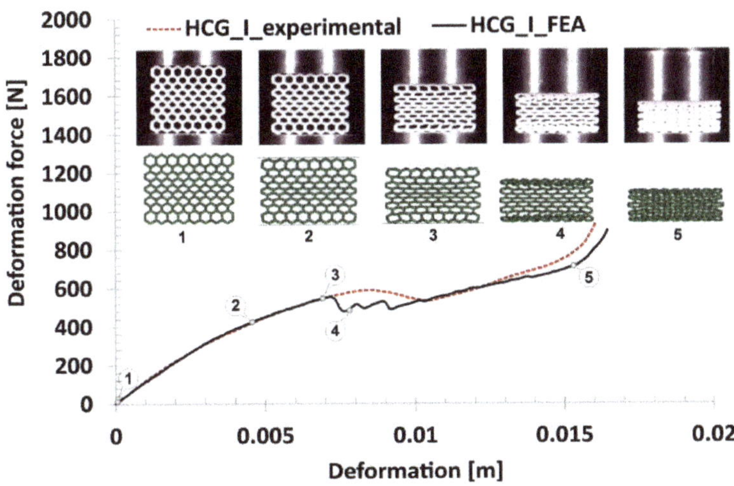

Figure 24. The results of gradually increasing honeycomb compression tests obtained based on the FE analysis.

The subsequent results shown in Figure 25 correspond to the honeycomb topology with discrete gradients (HCG_Ds). The first linear compression stage agrees well with the experimental data. The visible difference in the deformation force value in the plateau range was likely caused by the initially assumed wall thickness of the smaller singular cells of the FE model. In reference to the observations made during the technological studies, the adopted 3D printing technique as well as the TPU 95 filament requires individual 3D printing parameters depending on the type of objects printed. A large number of geometric elements of small dimensions require different 3D printing parameters. Generally, in this case, it is recommended to reduce the rotation velocity of the stepper motors in order to reduce the material flow. By using the same parameters for the HCG_Ds topology as for the referential HC, the dimensional deviations of the small singular cell thickness for HCG_Ds are greater. This situation explains the difference between the numerical results and the experimental data. Referring to the main mechanisms arriving during structure compression tests, similar to experimental results, only buckling and bending mechanisms could be observed.

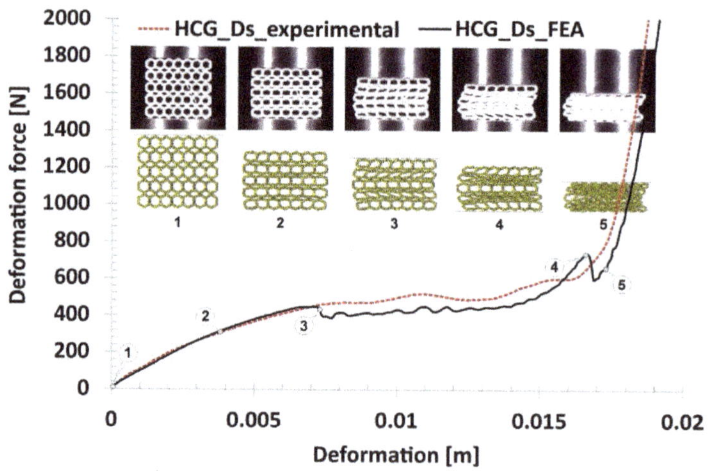

Figure 25. The results of discrete gradient honeycomb compression tests obtained based on the FE analysis.

The last considered topology refers to the bidirectional modification of the honeycomb gradient. Figure 26 shows the results obtained from the FE analysis. When comparing the quality of the proposed numerical model and the assumed initial boundary conditions, a conclusion similar to that of the of HCG_D topology variants can be drawn. The initial linear phase of the deformation has been reproduced accurately, however there is a visible difference in the deformation force history plot in a subsequent stage where some of the layers of the structure have collapsed. These differences are mainly due to the dimensional deviations of the small singular cells the topology of the structure consists of. Referring to the main mechanisms responsible for specimens collapse, the buckling and bending mechanisms can be distinguished.

Numerical investigations carried out have made it possible to predict the mechanical behaviour of the topologies proposed during quasi-static compression tests. The presented results of deformation history plots generally are in a good agreement with the data obtained during the experimental tests. In the case of gradient topologies, however, certain differences could be observed in the history plots obtained. This situation was mainly caused by dimensional deviations of actual structure specimens where small-sized singular cells were used. This characteristic can also be observed in Figure 27, where the comparison of deformation energy history plots was presented. The solution that can reduce this discrepancy requires an individual evaluation during the particular FE model design process.

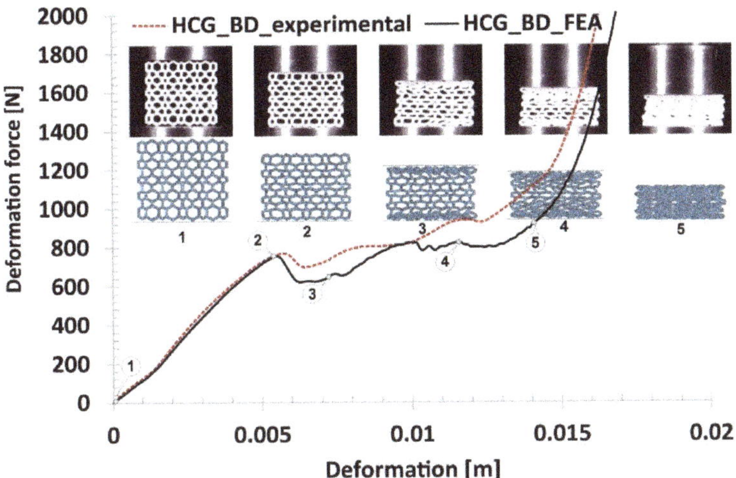

Figure 26. The results of bidirectional gradient honeycomb compression tests obtained based on the FE analysis.

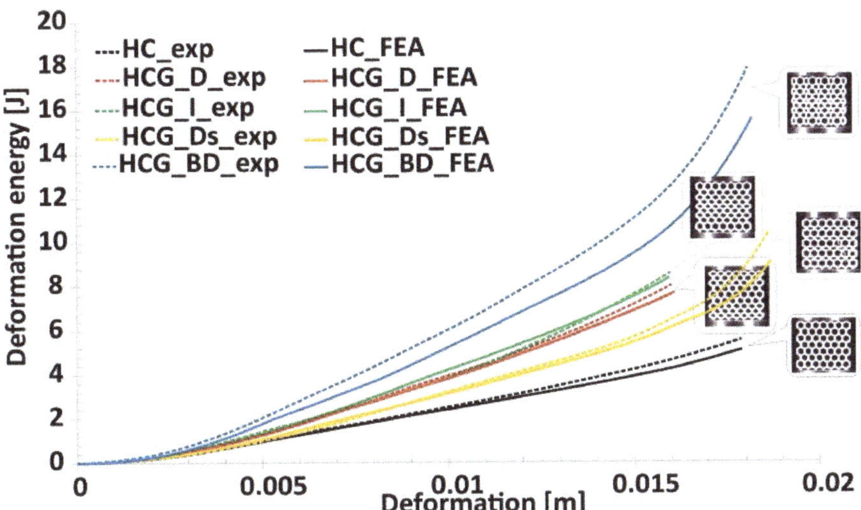

Figure 27. Comparison of deformation energy plots obtained from the FE analysis and experimental tests.

7. Conclusions

Based on the research conducted regarding the compression tests of regular and gradual honeycomb structure specimens, manufactured additively from flexible material, the following conclusions were made:

- Fused filament fabrication is a cost-effective 3D printing process that allows objects to be fabricated using specific filaments such as thermoplastic polyurethane (TPU 95) with unique mechanical properties. Nevertheless, the material proposed by the authors requires a direct-type filament feeding mechanism and it is recommended to conduct further technological studies in order

- to define the appropriate 3D printing parameters, which guarantee a structural and geometric quality of the manufactured objects.
- The conducted characterisation of the mechanical properties of the TPU 95 material allows to confirm that it exhibits hyperelastic properties with a high deformation range. For this reason, it is necessary to use a suitable constitutive material model in computer simulations.
- The experimental compression tests have shown a linear relationship between the relative density of the applied topology and the deformation energy value. Specimens with gradually changing topologies showed a higher value of the deformation energy compared to the reference honeycomb structure.
- By analysing the history plots of the deformation processes of the specimens, a lack of crack damage mechanism can be observed due to the high flexibility of the applied TPU 95 filament. The main mechanisms occurring during the compression test were buckling and bending.
- The proposed numerical approach to the investigations made it possible to predict the structural deformation process. The results obtained agree well with the data recorded during the experimental tests.
- The adopted simplified rubber material constitutive model, defined on the basis of experimental compression and tensile tests, enables correct reproduction of the mechanical response of the structural specimens made with the TPU 95 filament.
- The developed models will be used in further numerical investigations conducted under dynamic load conditions.
- Planned investigations taking into account dynamic loading conditions will be carried out experimentally as well as numerically. The experimental approach will be performed with the Split Hopkinson Pressure Bar stand implementation in a direct impact configuration. The numerical approach will allow for verification of the proposed numerical model. A good correlation between results enables us to perform further optimisation studies.
- Obtained results of the dynamic tests offer the chance for evaluation of developed structures as well as highly flexible Polyflex TPU 95a material in terms of energy absorption.

Author Contributions: P.P.—conceptualisation and methodology, computer simulations, original draft preparation, K.R. and K.C.—3D printing process, quality control, experimental tests, M.S.—characterisation of mechanical properties of the TPU 95 filament, experimental tests, J.M.—conceptualisation, computer simulations results consultation, writing—review and editing, R.W.—writing—review and editing, J.J.—conceptualisation, analysis of obtained results, writing—review and editing. All authors have read and agreed to the published version of the manuscript.

Funding: The research was supported by the Interdisciplinary Centre for Mathematical and Computational Modelling (ICM), University of Warsaw, under grant no GA73-30.

Conflicts of Interest: The authors declare no conflict of interest.

References

1. Li, N.; Huang, S.; Zhang, G.; Qin, R.; Liu, W.; Xiong, H.; Shi, G.; Blackburn, J. Progress in additive manufacturing on new materials: A review. *J. Mater. Sci. Technol.* **2019**, *35*, 242–269. [CrossRef]
2. Strong, D.; Kay, M.; Conner, B.; Wakefield, T.; Manogharan, G. Hybrid manufacturing—Integrating traditional manufacturers with additive manufacturing (AM) supply chain. *Addit. Manuf.* **2018**, *21*, 159–173. [CrossRef]
3. Prakash, K.S.; Nancharaih, T.; Rao, V.S. Additive manufacturing techniques in manufacturing —An Overview. *Mater. Today Proc.* **2018**, *5*, 3873–3882. [CrossRef]
4. Murr, L. Frontiers of 3D printing/additive manufacturing: From human organs to aircraft fabrication. *J. Mater. Sci. Technol.* **2016**, *32*, 987–995. [CrossRef]
5. Ngo, T.; Kashani, A.; Imbalzano, G.; Nguyen, Q.; Hui, D. Additive manufacturing (3D printing): A review of materials, methods, applications and challenges. *Compos. Part B Eng.* **2018**, *143*, 172–196. [CrossRef]
6. Craveiro, F.; Duarte, J.P.; Bartolo, P.; Bartolo, P. Additive manufacturing as an enabling technology for digital construction: A perspective on Construction 4.0. *Autom. Constr.* **2019**, *103*, 251–267. [CrossRef]

7. Du Plessis, A.; Broeckhoven, C.; Yadroitsava, I.; Yadroitsev, I.; Hands, C.H.; Kunju, R.; Bhate, D. Beautiful and functional: A review of biomimetic design in additive manufacturing. *Addit. Manuf.* **2019**, *27*, 408–427. [CrossRef]

8. Dizon, J.R.C.; Espera, A.H.; Chen, Q.; Advincula, R.C. Mechanical characterization of 3D-printed polymers. *Addit. Manuf.* **2018**, *20*, 44–67. [CrossRef]

9. Kumar, S.; Czekanski, A. Roadmap to sustainable plastic additive manufacturing. *Mater. Today Commun.* **2018**, *15*, 109–113. [CrossRef]

10. Yuan, S.; Shen, F.; Chua, C.K.; Zhou, K. Polymeric composites for powder-based additive manufacturing: Materials and applications. *Prog. Polym. Sci.* **2019**, *91*, 141–168. [CrossRef]

11. Azarov, A.V.; Antonov, F.K.; Golubev, M.; Khaziev, A.; Ushanov, S.A. Composite 3D printing for the small size unmanned aerial vehicle structure. *Compos. Part B Eng.* **2019**, *169*, 157–163. [CrossRef]

12. Valino, A.D.; Dizon, J.R.C.; Espera, A.H., Jr.; Chen, Q.; Messman, J.; Advincula, R.C. Advances in 3D printing of thermoplastic polymer composites and nanocomposites. *Prog. Polym. Sci.* **2019**, *98*, 101162. [CrossRef]

13. Zolfagharian, A.; Khosravani, M.R.; Kaynak, A. Fracture resistance analysis of 3D-printed polymers. *Polymers* **2020**, *12*, 302. [CrossRef] [PubMed]

14. Zolfagharian, A.; Kaynak, A.; Bodaghi, M.; Kouzani, A.Z.; Gharaie, D.W.P.; Nahavandi, S. Control-based 4D printing: Adaptive 4D-printed systems. *Appl. Sci.* **2020**, *10*, 3020. [CrossRef]

15. Kluczyński, J.; Śnieżek, L.; Kravcov, A.; Grzelak, K.; Svoboda, P.; Szachogłuchowicz, I.; Franek, O.; Morozov, N.; Torzewski, J.; Kubeček, P. The examination of restrained joints created in the process of multi-material FFF additive manufacturing technology. *Materials* **2020**, *13*, 903. [CrossRef]

16. Wang, Q.; Tian, X.; Huang, L.; Li, D.; Malakhov, A.V.; Polilov, A.N. Programmable morphing composites with embedded continuous fibers by 4D printing. *Mater. Des.* **2018**, *155*, 404–413. [CrossRef]

17. Mitchell, A.; Lafont, U.; Hołyńska, M.; Semprimoschnig, C. Additive manufacturing—A review of 4D printing and future applications. *Addit. Manuf.* **2018**, *24*, 606–626. [CrossRef]

18. Rayate, A.; Jain, P.K. A Review on 4D printing material composites and their applications. *Mater. Today Proc.* **2018**, *5*, 20474–20484. [CrossRef]

19. Wickramasinghe, S.; Do, T.; Tran, P. FDM-based 3D printing of polymer and associated composite: A review on mechanical properties, defects and treatments. *Polymers* **2020**, *12*, 1529. [CrossRef]

20. Kashyap, D.; Kumar, P.K.; Kanagaraj, S. 4D printed porous radiopaque shape memory polyurethane for endovascular embolization. *Addit. Manuf.* **2018**, *24*, 687–695. [CrossRef]

21. Haryńska, A.; Carayon, I.; Kucińska-Lipka, J.; Janik, H. Fabrication and characterization of flexible medical-grade TPU filament for fused deposition modeling 3DP technology. *Polymers* **2018**, *10*, 1304. [CrossRef]

22. Blok, L.; Longana, M.L.; Yu, H.; Woods, B. An investigation into 3D printing of fibre reinforced thermoplastic composites. *Addit. Manuf.* **2018**, *22*, 176–186. [CrossRef]

23. Sang, L.; Han, S.; Li, Z.; Yang, X.; Hou, W. Development of short basalt fiber reinforced polylactide composites and their feasible evaluation for 3D printing applications. *Compos. Part B Eng.* **2019**, *164*, 629–639. [CrossRef]

24. Soltani, A.; Noroozi, R.; Bodaghi, M.; Zolfagharian, A.; Hedayati, R. 3D printing on-water sports boards with bio-inspired core designs. *Polymers* **2020**, *12*, 250. [CrossRef]

25. Herzberger, J.; Sirrine, J.M.; Williams, C.B.; Long, T.E. Polymer design for 3D printing elastomers: Recent advances in structure, properties, and printing. *Prog. Polym. Sci.* **2019**, *97*, 101144. [CrossRef]

26. Robinson, M.; Soe, S.P.; Johnston, R.; Adams, R.; Hanna, B.; Burek, R.; McShane, G.; Celeghini, R.; Alves, M.; Theobald, P. Mechanical characterisation of additively manufactured elastomeric structures for variable strain rate applications. *Addit. Manuf.* **2019**, *27*, 398–407. [CrossRef]

27. Antolak-Dudka, A.; Płatek, P.; Durejko, T.; Baranowski, P.; Małachowski, J.; Sarzyński, M.; Czujko, T. Static and dynamic loading behavior of Ti6Al4V Honeycomb Structures Manufactured by Laser Engineered Net Shaping (LENSTM) technology. *Materials* **2019**, *12*, 1225. [CrossRef]

28. Delcuse, L.; Bahi, S.; Gunputh, U.; Rusinek, A.; Wood, P.; Miguelez, M. Effect of powder bed fusion laser melting process parameters, build orientation and strut thickness on porosity, accuracy and tensile properties of an auxetic structure in IN718 alloy. *Addit. Manuf.* **2020**, *36*, 101339. [CrossRef]

29. Stanczak, M.; Fras, T.; Blanc, L.; Pawlowski, P.; Rusinek, A. Blast-induced compression of a thin-walled aluminum honeycomb structure—Experiment and modeling. *Metals* **2019**, *9*, 1350. [CrossRef]

30. Kurzawa, A.; Pyka, D.; Jamroziak, K.; Bajkowski, M.; Bocian, M.; Magier, M.; Koch, J. Assessment of the impact resistance of a composite material with EN AW-7075 matrix reinforced with α-Al2O3 particles using a 7.62 × 39 mm projectile. *Materials* **2020**, *13*, 769. [CrossRef]

31. Jamroziak, K.; Bajkowski, M.; Bocian, M.; Polak, S.; Magier, M.; Kosobudzki, M.; Stepien, R. Ballistic head protection in the light of injury criteria in the case of the Wz.93 combat helmet. *Appl. Sci.* **2019**, *9*, 2702. [CrossRef]

32. Arkusz, K.; Klekiel, T.; Slawinski, G.; Będziński, R. Influence of energy absorbers on Malgaigne fracture mechanism in lumbar-pelvic system under vertical impact load. *Comput. Methods Biomech. Biomed. Eng.* **2019**, *22*, 313–323. [CrossRef] [PubMed]

33. Jiang, H.; Le Barbenchon, L.; Bednarcyk, B.A.; Scarpa, F.; Chen, Y. Bioinspired multilayered cellular composites with enhanced energy absorption and shape recovery. *Addit. Manuf.* **2020**, *36*, 101430. [CrossRef]

34. Hedayati, R.; Sadighi, M.; Aghdam, M.; Zadpoor, A.A. Mechanical properties of additively manufactured thick honeycombs. *Materials* **2016**, *9*, 613. [CrossRef]

35. Lopez, D.M.B.; Ahmad, R. Tensile mechanical behaviour of multi-polymer sandwich structures via fused deposition modelling. *Polymers* **2020**, *12*, 651. [CrossRef]

36. Velasco-Hogan, A.; Xu, J.; Meyers, M.A. Additive manufacturing as a method to design and optimize bioinspired structures. *Adv. Mater.* **2018**, *30*. [CrossRef]

37. Huang, J.; Zhang, Q.; Scarpa, F.; Liu, Y.; Leng, J. In-plane elasticity of a novel auxetic honeycomb design. *Compos. Part B Eng.* **2017**, *110*, 72–82. [CrossRef]

38. Huang, J.; Gong, X.; Zhang, Q.; Scarpa, F.; Liu, Y.; Leng, J. In-plane mechanics of a novel zero Poisson's ratio honeycomb core. *Compos. Part B Eng.* **2016**, *89*, 67–76. [CrossRef]

39. Boldrin, L.; Hummel, S.; Scarpa, F.; Di Maio, D.; Lira, C.; Ruzzene, M.; Remillat, C.; Lim, T.-C.; Rajasekaran, R.; Patsias, S. Dynamic behaviour of auxetic gradient composite hexagonal honeycombs. *Compos. Struct.* **2016**, *149*, 114–124. [CrossRef]

40. Zaharia, S.M.; Enescu, L.A.; Pop, M.A. Mechanical Performances of lightweight sandwich structures produced by material extrusion-based additive manufacturing. *Polymers* **2020**, *12*, 1740. [CrossRef]

41. Bates, S.R.; Farrow, I.R.; Trask, R.S. Compressive behaviour of 3D printed thermoplastic polyurethane honeycombs with graded densities. *Mater. Des.* **2019**, *162*, 130–142. [CrossRef]

42. Kucewicz, M.; Baranowski, P.; Małachowski, J.; Popławski, A.; Płatek, P. Modelling, and characterization of 3D printed cellular structures. *Mater. Des.* **2018**, *142*, 177–189. [CrossRef]

43. Kucewicz, M.; Baranowski, P.; Małachowski, J. A method of failure modeling for 3D printed cellular structures. *Mater. Des.* **2019**, *174*, 107802. [CrossRef]

44. Yang, C.; Vora, H.D.; Chang, Y. Behavior of auxetic structures under compression and impact forces. *Smart Mater. Struct.* **2018**, *27*, 025012. [CrossRef]

45. Qi, C.; Jiang, F.; Yu, C.; Yang, S. In-plane crushing response of tetra-chiral honeycombs. *Int. J. Impact Eng.* **2019**, *130*, 247–265. [CrossRef]

46. Tabacu, S.; Ducu, C. Experimental testing and numerical analysis of FDM multi-cell inserts and hybrid structures. *Thin-Walled Struct.* **2018**, *129*, 197–212. [CrossRef]

47. Bodaghi, M.; Damanpack, A.; Hu, G.; Liao, W.-H. Large deformations of soft metamaterials fabricated by 3D printing. *Mater. Des.* **2017**, *131*, 81–91. [CrossRef]

48. Hallquist, J. *LS-DYNA® Theory Manual*; Livermore Software Technology Corporation: Livermore, CA, USA, 2006; ISBN 9254492507.

49. Kucewicz, M.; Baranowski, P.; Stankiewicz, M.; Konarzewski, M.; Płatek, P.; Małachowski, J. Modelling and testing of 3D printed cellular structures under quasi-static and dynamic conditions. *Thin-Walled Struct.* **2019**, *145*, 106385. [CrossRef]

50. Baranowski, P.; Płatek, P.; Antolak-Dudka, A.; Sarzyński, M.; Kucewicz, M.; Durejko, T.; Małachowski, J.; Janiszewski, J.; Czujko, T. Deformation of honeycomb cellular structures manufactured with Laser Engineered Net Shaping (LENS) technology under quasi-static loading: Experimental testing and simulation. *Addit. Manuf.* **2019**, *25*, 307–316. [CrossRef]

51. Hanssen, A.; Hopperstad, O.; Langseth, M.; Ilstad, H. Validation of constitutive models applicable to aluminium foams. *Int. J. Mech. Sci.* **2002**, *44*, 359–406. [CrossRef]

52. Baranowski, P.; Małachowski, J.; Mazurkiewicz, Ł. Numerical and experimental testing of vehicle tyre under impulse loading conditions. *Int. J. Mech. Sci.* **2016**, *106*, 346–356. [CrossRef]

53. Baranowski, P.; Małachowski, J.; Janiszewski, J.; Wekezer, J. Detailed tyre FE modelling with multistage validation for dynamic analysis. *Mater. Des.* **2016**, *96*, 68–79. [CrossRef]
54. Freidenberg, A.; Lee, C.; Durant, B.; Nesterenko, V.; Stewart, L.; Hegemier, G. Characterization of the Blast Simulator elastomer material using a pseudo-elastic rubber model. *Int. J. Impact Eng.* **2013**, *60*, 58–66. [CrossRef]
55. Karagiozova, D.; Mines, R. Impact of aircraft rubber tyre fragments on aluminium alloy plates: II—Numerical simulation using LS-DYNA. *Int. J. Impact Eng.* **2007**, *34*, 647–667. [CrossRef]

Article

System Performance and Process Capability in Additive Manufacturing: Quality Control for Polymer Jetting

Razvan Udroiu * and Ion Cristian Braga

Department of Manufacturing Engineering, Transilvania University of Brasov, 29 Eroilor Boulevard, 500036 Brasov, Romania; braga.ion.cristian@unitbv.ro
* Correspondence: udroiu.r@unitbv.ro; Tel.: +40-268-421-318

Received: 19 May 2020; Accepted: 3 June 2020; Published: 4 June 2020

Abstract: Polymer-based additive manufacturing (AM) gathers a great deal of interest with regard to standardization and implementation in mass production. A new methodology for the system and process capabilities analysis in additive manufacturing, using statistical quality tools for production management, is proposed. A large sample of small specimens of circular shape was manufactured of photopolymer resins using polymer jetting (PolyJet) technology. Two critical geometrical features of the specimen were investigated. The variability of the measurement system was determined by Gage repeatability and reproducibility (Gage R&R) methodology. Machine and process capabilities were performed in relation to the defined tolerance limits and the results were analyzed based on the requirements from the statistical process control. The results showed that the EDEN 350 system capability and PolyJet process capability enables obtaining capability indices over 1.67 within the capable tolerance interval of 0.22 mm. Furthermore, PolyJet technology depositing thin layers of resins droplets of 0.016 mm allows for manufacturing in a short time of a high volume of parts for mass production with a tolerance matching the ISO 286 IT9 grade for radial dimension and IT10 grade for linear dimensions on the Z-axis, respectively. Using microscopy analysis some results were explained and validated from the capability study.

Keywords: additive manufacturing; material jetting; polymer; machine capability; process capability; statistical process control; quality; variability; tolerance grade

1. Introduction

The applications of additive manufacturing (AM) to industry have developed from rapid prototyping (RP) and rapid tooling to rapid manufacturing (RM). Additive manufacturing will revolutionize future manufacturing as a key technology in the implementation of the new industrial revolution, Industry 4.0 [1].

Nowadays, the AM processes defined by ISO/ASTM 52900-15 [2] standard are starting to find applications in industry. An industrial additive manufacturing system [3] should have six main components: design, pre-processing, manufacture, post-processing, quality control, and maintenance. The performance of AM systems is an important task to be estimated for the production of parts in an industrial process. There are many AM processes [2] and technologies associated with them, as follows:

- Vat photo-polymerization (VP) process with the stereolithography (SLA) technology;
- Binder jetting (BJ) process with 3D inkjet printing (3DP) technology;
- Material extrusion (ME) process with the fused deposition modeling (FDM) technology;
- Material jetting (MJ) process with polymer jetting (PolyJet) and multi-jet printing (MJM) technologies;
- Sheet lamination (SL) process with the laminated object manufacturing (LOM) technology;

- Powder bed fusion (PBF) process with selective laser sintering/melting (SLS/SLM) and electron beam melting (EBM) technologies; and
- Directed energy deposition (DED) process with laser engineered net shaping (LENS) technology.

Polymers have become very popular as materials for AM, being used in most of the AM processes and targeting a variety of applications [4]. The performance of all the AM systems that are connected to the mentioned AM processes should be analyzed in order to determine their capability to produce parts for the industry. The artifacts or test pieces are primarily used to quantitatively assess the geometric performance of AM systems [5]. Additionally, the AM product characterization needs other tests such as feedstock materials characterization, mechanical tests [6,7], and surface texture characterization [8–10]. The test artifacts are intended to reveal the strengths and weaknesses of different additive manufacturing techniques. Furthermore, they allow the comparison of the performances of different AM systems and the same AM system over time [11]. According to [5], three main characteristics, accuracy, resolution, and surface texture, of the AM systems can be estimated based on some standardized artifacts. Thus, seven artifact geometries have been proposed as follows: linear and circular artifact artifacts to test the accuracy, pins, holes, ribs, and slots artifacts to test the resolution, and the surface texture artifact to test the texture of the surfaces.

Current geometric dimensioning and tolerancing (GD&T) standards have been developed based on the capabilities of traditional manufacturing processes as subtractive manufacturing and formative manufacturing methodologies [12]. New GD&T standards need to be implemented for the different AM processes that use a large variety of materials (plastics, metals, composites, ceramics etc.). The effect of process parameters on the mechanical and geometric performances of polylactic acid (PLA) based composite materials was investigated in [13–15] and the results show the great potential of 3D printed composites in different applications. The physical and chemical properties of polymers relevant to dimensional accuracy require different evaluation and quantification of geometrical tolerances in comparison to metal materials. The tolerance standards applicable for metal parts, therefore, cannot be adopted for plastic structures or can only be applied to a very limited extent.

In the production process, the variations and fluctuations in the manufacturing accuracy are influenced by many factors such as machines, workpiece, methods, people, and environment, etc. The inherent fluctuations have less impact on product quality [16]. The abnormal variations have a large impact on product quality [17]. The most known methods used to control and reduce the manufacturing process variation are the statistical process control, measurement system analysis, six sigma method, and Taguchi's design of experiments [18].

Statistical process control (SPC) uses statistical methods in quality control to monitor, maintain, and improve the capability of manufacturing processes to assure product conformance [16,19]. Akande et al. [20] analyzed quality characteristics of strength, bending stiffness, density, and dimensional accuracy of parts built by the SLS process using SPC control charts. They concluded that SPC ensures consistency in product quality for long term production.

Any quality control process needs to quantify, first, the machine capability (short-term study or machine performance) in one continuous production run and manufacturing process capability (long-term study) in series production [19,21]. Measurement process capability provides the evidence for conformity or nonconformity with specification according to ISO 14253:2017 [22].

Experimental and theoretical studies have been developed in order to characterize the performance of AM processes and have particularly focused on quality control in additive manufacturing. Additionally, the standards focused on AM systems are under development. The use of AM processes in mass production depends on the part quality. Some issues are the inconsistency of AM repeatability and reproducibility that have not been solved yet for all the AM processes. Singh et al. [18] analyzed the repeatability of acrylonitrile butadiene styrene (ABS) replicas built by the FDM process and chemical vapor smoothing, but the repeatability variation and the appraiser variation were not calculated. Baturynska [23], using statistical analysis, attempted to improve the dimensional accuracy of the parts built by polymer powder bed fusion. She developed linear regression models to predict the value of the thickness, width, and length of rectangular specimens and to compensate for the shrinkage effect. The material jetting process allows time between the jetting of each layer of material to relieve internal stresses [24]. George et al. [24] reviewed the accuracy and reproducibility of 3D printed medical models from polymers, using material extrusion (FDM), powder bed fusion (SLS), binder jetting, and material jetting. They concluded that regular testing of the accuracy of AM systems and preventive maintenance are necessary steps for quality assurance. Preißler et al. [25] investigated a process capability for a fused filament fabrication (FFF) process using PLA material, based on a customized pyramid object manufactured in 25 samples. The results for a 30 mm dimension and tolerance of ±0.2 mm through the quality control chart shows that the process was not in the statistical control.

Singh [26] investigated the process capability of the linear dimensions of a prismatic component built by PolyJet technology from an EDEN 260 machine. The results of this study suggested that that process lies in the ±4.5 sigma limit with regard to the dimensional accuracy of the chosen specimen. However, the variability of the measurement system was not performed and the number of 16 parts used to determine the process capability was too low, according to the capability standards [27]. Kitsakis et al. [28] investigated the IT (International Tolerance) grades for the dimensions of eight samples printed with deposition layers of 30 microns on the Objet Eden 250 3D printer, and they assigned the IT11 grade for it. The variability of the measurement system used in this study was not accomplished. Yap et al. [29] investigated the design capability and manufacturing accuracy of the PolyJet 3D printing process on an Objet500 Connex3 PolyJet printer using artifacts with customized features and concluded that the accuracy of the parts printed in glossy mode was better than that in matte finishing, but the minimum clearance gap for parts was obtained in a matte finish. Minetola et al. [30] evaluated the dimensional accuracy of three AM systems for polymeric materials using the ISO IT grades of an artifact from the GrabCAD library, building two replicas. They concluded that a smaller layer thickness provided higher dimensional accuracy of the part dimensions.

From the literature survey, the results are as follows:

- The AM artifacts are intended to investigate the strengths and weaknesses of additive manufacturing processes and they allow the comparison of the performances of different AM systems.
- The AM process control has an important role on the part quality, but there is a lack of adequate AM control methods and standards. There is still no AM standard for machine performance and process capability determination in mass production.
- Only a few research studies have focused on the repeatability, ISO IT grades, and process capability of polymer based AM systems.

Having benefits in terms of cost reduction and shorten of the time-to-market in products, the implementation of polymer-based AM technologies within production depends on the process capability and control.

The main aim of this article was to define a methodology for statistically analyzing the AM system performances and AM process control. A case study regarding the EDEN 350 AM system and polymer jetting process was conducted to validate the proposed basic methodology.

2. Materials and Methods

2.1. New Methodology for Statistical Quality Tools in AM Production

The main objectives of the new methodology in AM are to define statistical quality tools based on standards for the assessment of the variability of the measurement system, the additive manufacturing repeatability, AM system capability or AM system performance, and AM process capability. This methodology includes experiments, statistical analysis, and results interpretation. SPC tools are used to provide the mean of identifying possible changes in the process [16].

The new methodology in AM consists of a preparatory step, followed by six main steps, as shown in Figure 1.

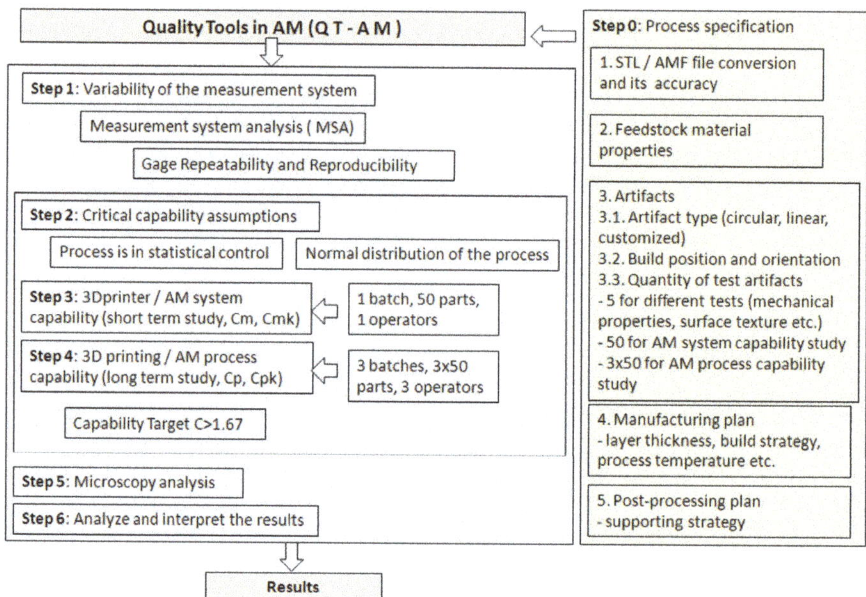

Figure 1. Flowchart of the proposed methodology/procedure named Quality Tools in AM (QT-AM).

The preparatory step defines the AM process specification as follows: STL (Standard Triangulation Language) or AMF (Additive manufacturing file) file conversion and its accuracy, feedstock material properties, artifact type, build orientation and position, the sample size of specimens, and the manufacturing and the post-processing plan. According to the ISO/ASTM 52901:2017 standard [31], the part definition made by AM, for a purchase purpose, should include the following characteristics: part geometry, tolerances, surface texture, feedstock material, build orientation, acceptable imperfections or deviations, and process control information (e.g., repeatability). The main characteristics of part geometry can be defined as a digital file containing the 3D model and a part engineering drawing.

In traditional manufacturing, the specific requirements (dimensions, tolerances, surface finish, material, etc.) of the 3D model and drawings are set based on standards according to product material. Thus, ISO 286 is usually used for parts made of metal [32] and DIN 16742 for plastic parts [33]. In AM, the general tolerances for linear dimensions are specified according to the general standard ISO 2768-1 [34], based on the ISO/ASTM 52901:2017 recommendation. The surface texture or surface finish of the part should be specified by a maximum value.

Feedstock material properties need to conform to the suppliers' specifications. Artifact manufacturing should be undertaken according to a manufacturing plan (layer thickness, build strategy, process temperature). The CAD model of the artifact is converted to a STL file format. The conversion parameters used within different CAD software as well as any maximum deviation (chord height and angular tolerance) should be chosen correlated to the 3D printing layer thickness. Where supports cannot be avoided, a supporting strategy should be documented. It includes the support geometry, support material, the removal technique, and the specific post-processing treatments. The support material can be made from the same material as the artifact (model) material or can be different. The application of support structures or support material should be minimized on the critical features of the part.

The amount of variability induced in measurements by the measurement system itself should be determined before any capability study is performed (Figure 1). The measurements were performed using a grade "A" measurement method according to the ASTM 52902-19 standard [5]. Therefore, for simple and inexpensive measurements commonly available in a shop floor, a digital caliper was used.

In the second phase of the methodology, the critical capability assumptions are analyzed, as shown in Figure 1. AM machine capability or AM system performance has the main purpose of the checking of existing 3D printers, objective arguments in case of 3D printer defects, and findings for target specifications when purchasing a new 3D printer. 3D printer/AM system capability and 3D printing/AM process capability studies are determined within the third and fourth steps. Capability is the ability of a system, or process, to realize a product that will fulfill the requirements for that product. Capability conditions under which the process is evaluated include the following, according to the ISO 22514-1:2014 standard [35]:

- Methods applied to demonstrate that the process is in control;
- Technical conditions (input batches, operators, tools, etc.);
- Measurement process (resolution, repeatability, reproducibility, etc.); and
- Data collection (duration, frequency).

Capability analysis should be carried out for a new or changed production process and then over time to control the process according to the standard ISO/TS 16949 [36]. Capability analysis is summarized in indices that show the system's ability to meet its requirements. Machine and process capabilities provide results on how well a machine and a process performs in relation to defined tolerance limits. These two branches differ because they are determined in different conditions, but principally similar indices are calculated. The target capability indices commonly used in the automotive industry are greater than 1.67, which corresponds to a safety or critical parameter for a new process [16]. The quality condition is excellent if the capability indices are between 1.67 and 2 [16,37].

A quality inspection through a microscopy study is performed in the fifth step of the methodology. Optical micrographs were performed using a Zeiss O-Inspect (Carl Zeiss Industrielle Messtechnik, Oberkochen, Germany) multi-sensor measuring machines.

2.2. Process Specifications. Materials, Artifact, and Manufacturing Method

In this work, a part used in the pre-production of plastic parts has been selected as the benchmark. The part presents similar geometric basic features as the circular artifact shown in Figure 2. The circular artifact consists of a circular upper surface and a steep lower surface ending with a sharp edge (Figure 2). Two critical dimensions in terms of assembly and functionality of the artifact, the height H = 12 mm, and the diameter D = 14.5 mm, have been selected for the machine and process capability study. A fine tolerance class of ±0.1 mm was selected, taking into account the ranges of nominal lengths between 6–30 mm according to the ISO 2768-1 standard [34].

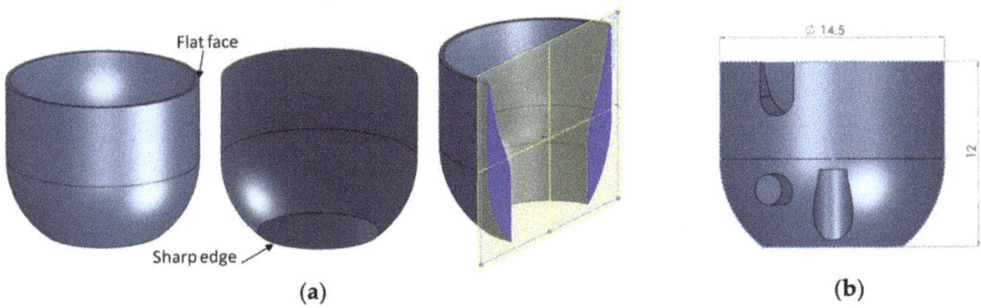

Figure 2. (**a**) Views and section view of the artifact. (**b**) The part used in the pre-production.

SolidWorks version 2013 software (Dassault Systèmes, Massachusetts, MA, USA) was used to design the 3D model and to generate the STL file. The 3D model of the part was converted into a STL file, which is the input file format of the Objet EDEN 350 PolyJet machine (Stratasys, Rehovot, Israel) [38]. The STL file conversion tolerances were set to a deviation of 0.01 mm and an angular tolerance of 4 degrees.

Feedstock materials used in this study were Objet VeroBlue RGD840 resin used as the model material and FullCure 705 as the support material [39]. The composition of the Objet VeroBlue RGD840 resin consists of an acrylic monomer, urethane acrylate oligomer, epoxy acrylate, and photo-initiator. FullCure 705 resin is made of an acrylic monomer, polyethylene glycol 400, propane-1, 2-diol, glycerol, and photo-initiator. The main properties of the Objet VeroBlue RGD840 material are shown in Table 1 [39]. Characteristics may vary if different orientations of specimens and test conditions are applied [6,7].

Table 1. Objet VeroBlue RGD840 properties [39].

Property	ASTM	Metric
Tensile Strength	D-638-03	50–60 MPa
Elongation at Break	D-638-05	15–25%
Flexural Strength	D-790-03	60–70 MPa
Rockwell Hardness	Scale M	73–76 Scale M
Water Absorption	D-570-98 24h	1.5–2.2%

The orientation of the specimens on the build tray affects how quickly, efficiently, and qualitatively they will be manufactured by the AM system [40]. Additionally, within the PolyJet process, the orientation of parts has an influence on the quantity and where the support material is used. The circular specimens were printed in a standing up position on the build platform, as shown in Figure 3. It is advantageous to print a circular model that has holes standing up on the build platform, so support material does not fill the holes [38]. Additionally, if a circular model is lying down on the build platform and printed in glossy printing mode, then the surface quality is affected by some errors [41]. The experimental roughness (Ra) values for the PolyJet material jetting process are specified according to the finish type as follows: for matte finish in the range of 0.5–15 µm, and for the glossy finish in the range of 0.5–4 µm [42]. The dimensional accuracy and the quality of the surface of a circular artifact built in standing up position are not significantly influenced by the orientation and positioning on the build platform.

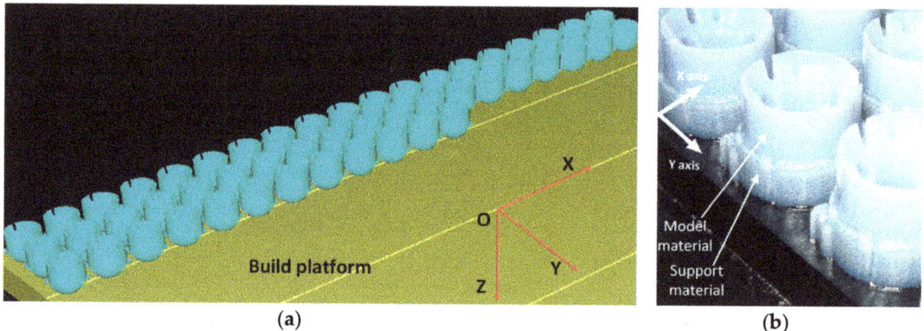

Figure 3. (**a**) Layout of the EDEN 350 build platform illustrating the 50 parts patterned in an array. (**b**) Detail of the printed specimens on the build platform.

An Objet EDEN 350 PolyJet system was used to manufacture the specimens. Based on drop-on-demand (DOD) inkjet technology [43], the PolyJet system deposits layers of resin droplets of 0.016 mm thick. It levels each deposited resin layer and hardens it using ultraviolet (UV) light. During the process, the print heads and the photopolymer resins are heated at around 72 °C. The print heads were vacuumed at 6.2 atm. The experiments were performed under a controlled laboratory temperature of 20 °C and relative humidity of 30%. PolyJet 3D printers only use a solid infill pattern on parts. A different infill type can be added in the design stage of the CAD model, but the part's interior will likely be filled with support material in the printing process. A solid infill pattern was used for all of the samples.

The build platform preparation (Figure 3), STL model slicing, and G-code generation were performed using the Objet Studio client/server software (Objet Geometries, Rehovot, Israel). The specimens were 3D printed in a glossy finish style. Only the bottom surfaces of the specimen were affected by the support material. The support material was removed with a pressure water jet from the bottom surface of the 3D printed specimens.

The density of the printed material was determined using the Archimedes density method [44,45] by calculating the volume of five specimens, in addition to determining the mass of the parts using a precision scale. The results showed a mean density measured of the printed material of 1.15 g/cm^3.

One batch of 50 parts was 3D printed for the AM system capability study, and three batches each containing 50 parts for the AM process capability study. A batch of 50 artifacts was manufactured in 1 h and 40 min, using 78 g of model material and 54 g of support material.

2.3. The Variability of the Measurement System

Within both manufacturing processes and quality systems, there is variation. All measurement data had some degree of variance or errors. A robust statistical process control (SPC) process requires accurate data to have the greatest impact on product quality. The percentage of variance due to the measurement system has to be determined. The measuring system can be affected by various sources of variation, called factors [46]: measuring instruments, operators, measuring method, specifications (the engineering tolerance), and parts or specimens.

The variability of the measurement system was determined by Gage repeatability and reproducibility methodology. Repeatability is due to measuring instrument variation and reproducibility is due to operator (appraiser) variation. Gage R&R study was performed using the analysis of variance (ANOVA) method [47]. The ANOVA Gage R&R method estimates:

- The amount of measurement system variation compared with the process variation;
- The amount of variation in the measurement system that is due to operator influence; and
- The measurement system's capability to discriminate between different parts.

The measurement system used in the analysis included:

- A Mitutoyo 500-196-30 digital scale caliper with advanced onsite sensor (AOS), a measuring range from 0 to 150 mm, and resolution 0.001 mm was used;
- The method that describes the way to keep the part in hand and the area to be measured for the height and for the diameter;
- A sample of 10 parts was used to be measured by three operators, twice, for each characteristic, the height, and the diameter. The parts were measured randomly;
- Circular 3D printed parts were made of polymers; and
- A controlled laboratory temperature of 20 °C and relative humidity of 30%.

Using Minitab 19 software (Minitab, Ltd., Coventry, United Kingdom) [48], a worksheet for Gage R&R analysis was created and the order of the measurements for each operator was imposed. The total sample size was 60 measurements.

2.4. System and Process Capability for PolyJet Technology

The Gauge R&R should be proven before the capability analysis. Two critical assumptions need to be considered when performing the machine and process capability analyses with continuous data, namely, the process is in statistical control, and a normal distribution of the process is required. A process is considered stable if its output is within the predictable limits. In order to assess whether or not a process is in statistical control, it uses control charts [16,19].

Short-term performance studies are typically performed on machines where parts are produced consecutively under repeatability conditions and the sample size produced is at least 50 workpieces to be manufactured in one shift [27]. 3D printer capability or AM system capability is used to assess the quality and performance of a single AM machine. The AM system capability was evaluated within the following conditions:

- 50 parts are printed at once;
- One operator manages the 3D printing process;
- The variation of the material batch or the printer user variation is not included in the total variation of the process; and
- The parts are measured and the data statistically analyzed.

The quality of the production processes is measured by establishing some characteristics and monitoring the long-term capability of its 3D printing process capability or AM process capability is a long-term study on a stable process that indicates the performance quality of the 3D printing process. The AM process capability can be evaluated within the following conditions:

- The specimens are 3D printed in three batches, each batch containing 50 specimens;
- Different operators manage the 3D printing process of the three batches based on the established parameters; and
- The parts are measured and the data transposed into the Destra software (Q-DAS GmbH, Weinheim, Germany), with the order of the measurements not being important.

D and H dimensions of the parts were measured using the Mitutoyo 500-196-30 digital scale caliper (Mitutoyo Corporation, Kawasaki, Japan). System and process capability is determined by calculating the capability coefficients described in Equation (1). The lower specification limit (LSL) and upper specification limit (USL) are the targets set for the process. The potential machine and process capability indices (C_m, C_p) represent the number of times the process spread fits into the tolerance interval. A high potential capability index does not guarantee that the process is close to the target value, which is why the position of the process spread in relation to the tolerance interval is determined by calculating the critical capability machine/process index (C_{mk}/C_{pk}).

$$\left\{ \begin{array}{l} C_i = \dfrac{USL-LSL}{x_{i\_99.865\%}-x_{i\_0.135\%}} \\[2mm] C_{ik} = min\left\{ \dfrac{USL-x_{i_{50}\%}}{x_{i_{99.865}\%}-x_{i_{50}\%}}, \dfrac{x_{i_{50}\%}-LSL}{x_{i_{50}\%}-x_{i_{0.135}\%}} \right\} \quad ,i = \{m - machine, p - process\} \\[2mm] C_{target} = 1.67(1.33) \end{array} \right. \tag{1}$$

The location and dispersion were calculated using the $M1_{1,6}$ method according to the ISO 22514-2: 2017 standard [21]. Subscripts 1 and 6 refer to equations for calculating the estimator for the location and dispersion, respectively. This means that the arithmetic mean of the values is used for the location being assumed, and externally tested the normal distribution, and the distance between the edges 0.135% and 99.865% for dispersion. The reference interval of the product characteristic is bounded by the 99.865% distribution quantile, and the 0.135% distribution quantile. The length of the interval is $X_{99,865\%}-X_{0,135\%}$ [35]. $X_{50\%}$ represents the 50% distribution quantile.

The results of the short-term and long-term capabilities were analyzed based on the requirements from the SPC Reference Manual, from Automotive Industry Action Group (AIAG) [19]. Destra software [49] was used to perform the capability study. The capability can be evaluated graphically by drawing capability histograms and capability plots. The requirements for indices C_m, C_{mk}, C_p, and C_{pk} demand a minimum value of 1.67 for all of them.

2.5. Capable Tolerance Specification for PolyJet Technology

Tolerance specification (tolerance, lower and upper limits) for the dimensions of the 3D printed circular part was chosen based on the general tolerances standards [34] and plastics molded parts tolerances [33]. A tolerance of ±0.1 mm was selected. Based on this specification, the AM system and process capability for PolyJet technology was calculated. The capability indices were compared with a capability target index of 1.67.

Rather than estimating the process capability for a particular tolerance, a capable tolerance and its limit deviations were calculated based on a target capability index. The target capability index was set to 1.67. The index of the process K was calculated using Equation (2) and describes the level by which the process is off target value and represents an appropriate measure of process centering [37,50]. The lower (LSL$_T$) and upper (USL$_T$) specification limits of the capable tolerance were calculated based on Equation (3). Upper limit deviation (ULD) and lower limit deviation (LLD) from nominal size were then determined based on Equation (3).

$$K = \frac{(USL + LSL) - 2x_{mean}}{USL - LSL} \tag{2}$$

The process mean is positioned between the midpoint of the specifications and one of the required limits if $0 < |K| < 1$. $|K| > 1$ indicates that the process mean is situated outside the required limits.

$$
\begin{cases}
\textit{if } K > 0 \textit{ then}
\begin{cases}
LSL_T = X_{50\%} - C_{Pk}(X_{50\%} - X_{0.135\%}) \\
LLD = T_m - LSL_T \\
USL_T = LSL + C_P(X_{99.865\%} - X_{0.135\%}) \\
ULD = USL_T - T_m
\end{cases} \\
\textit{if } K < 0 \textit{ then}
\begin{cases}
USL_T = X_{50\%} + C_{Pk}(X_{99.865\%} - X_{50\%}) \\
ULD = USL_T - T_m \\
LSL_T = USL - C_P(X_{99.865\%} - X_{0.135\%}) \\
LLD = T_m - LSL_T
\end{cases}
\end{cases}
\tag{3}
$$

The capable lower limit deviation and capable upper limit deviation were determined based on the relations LLD$_C$ < LLD$_T$ and ULD$_C$ > ULD$_T$. The capable tolerance is calculated as follows: T_c = ULD$_C$ − LLD$_C$. A confirmatory analysis of AM process capability was performed using the determined capable tolerance.

3. Results and Discussion

3.1. The Variability of the Measurement System

The variance components (VarComp) compare the variation from each source of measurement error to the total variation. In these results, the %Contribution column (Table 2) shows that the variation from Part-To-Part for H and D dimension was 99.46%/99.12%, which is much larger than the total Gage R&R, which was 0.54%/0.88%. Thus, the largest part of the variation was due to the differences between parts. This means that the measurement system can reliably distinguish between parts.

Table 2. Variance components for the characteristic diameter and height.

Source	VarComp [1]	Contribution [1]	VarComp [2]	Contribution [2]
Total Gage R&R	0.0000032	0.88%	0.0000029	0.54%
Repeatability	0.0000031	0.86%	0.0000028	0.52%
Reproducibility	0.0000001	0.02%	0.0000001	0.01%
Operators	0.0000001	0.02%	0.0000001	0.01%
Part-To-Part	0.0003585	99.12%	0.0005336	99.46%
Total Variation	0.0003617	100%	0.0005365	100%

[1] Diameter, [2] Height.

The measurement system variation compared to the total variation is shown in Tables 3 and 4. The total Gage R&R equaled 7.33%/9.37% of the study variation for the H and D dimensions. In order to evaluate the capability of the measurement system to evaluate parts versus specification, the values %Tolerance are used, these values being calculated for each characteristic as the ratio between the study variation for each source and the process tolerance.

Table 3. Gage evaluation for diameter (D).

		Study Var	%Study Var	%Tolerance
Source	StdDev (SD)	(6 × SD)	(SV)	(SV/Toler)
Total Gage R&R	0.0017829	0.010698	9.37%	5.35
Repeatability	0.0017611	0.010566	9.26%	5.28
Reproducibility	0.0002783	0.00167	1.46%	0.83
Operators	0.0002783	0.00167	1.46%	0.83
Part-To-Part	0.0189344	0.113606	99.56%	56.8
Total Variation	0.0190181	0.114109	100%	57.05
Number of Distinct Categories = 14				

Table 4. Gage evaluation for height (H).

		Study Var	%Study Var	%Tolerance
Source	StdDev (SD)	(6 × SD)	(SV)	(SV/Toler)
Total Gage R&R	0.0016968	0.010181	7.33%	5.09
Repeatability	0.0016783	0.01007	7.25%	5.03
Reproducibility	0.00025	0.0015	1.08%	0.75
Operators	0.00025	0.0015	1.08%	0.75
Part-To-Part	0.0231008	0.138605	99.73%	69.3
Total Variation	0.0231631	0.138978	100%	69.49
Number of Distinct Categories = 16				

The repeatability variation and the reproducibility variation, which shows the equipment variation (EV) and the appraiser variation (AV), respectively, were lower than 10%. Based on the requirements specified in the MSA 4 [46], the measurement system can be accepted. The number of distinct categories was greater than five (Tables 3 and 4), resulting in an acceptable measurement system [46].

The variability results of the measurement system are graphically provided in Figures 4 and 5. In the Components of Variation graph, the %Contribution from Part-To-Part is larger than that of the total Gage R&R. Thus, much of the variation is due to differences between parts. The R Chart by Operator shows that Operators measured parts consistently. In the Xbar Chart by Operator, most of the points were outside the control limits. Thus, much of the variation is due to differences between parts.

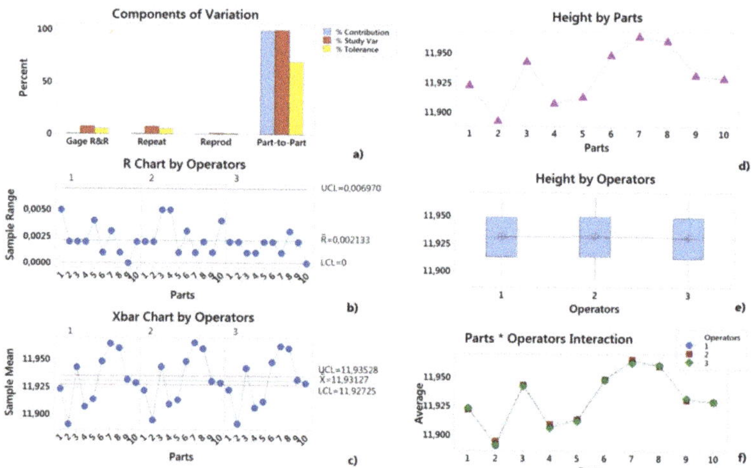

Figure 4. Gage R&R (ANOVA) report for height (H): (**a**) Components of Variation graph; (**b**) R Chart by Operator; (**c**) Xbar Chart by Operators; (**d**) By Parts graph; (**e**) By Operators graph; (**f**) Parts * Operators interaction.

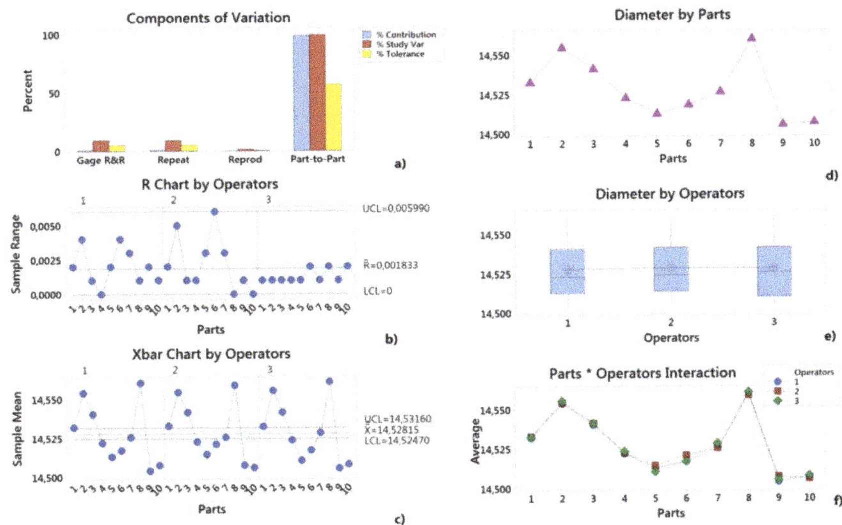

Figure 5. Gage R&R (ANOVA) report for diameter (D): (**a**) Components of Variation graph; (**b**) R Chart by Operator; (**c**) Xbar Chart by Operators; (**d**) By Parts graph; (**e**) By Operators graph; (**f**) Parts * Operators interaction.

The By Operator graphs (Figures 4e and 5e) show that the differences between operators were smaller than the differences between parts. In the Parts * Operators Interaction graphs (Figures 4f and 5f), the lines were approximately parallel and the p-value for the Parts * Operators interaction was 0.779/0.195 for the H and D dimensions. This indicates that no significant interaction between each Parts and Operators exists.

The Gage R&R result shows that for the height as well as for the diameter, a variation due to the measurement system was much lower than the part-to-part variation, as a result, the next studies could be based on measurements.

3.2. System Performance of Objet EDEN 350 PolyJet

First, both critical assumptions for performing the machine capability (system performance) analyses were graphically checked. The control charts from Figures 6 and 7 show the manufacturing process information for all 50 measurements of the D and H measured dimensions. The distributions were stable over the period of study, as shown in Figures 6 and 7.

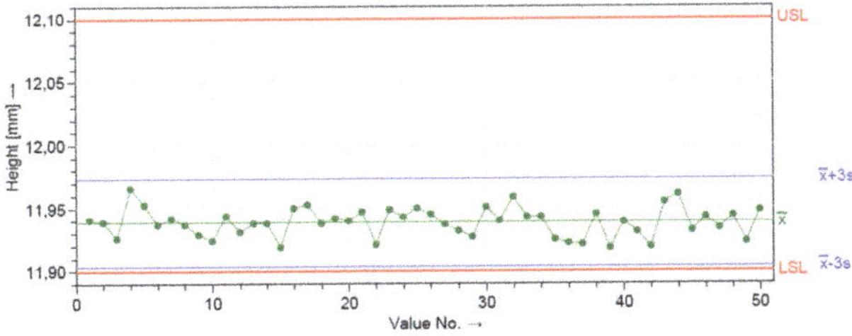

Figure 6. Control chart for the short-term capability study of the height dimension (H).

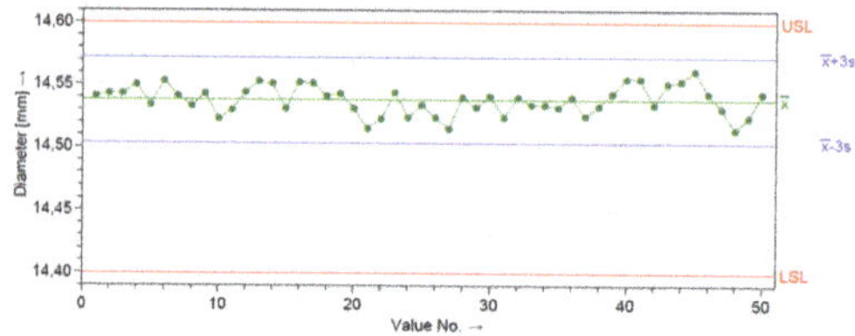

Figure 7. Control chart for the short-term capability study of the diameter dimension (D).

The dimensional values lay within the LSL and USL, indicating that the process is in statistical control for both dimensions. A normal distribution was detected based on the Anderson–Darling normality test. Figure 8 shows the histograms of the individuals and the distribution models.

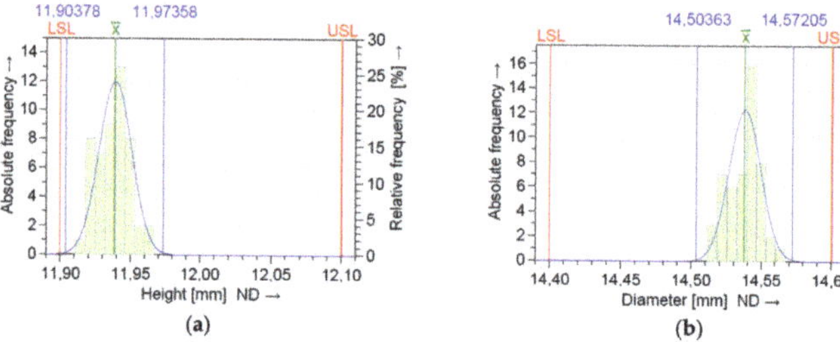

Figure 8. Histogram of the individuals and the distribution model for short term capability study: (**a**) height (H); (**b**) diameter (D).

The measurements were located near the upper specification limit (USL) of the diameter (D) and near the lower specification limits (LSL) of the height (H), respectively. This graphics show the shape of the subgroup frequencies.

The numerical results of the machine capability analysis are shown in Tables 5 and 6 for both dimensions of the circular specimen, where Tm is the tolerance center, T is the tolerance of the characteristic, n is the sample size, x_{min} the minimum value of the characteristic, x_{max} is the maximum value of the characteristic, x_{mean} is the median of all values, StDev is the standard deviation of all individuals, $X_{0.135\%}$ is the 0.135% distribution quantile, $X_{50\%}$ is the 50% distribution quantile, and $X_{99.865\%}$ is the 99.865% distribution quantile. The potential and the critical capability index both showed three values (Figure 9) that specify the two-sided 95% confidence interval for the respective capability index: lower confidence limit, estimator, and upper confidence limit.

Table 5. Machine capability analysis for the H dimension.

Drawing Values		Collected Values		Statistics	
Tm	12	n	50	StDev	0.0116
LSL	11.9	x_{min}	11.918	$X_{0.135\%}$	11.90378
USL	12.1	x_{max}	11.966	$X_{99.865\%}$	11.97358
T	0.2	x_{mean}	11.939	$X_{50\%}$	11.93868

Table 6. Machine capability analysis for the D dimension.

Drawing Values		Collected Values		Statistics	
Tm	14.5	n	50	StDev	0.0114
LSL	14.4	x_{min}	14.514	$X_{0.135\%}$	14.50363
USL	14.6	x_{max}	11.561	$X_{99.865\%}$	14.57205
T	0.2	x_{mean}	14.54	$X_{50\%}$	14.53784

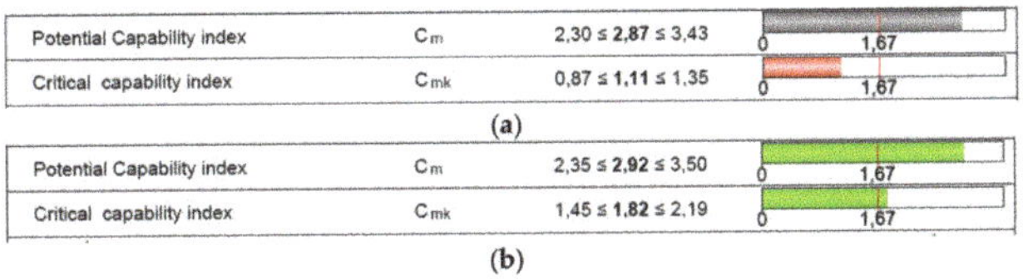

Potential Capability index	C_m	$2,30 \leq 2,87 \leq 3,43$	
Critical capability index	C_{mk}	$0,87 \leq 1,11 \leq 1,35$	

(a)

Potential Capability index	C_m	$2,35 \leq 2,92 \leq 3,50$	
Critical capability index	C_{mk}	$1,45 \leq 1,82 \leq 2,19$	

(b)

Figure 9. Machine capability analysis report for: (**a**) dimension H; (**b**) dimension D.

The requirements for indices C_m and C_{mk} were met for the D dimension (Figure 9b). Based on the measured parts, the critical capability index was lower than the target for the characteristic H. Therefore, the 3D printer capability was not proven (Figure 9a).

3.3. Process Capability of PolyJet

The control charts of the process capability for both dimensions of the diameter and height of the circular specimen are shown in Figures 10 and 11. The Xbar-S control charts for the subgroups with the sample size of five pieces were chosen to check if the process variation was in control. The mean data and standard deviation data showed that none of the points were outside the control limits (UCL, upper control limit, LCL, lower control limit), and the points displayed a random pattern. Thus, the process variation was in control.

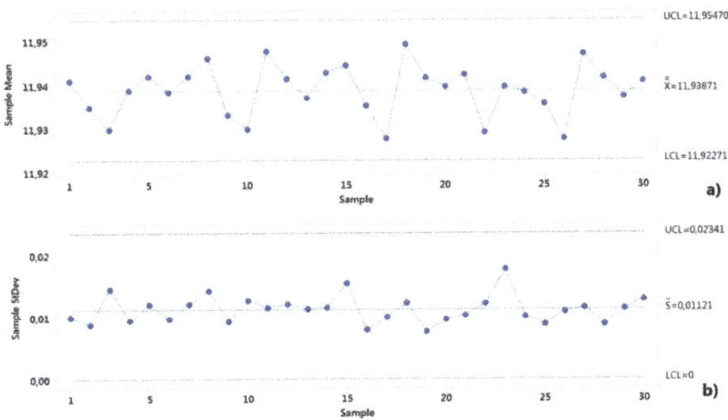

Figure 10. The control charts for the long-term capability study of height dimension (H): (**a**) mean data; (**b**) standard deviation data.

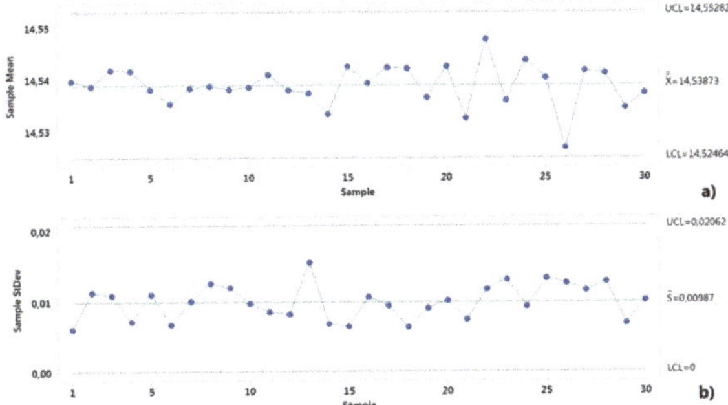

Figure 11. The control charts for the long-term capability study of diameter dimension (D): (**a**) mean data; (**b**) standard deviation data.

The model distribution of the data for the dimensions H and D showed a normal distribution, as shown in Figure 12. The entire production process was stable and controllable.

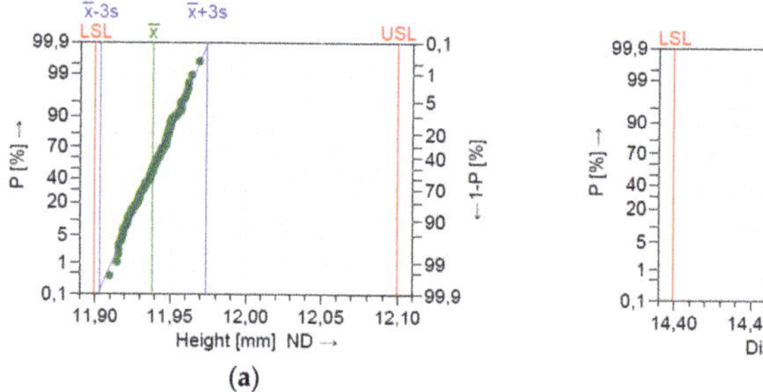

Figure 12. Normal probability plot graph for the long-term capability study: (**a**) height dimension (H); (**b**) diameter dimension (D).

The location of the process distribution (Figure 13) was near the upper tolerance limits for the dimension of diameter (D) and near the lower tolerance limits for the dimension of height (H), respectively.

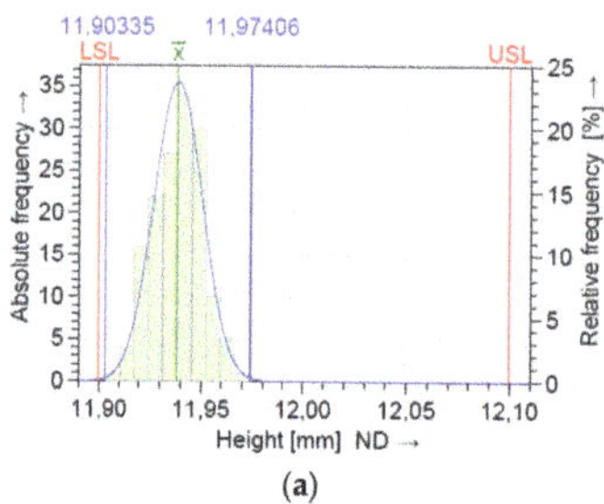
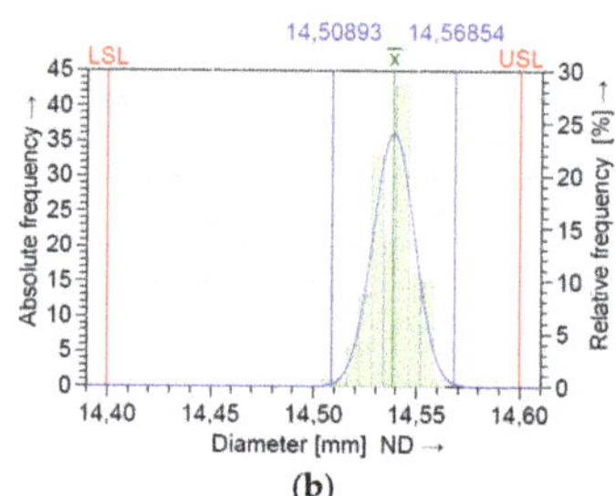

Figure 13. Histogram of individuals and the distribution model for the long-term capability study: (**a**) height dimension (H); (**b**) diameter dimension (D).

The numerical results of the process capability analysis are shown in Tables 7 and 8 for both dimensions of diameter and height. The standard deviation of height was slightly larger than that of diameter.

Table 7. Process capability analysis for the H dimension.

Drawing Values		Collected Values		Statistics	
Tm	12	n	150	StDev	0.0118
LSL	11.9	x_{min}	11.910	$X_{0.135\%}$	11.90335
USL	12.1	x_{max}	11.969	$X_{99.865\%}$	11.97406
T	0.2	x_{mean}	11.939	$X_{50\%}$	11.93871

Table 8. Process capability analysis for the D dimension.

Drawing Values		Collected Values		Statistics	
Tm	14.5	n	150	StDev	0.00994
LSL	14.4	x_{min}	14.510	$X_{0.135\%}$	14.50893
USL	14.6	x_{max}	14.562	$X_{99.865\%}$	14.56854
T	0.2	x_{mean}	14.540	$X_{50\%}$	14.53873

Based on the requirements, the target for C_{pk} is very often established at a minimum of 1.67. Some of the industry manufacturers accept even lower values of 1.33 for C_p and C_{pk} [51]. Even so, the result for the height characteristic in terms of C_{pk} was lower than 1.67 or 1.33. The requirements for indices C_{pm} and C_{pk} were met for the diameter dimension (Figure 14), but were not met for the height dimension.

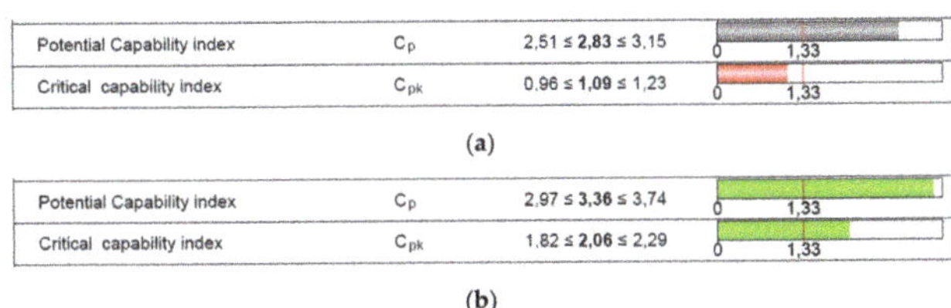

Potential Capability index	C_p	$2{,}51 \le 2{,}83 \le 3{,}15$
Critical capability index	C_{pk}	$0{,}96 \le 1{,}09 \le 1{,}23$

(a)

Potential Capability index	C_p	$2{,}97 \le 3{,}36 \le 3{,}74$
Critical capability index	C_{pk}	$1{,}82 \le 2{,}06 \le 2{,}29$

(b)

Figure 14. Process capability analysis report: (**a**) height dimension (H); (**b**) diameter dimension (D).

3.4. Capable Tolerance and Its Limits Deviation for PolyJet Process

The calculation of the capability indices was based on the location and dispersion of the characteristic value with respect to the specified tolerance. x_{mean} indicates the location of the process. It can be observed from the process capability graphics (Figure 13) that the x_{mean} was lower than the nominal value for the H characteristic and higher for the D characteristic, respectively.

Capable tolerance and its limit deviations were calculated based on a target capability index of 1.67 for both dimensions of height and diameter. The index of process K was calculated, and the results showed the value of 0.62 for the height and −0.38 for the diameter, respectively. The capable lower limit deviation and capable upper limit deviation were determined for both dimensions.

The capable limit deviations of the circular specimen were found as ULD = max{ULD$_D$, ULD$_H$} and LLD = min{LLD$_D$, LLD$_H$}, where the subscripts H and D represent the characteristic height and diameter, respectively. The results show that the capable lower limit deviation and capable upper limit deviation of the circular artifact were LLD = −0.13 mm and ULD = +0.09 mm, respectively. The capable tolerance interval of the circular artifact was T_C = 0.22 mm.

A confirmatory analysis of AM process capability was performed using the determined capable tolerance of the circular artifact. The process capability result was "too high" (C_{pk}>1.67), as shown in Figure 15. Thus, the requirements were met.

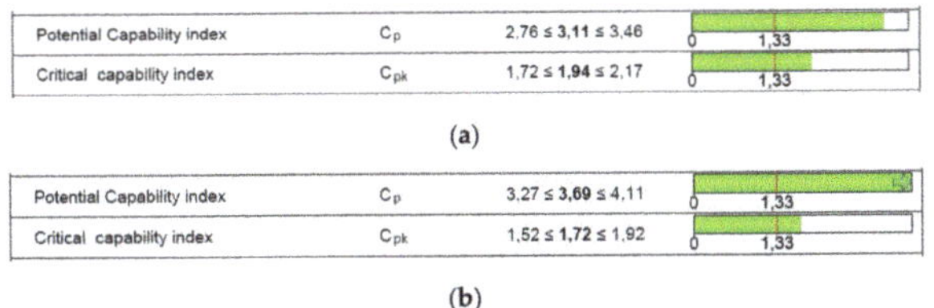

Potential Capability index	C_p	$2{,}76 \le 3{,}11 \le 3{,}46$
Critical capability index	C_{pk}	$1{,}72 \le 1{,}94 \le 2{,}17$

(a)

Potential Capability index	C_p	$3{,}27 \le 3{,}69 \le 4{,}11$
Critical capability index	C_{pk}	$1{,}52 \le 1{,}72 \le 1{,}92$

(b)

Figure 15. Process capability analysis report based on the capable tolerance of the circular artifact: (**a**) height dimension (H); (**b**) diameter dimension (D).

3.5. Determination of Tolerance Grade (ISO IT grade)

Tolerance grades indicate the degree of accuracy of manufacture. Since IT grades provide guidance on how precise a manufactured feature of a particular size should be, they can be used to compare different manufacturing processes [52]. The lower value of IT Grade implies a better dimensional accuracy. The IT Grade was calculated for 50 specimens of the circular artifact, based on the standard ISO 286 specifications [32]. The dimensional accuracy and IT Grade depend on the size of the feature.

Two dimensions of the circular artifact, the height and the diameter were analyzed. These sizes were within the ISO basic size range of (10–18 mm).

$$n_i = \frac{|D_N - D_{Mi}|}{0.45\left(\sqrt{D_{min}D_{max}}\right)^{\frac{1}{3}} + 0.001\sqrt{D_{min}D_{max}}}, i = \{1, \ldots, 50\} \tag{4}$$

The relative magnitude of each IT (International Tolerance) Grade is calculated relative to the standard tolerance unit i. The standard tolerance unit is i = 1.083 μm for the ISO basic size range of D_{min} = 10 mm to D_{max} = 18 mm. The tolerance unit "n" was calculated using Equation (4), where 'D' is the geometric mean of the ISO basic size range; D_N is the nominal dimension; and D_M is the measured dimension. Table 9 shows the IT grades for the height (H), and the diameter (D) of the circular specimen. The tolerance grades were determined based on the tolerance unit n.

Table 9. International tolerance grades for the circular artifact.

ISO 286 Standard Requirements		IT8	IT9	IT10	IT11
Max magnitude of the tolerance zone		25 i	40 i	64 i	100 i
Size range (10–18 mm), i = 1.083 μm		27 μm	43 μm	70 μm	109 μm
Collected values					
n	Linear dimension (Height = 12 mm)	-	(32–42) μm	(44–69) μm	(73–75) μm
	Radial dimension (Diameter = 14.5 mm)	(19–24) μm	(28–41) μm	(47–57) μm	-

The results show that the International Tolerance Grade of the height dimension was IT10 for 86% of specimens (Figure 16). A significant variation in the IT Grade percent was detected for the diameter dimension with a 58% IT10 distribution, as shown in Figure 16. The IT Grade, which represents the dimensional accuracy of the AM systems for each interval of ISO basic sizes can be determined using the same procedure.

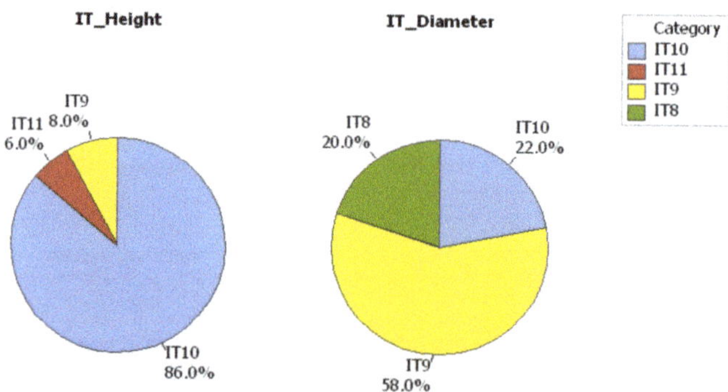

Figure 16. IT grades for the characteristics of height and diameter of the circular artifact within the size range (10–18 mm).

3.6. Quality Inspection through Microscopy Analysis

A microscopy analysis study was performed to conduct a quality inspection of the critical features of the circular workpiece. The dimension H was measured between the upper and lower surface of the specimen, and dimension D on the upper surface of the workpiece, respectively. The quality of these surfaces should be investigated. There was no support material deposited on the upper surfaces of the model printed in glossy mode, only on the bottom surfaces, as shown in Figures 17 and 18.

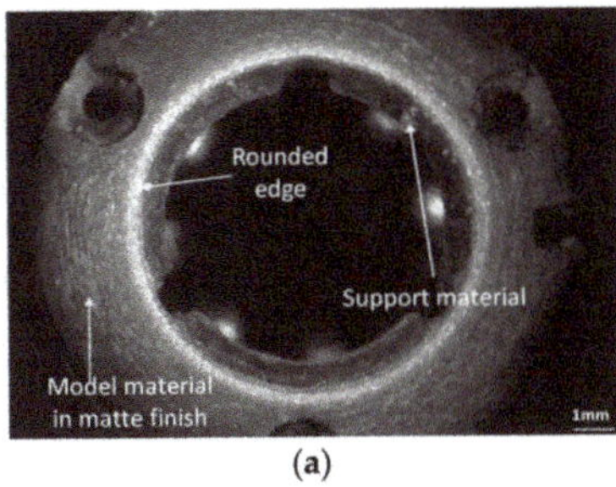

(a)

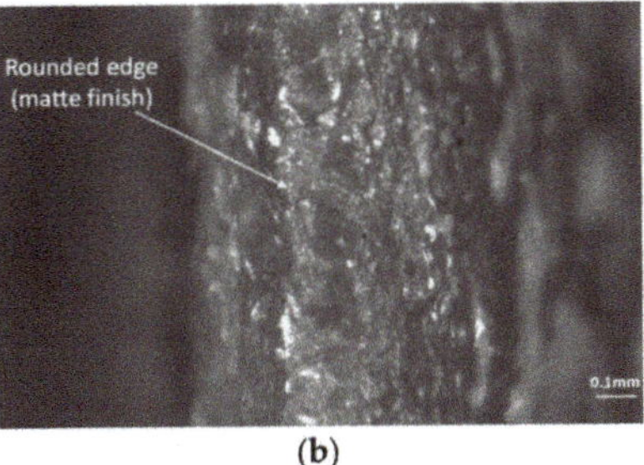

(b)

Figure 17. The lower surface of the circular artifact manufactured by the Objet EDEN 350 PolyJet. (a) Lower surface detail affected by support material; (b) lower edge detail in a matte finish.

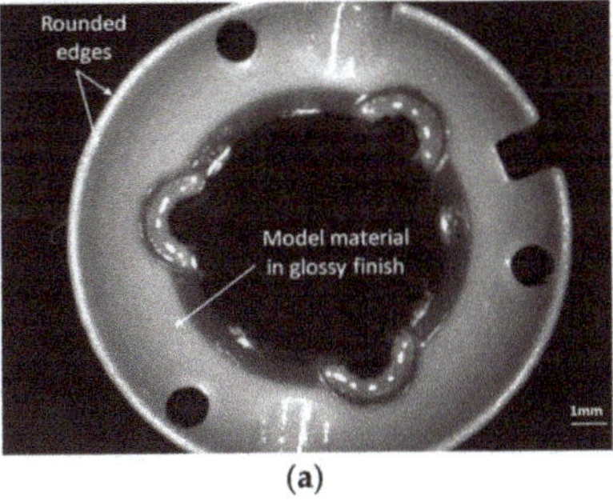

(a)

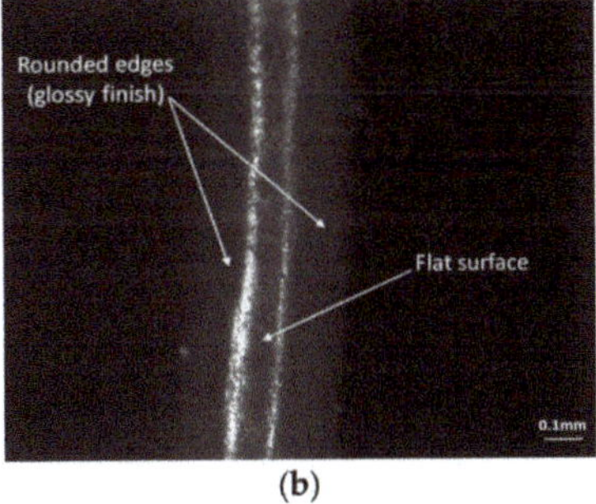

(b)

Figure 18. The upper surface of the circular artifact manufactured by the Objet EDEN 350 PolyJet. (a) Upper surface in glossy mode; (b) upper edge detail in a glossy finish.

The lower surface of the circular specimen was affected by the material support. Small pieces of support material were detected on the lower surface of the specimen, even if the specimen was cleaned with a pressure water jet after 3D printing (Figure 17).

The quality of the lower and upper edges of the circular artifact may influence the artifact height dimension. A good quality surface without material defects was detected on the upper edge of the specimen (Figure 18b). The edges of the upper surface printed in glossy mode were rounded, as shown in Figure 18b. Additionally, the sharp edge of the lower surface affected by the support material was rounded, as shown in Figure 17. This edge roundness can explain why the distribution of the height measurements was located near the lower tolerance limits and was lower than the nominal value.

For both the glossy and matte finishes, microscopic investigations on the lateral surface of the circular artifact were conducted on a perpendicular and parallel direction to the X-axis (Figures 19 and 20). It can be seen as a clean and smooth surface in the X-axis direction (Figure 19a) for a glossy finish. Rough areas were detected on the perpendicular direction to the X-axis (Figure 19b). The steep surface affected by the support material indicates a homogenous material that contained small inclusions of the FullCure 705 support material (Figure 20).

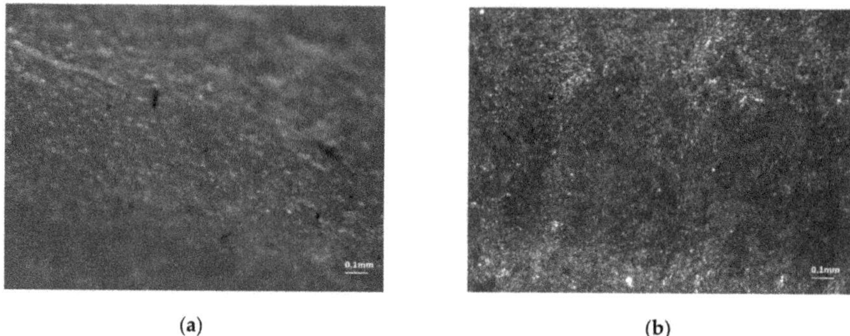

(a) (b)

Figure 19. Microscopic views (1:1 × 10^{-4} m scale) of the lateral surface of the artifact in the glossy finish area located: (**a**) parallel to the X-axis (0°); (**b**) parallel to the Y-axis (90°).

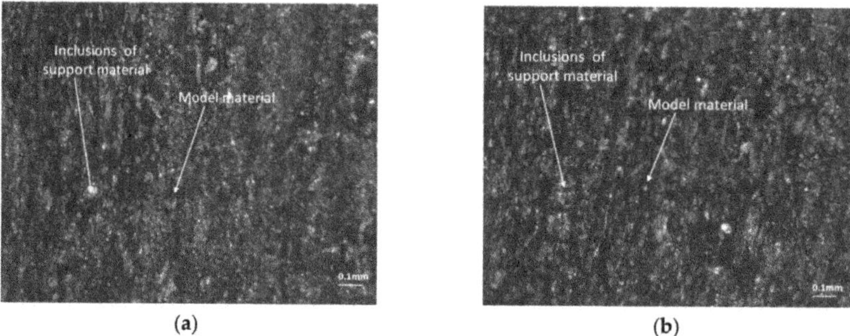

(a) (b)

Figure 20. Microscopic views (1:1 × 10^{-4} m scale) on the lateral surface of the artifact, in the matte finish area affected by the support material located: (**a**) parallel to the X-axis (0°); (**b**) parallel to the Y-axis (90°).

4. Conclusions

This paper contributes to the characterization of the dimensional accuracy, repeatability, system performance, and process capability of polymer-based AM processes and systems. The methodology used for quality control in additive manufacturing allows the polymer-based AM processes to be implemented in production. Additionally, this methodology can be used as the AM's machine monitoring technique.

The following conclusions are drawn:

- The properties of the polymers used in additive manufacturing processes are relevant to the dimensional accuracy of the parts and require different evaluation and quantification of geometrical tolerances in comparison to metal materials and other plastics.
- The implementation of AM for pre-production series and short series production mainly depends on the repeatability, machine capability, and process capability.
- The values of the system and process capability indices (C_m, C_{mk}, C_p, and C_{pk}) of the circular parts produced with Objet VeroBlue RGD840 material by PolyJet technology were greater than 1.67 within the capable tolerance interval of 0.22 mm. The capable lower limit deviation and capable upper limit deviation of the circular artifact were −0.13 mm and +0.09 mm, respectively.
- From the statistical analysis conducted on the geometrical dimensions of the circular parts, the distribution of the measurements showed that they were not centered on the nominal value.

These were located near the upper tolerance limits for the dimension of diameter (D) and near the lower tolerance limits for the dimension of height (H), respectively.

- The roundness of the artifact edges detected through the microscopy investigations explains why the distribution of the height measurements was located near the lower tolerance limits and was lower than the nominal value. Additionally, the height values resulting from the measurements were lower than the nominal value.
- The International Tolerance Grade for polymer manufactured circular parts was found to be between IT8 to IT10, which is in-line as per the ISO-286 for materials. The IT Grade of the height dimension was IT10 for 86% of specimens and 58% for the diameter dimension, respectively.
- A small size specimen, built in a minimum of 50 pieces, should be used for AM system capability determination to minimize material consumption and related costs. Three batches of 50 specimens should be built for the process capability study.

Further research is required for the capability characterization of other AM machines and processes using different types of materials.

Author Contributions: Conceptualization, R.U.; Data curation, R.U.; Formal analysis, R.U. and I.C.B.; Investigation, R.U. and I.C.B.; Methodology, R.U.; Project administration, R.U.; Writing—original draft, R.U. and I.C.B.; Writing—review & editing, R.U. All authors have read and agreed to the published version of the manuscript.

Funding: This research received no external funding.

Acknowledgments: We hereby acknowledge the research platform PLADETINO (Platform for Innovative Technological Development) from Transilvania University of Brasov (grant no. 13/2008), and CNCSIS 78 for partly providing the infrastructure used in this work.

Conflicts of Interest: The authors declare no conflicts of interest.

References

1. Kang, H.S.; Lee, J.-Y.; Choi, S.; Kim, H.; Park, J.H.; Son, J.Y.; Kim, B.H.; Noh, S.D. Smart manufacturing: Past research, present findings, and future directions. *Int. J. Precis. Eng. Manuf. Technol.* **2016**, *3*, 111–128. [CrossRef]
2. International Organization for Standardization. *Standard Terminology for Additive Manufacturing—General Principles—Terminology*; ISO/ASTM 52900-15; ISO/ASME International: Geneva, Switzerland, 2015.
3. Eyers, D.; Potter, A. Industrial Additive Manufacturing: A manufacturing systems perspective. *Comput. Ind.* **2017**, *92*, 208–218. [CrossRef]
4. Tofail, S.A.; Koumoulos, E.P.; Bandyopadhyay, A.; Bose, S.; O'Donoghue, L.; Charitidis, C.A. Additive manufacturing: Scientific and technological challenges, market uptake and opportunities. *Mater. Today* **2018**, *21*, 22–37. [CrossRef]
5. International Organization for Standardization. *Additive Manufacturing—Test Artifacts—Geometric Capability Assessment of Additive Manufacturing Systems*; ISO/ASTM 52902-19; ASTM International: West Conshohocken, PA, USA, 2019. [CrossRef]
6. Cazón-Martín, A.; Morer, P.; Matey, L. PolyJet technology for product prototyping: Tensile strength and surface roughness properties. *Proc. Inst. Mech. Eng. Part B J. Eng. Manuf.* **2014**, *228*, 1664–1675. [CrossRef]
7. Mueller, J.; Shea, K.; Daraio, C. Mechanical properties of parts fabricated with inkjet 3D printing through efficient experimental design. *Mater. Des.* **2015**, *86*, 902–912. [CrossRef]
8. Leach, R. Metrology for Additive Manufacturing. *Meas. Control* **2016**, *49*, 132–135. [CrossRef]
9. Krolczyk, G.; Raos, P.; Legutko, S. Experimental analysis of surface roughness and surface texture of machined and fused deposition modelled parts. *Tehnički Vjesnik* **2014**, *21*, 217–221.
10. García-Plaza, E.; López, P.J.N.; Caminero, M.Á.; Muñoz, J.M.C. Analysis of PLA Geometric Properties Processed by FFF Additive Manufacturing: Effects of Process Parameters and Plate-Extruder Precision Motion. *Polymers* **2019**, *11*, 1581. [CrossRef]
11. Moylan, S. Progress Toward standardized additive manufacturing test artifacts. In Proceedings of the ASPE 2015 Spring Topical Meeting Achieving Precision Tolerances in Additive Manufacturing, Raleigh, NC, USA, 26–29 April 2015; pp. 100–105.

12. Ameta, G.; Lipman, R.; Moylan, S.P.; Witherell, P. Investigating the Role of Geometric Dimensioning and Tolerancing in Additive Manufacturing. *J. Mech. Des.* **2015**, *137*, 111401. [CrossRef]
13. Reverte, J.; Caminero, M.Á.; Chacón, J.; García-Plaza, E.; Núñez, P.; Becar, J. Mechanical and Geometric Performance of PLA-Based Polymer Composites Processed by the Fused Filament Fabrication Additive Manufacturing Technique. *Materials* **2020**, *13*, 1924. [CrossRef]
14. Chacón, J.; Caminero, M.Á.; Núñez, P.; García-Plaza, E.; García-Moreno, I.; Reverte, J. Additive manufacturing of continuous fibre reinforced thermoplastic composites using fused deposition modelling: Effect of process parameters on mechanical properties. *Compos. Sci. Technol.* **2019**, *181*, 107688. [CrossRef]
15. Caminero, M.Á.; Chacón, J.; García-Plaza, E.; Núñez, P.; Reverte, J.; Becar, J. Additive Manufacturing of PLA-Based Composites Using Fused Filament Fabrication: Effect of Graphene Nanoplatelet Reinforcement on Mechanical Properties, Dimensional Accuracy and Texture. *Polymers* **2019**, *11*, 799. [CrossRef] [PubMed]
16. Montgomery, D.C. *Introduction to Statistical Quality Control*, 6th ed.; John Wiley & Sons Inc.: New York, NY, USA, 2009.
17. Guoqing, W. Statistical Process Control Analysis Based on Software Q-Das. *Am. J. Theor. Appl. Stat.* **2014**, *3*, 90. [CrossRef]
18. Singh, J.; Singh, R.; Singh, H. Repeatability of linear and radial dimension of ABS replicas fabricated by fused deposition modelling and chemical vapor smoothing process: A case study. *Measurement* **2016**, *94*, 5–11. [CrossRef]
19. Automotive Industry Action Group. *Statistical Process Control, Reference Manual*, 2nd ed.; Automotive Industry Action Group: Detroit, MI, USA, 2005.
20. Akandeac, S.O.; Dalgarnoa, K.W.; Munguiaa, J.; Pallarib, J. Statistical Process Control Application to Polymer based SLS process. In Proceedings of the 26th Annual International Solid Freeform Fabrication Symposium, Austin, TX, USA, 10–12 August 2015; pp. 1634–1643.
21. International Organization for Standardization. *Statistical Methods in Process Management—Capability and Performance—Part 2: Process Capability and Performance of Time-Dependent Process Models*; ISO 22514-2; ISO: Geneva, Switzerland, 2017.
22. International Organization for Standardization. *Geometrical Product Specifications (GPS)—Inspection by Measurement of Workpieces and Measuring Equipment—Part 1: Decision Rules for Verifying Conformity or Nonconformity with Specifications*; ISO 14253-1; ISO: Geneva, Switzerland, 2017.
23. Baturynska, I. Statistical analysis of dimensional accuracy in additive manufacturing considering STL model properties. *Int. J. Adv. Manuf. Technol.* **2018**, *97*, 2835–2849. [CrossRef]
24. George, E.; Liacouras, P.; Rybicki, F.J.; Mitsouras, D. Measuring and Establishing the Accuracy and Reproducibility of 3D Printed Medical Models. *RadioGraphics* **2017**, *37*, 1424–1450. [CrossRef]
25. Preißler, M.; Rosenberger, M.; Notni, G. An Investigation for Process Capability in Additive Manufacturing. In Proceedings of the 59th Ilmenau Scientific Colloquium, Ilmenau, Germany, 11–15 September 2017.
26. Singh, R. Process capability study of polyjet printing for plastic components. *J. Mech. Sci. Technol.* **2011**, *25*, 1011–1015. [CrossRef]
27. International Organization for Standardization. *Statistical Methods in Process Management—Capability and Performance—Part 3: Machine Performance Studies for Measured Data on Discrete Parts*; ISO 22514-3; ISO: Geneva, Switzerland, 2008.
28. Kitsakis, K.; Kechagias, J.; Vaxevanidis, N.; Giagkopoulos, D. *Tolerance Analysis of 3d-MJM Parts According to IT Grade*; IOP Conf. Series: Materials Science and Engineering; IOP Science: Kozani, Greece, 2016; Volume 161. [CrossRef]
29. Yap, Y.L.; Wang, C.; Sing, S.; Dikshit, V.; Yeong, W.Y.; Wei, J. Material jetting additive manufacturing: An experimental study using designed metrological benchmarks. *Precis. Eng.* **2017**, *50*, 275–285. [CrossRef]
30. Minetola, P.; Calignano, F.; Galati, M. Comparing geometric tolerance capabilities of additive manufacturing systems for polymers. *Addit. Manuf.* **2020**, *32*, 101103. [CrossRef]
31. International Organization for Standardization. *Additive Manufacturing—General Principles—Requirements for Purchased AM Parts*; ISO/ASTM 52901-17; ISO/ASME International: Geneva, Switzerland, 2017.
32. International Organization for Standardization. *Geometrical Product Specifications (GPS)—ISO Code System for Tolerances on Linear Sizes—Part 1: Basis of Tolerances, Deviations and Fits*; ISO 286-1:2010; ISO: Geneva, Switzerland, 2010.
33. German Institute for Standardisation. *Plastics Moulded Parts—Tolerances and Acceptance Conditions*; DIN 16742-13; DIN: Berlin, Germany, 2013.

34. International Organization for Standardization. *General Tolerances—Part 1: Tolerances for Linear and Angular Dimensions without Individual Tolerance Indications*; ISO 2768-1:1989; ISO: Geneva, Switzerland, 2015.

35. International Organization for Standardization. *Statistical Methods in Process Management. Capability and Performance—Part 1: General Principles and Concepts*; ISO 22514-1; ISO: Geneva, Switzerland, 2014.

36. International Organization for Standardization. *Quality Management Systems. Particular Requirements for the Application of ISO 9001:2000 for Automotive Production and Relevant Service Part Organizations*; ISO/TS 16949; ISO: Geneva, Switzerland, 2009.

37. Arcidiacono, G.; Nuzzi, S. A Review of the Fundamentals on Process Capability, Process Performance, and Process Sigma, and an Introduction to Process Sigma Split. *Int. J. Appl. Eng. Res.* **2017**, *12*, 4556–4570.

38. Objet Geometries. *Eden 500V/350V/350 3-D Printer System. User Guide*; Objet Geometries Ltd.: Rehovot, Israel, 2007.

39. Stratasys. PolyJet Materials Data Sheet. Available online: http://www.stratasys.com (accessed on 10 September 2015).

40. Udroiu, R.; Nedelcu, A. Optimization of additive manufacturing processes focused on 3D Printing. In *Rapid Prototyping Technology—Principles and Functional Requirements*; Hoque, M.E., Ed.; InTechOpen: Rijeka, Croatia, 2011; pp. 1–28. [CrossRef]

41. Chen, Y.; Lü, J. RP part surface quality versus build orientation: When the layers are getting thinner. *Int. J. Adv. Manuf. Technol.* **2012**, *67*, 377–385. [CrossRef]

42. Udroiu, R.; Braga, I.C.; Nedelcu, A. Evaluating the Quality Surface Performance of Additive Manufacturing Systems: Methodology and a Material Jetting Case Study. *Materials* **2019**, *12*, 995. [CrossRef] [PubMed]

43. Derby, B. Inkjet Printing of Functional and Structural Materials: Fluid Property Requirements, Feature Stability, and Resolution. *Annu. Rev. Mater. Res.* **2010**, *40*, 395–414. [CrossRef]

44. Du Plessis, A.; Sperling, P.; Beerlink, A.; Tshabalala, L.; Hoosain, S.; Mathe, N.; Le Roux, S.G. Standard method for microCT-based additive manufacturing quality control 2: Density measurement. *MethodsX* **2018**, *5*, 1117–1123. [CrossRef] [PubMed]

45. Sabău, E.; Udroiu, R.; Bere, P.; Buranský, I.; Miron-Borzan, C.-Ş. A Novel Polymer Concrete Composite with GFRP Waste: Applications, Morphology, and Porosity Characterization. *Appl. Sci.* **2020**, *10*, 2060. [CrossRef]

46. Automotive Industry Action Group: MSA-4. *Measurement Systems Analysis—Reference Manual*, 4nd ed.; Automotive Industry Action Group: Southfield, MI, USA, 2010.

47. Zanobini, A.; Sereni, B.; Catelani, M.; Ciani, L. Repeatability and Reproducibility techniques for the analysis of measurement systems. *Measurement* **2016**, *86*, 125–132. [CrossRef]

48. Minitab. Getting Started with Minitab 19. Available online: https://www.minitab.com (accessed on 21 September 2019).

49. Q-DAS Statistical Software (Q-DAS). Available online: https://www.q-das.com/en/ (accessed on 10 February 2020).

50. Sambrani, V.N. Process Capability—A Managers Tool for 6 Sigma Quality Advantage. *Glob. J. Manag. Bus. Res. G Interdiscip.* **2016**, *16*, 63–69.

51. Galve, J.E.; Elduque, D.; Pina, C.; Clavería, I.; Acero, R.; Fernández, Á.; Javierre, C. Dimensional Stability and Process Capability of an Industrial Component Injected with Recycled Polypropylene. *Polymers* **2019**, *11*, 1063. [CrossRef]

52. Maurya, N.K.; Rastogi, V.; Singh, P. Comparative Study and Measurement of Form Errors for the Component Printed by FDM and PolyJet Process. *Instrum. Mes. Métrol.* **2019**, *18*, 353–359. [CrossRef]

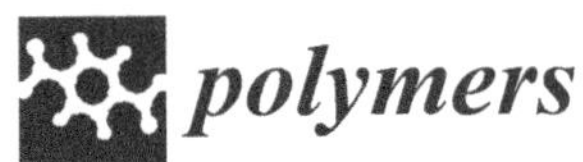 *polymers*

Article

3D Direct Printing of Silicone Meniscus Implant Using a Novel Heat-Cured Extrusion-Based Printer

Eric Luis [1], Houwen Matthew Pan [2], Swee Leong Sing [1], Ram Bajpai [3,4], Juha Song [2] and Wai Yee Yeong [1,*]

[1] Singapore Centre for 3D Printing, School of Mechanical and Aerospace Engineering, Nanyang Technological University, 50 Nanyang Avenue, Singapore 639798, Singapore; G001@ntu.edu.sg (E.L.); slsing@ntu.edu.sg (S.L.S.)

[2] School of Chemical and Biomedical Engineering, Nanyang Technological University, 70 Nanyang Avenue, Singapore 639798, Singapore; matthew.pan@u.nus.edu (H.M.P.); songjuha@ntu.edu.sg (J.S.)

[3] Center for Population Health Sciences, Lee Kong Chian School of Medicine, Nanyang Technological University, 11 Mandalay Road, Singapore 308232, Singapore; r.bajpai@keele.ac.uk

[4] School of Primary, Community and Social Care, Keele University, Keele ST5 5BG, UK

* Correspondence: wyyeong@ntu.edu.sg

Received: 30 March 2020; Accepted: 25 April 2020; Published: 1 May 2020

Abstract: The first successful direct 3D printing, or additive manufacturing (AM), of heat-cured silicone meniscal implants, using biocompatible and bio-implantable silicone resins is reported. Silicone implants have conventionally been manufactured by indirect silicone casting and molding methods which are expensive and time-consuming. A novel custom-made heat-curing extrusion-based silicone 3D printer which is capable of directly 3D printing medical silicone implants is introduced. The rheological study of silicone resins and the optimization of critical process parameters are described in detail. The surface and cross-sectional morphologies of the printed silicone meniscus implant were also included. A time-lapsed simulation study of the heated silicone resin within the nozzle using computational fluid dynamics (CFD) was done and the results obtained closely resembled real time 3D printing. Solidworks one-convection model simulation, when compared to the on-off model, more closely correlated with the actual probed temperature. Finally, comparative mechanical study between 3D printed and heat-molded meniscus is conducted. The novel 3D printing process opens up the opportunities for rapid 3D printing of various customizable medical silicone implants and devices for patients and fills the current gap in the additive manufacturing industry.

Keywords: 3D printing; additive manufacturing; material extrusion; silicone; meniscus implant

1. Introduction

The meniscus, which is a pair fibrocartilaginous cushion within the knee joint, acts as a weight bearing cushion, lubricating device and knee stabilizer. It is mainly C-shaped and triangular shaped in the horizontal and cross-sectional sections, respectively. Its relatively centralized acellular extracellular matrix and limited peripheral vascular supply from the meniscocapsular plexus severely restricts the meniscus regenerative capability when damaged or torn. The compressive modulus of 0.03 MPa is weakest at its supero-medial portion of the posterior horns and these regions are the commonest sites of tear and injury [1].

Treatment options for meniscus tears include direct repair, replacement with either scaffolds or implants and partial or total meniscectomy (removal of the damaged meniscus tissues). Repair methods for meniscal tears with sutures, pins and fibrin glue were used with limited success due to poor vascular supply. Replacement with meniscus scaffolds made of polycaprolactone (PCL) or collagen have produced equivocal results owing to inferior compressive mechanical properties. Partial meniscus implants which include collagen meniscus implant (Menaflex) and Actfit were used with limited success and have not yet gained approval from FDA. Total meniscus implant NuSurface is the only total meniscus implant currently undergoing Sun Clinical Trial 2 [2,3].

Silicone is used in biomedical manufacturing for biliary stents [4], cochlear implants [5], peripheral nerve sheaths [6], breast [7–10], chin, nose [11], testicular and hand implants [12], interphalangeal joint replacement implants [13], knee prosthesis [14], cardiopulmonary bypass tubing [15] and silicone membranes [16]. Recently, microfluidics [17], lab-on-a-chip [18] and biosensors [19] have also been manufactured with silicone. Despite these, the manufacturing of meniscus implants using silicone was limited. Meanwhile, most information in literature are on the general properties of silicone resins but little on the rheological properties. Takahashi et al. studied the rheological behavior of several silicone resins in the un-crosslinked state and found that the temperature dependence of the viscoelastic behavior can be described by a Williams–Landel–Ferry (WLF) equation [20].

Most of the implants are produced by indirect molding technique, whereby the mold is first 3D printed or machined and the silicone is subsequently poured into the mold and left to cure. This indirect molding technique, both time consuming and costly, also posed technical challenges when customized implants of different geometries or hollow silicone parts need to be manufactured.

Industrial silicone 3D printing began in 2015. The current state-of-the art in silicone 3D printing includes (a) using UV light of 365 nm to cure UV-sensitive silicone, as they are being deposited layer-by-layer by an inkjet printer (ACEO, Wacker, Munich, Germany), (b) freeform reversible embedding technique using a 2-part A/B silicone where the part A-silicone catalyst-cross-linker is extruded into a bath of part B-silicone (Fripp Design, Sheffield, UK) [21], (c) using a progressive cavity pump to extrude moisture-cured one part Oxime silicone elastomer [22], (d) a hybrid ink-jetting and UV-extrusion techniques with a printing speed which is 20 times faster [23], (e) a combination stereolithographic with low-one-photon-polymerization (LOPP) [24] and (f) 3D printing of PDMS elastomer in a hydrophilic Carbapol bath support [25].

The first successful direct 3D printing of heat-cured medical grade silicone meniscus implants is reported in this study. To date, no direct 3D printing of medical silicone implants has been described. First, optimum curing temperatures and times are obtained from the rheological characterization of the silicone resins [26,27]. This is followed by parametric optimization of nozzle diameters, nozzle temperatures and bed temperatures, using a range of parametric values, to print out standard ASTM cylindrical blocks, T-bones and finally meniscus implants. Computational fluid dynamics (CFD) simulation is used to study the heat transfer within the silicone resin across the heated nozzle and Solidworks is used to study the heat gradient distribution across the silicone meniscus implant. The results were validated against experimental values using standard heat probes. The new direct 3D printing technology provides an avenue for rapid 3D printing of various customizable medical silicone implants and devices for patients and fills the current gap in the additive manufacturing (AM) industry.

2. Materials and Methods

2.1. Silicone Material Specification

The silicone material used in this study is a water-white translucent, two-part platinum-catalyzed silicone elastomer that cure upon exposure to room temperature or can be accelerated by heat (Smooth-On®, Smooth-On Inc, Macungie, PA, USA). Ecoflex30 and Ecoflex50 are chosen for its safety and biocompatible profiles. They are mixed 1A:1B by weight or volume and cured at room temperature with negligible shrinkage. Their low viscosities permit easy mixing and de-airing. The properties of Ecoflex30 and Ecoflex50 are shown in Table 1.

Table 1. Ecoflex sample technical data (from safety data sheet provided by Smooth-On Inc).

	Mixed Viscosity (cp)	Specific Gravity (g/cc)	Specific Vol. (cu. in./lb.)	Pot Life (min)	Cure Time (h)	Shore Hardness	Tensile Strength (psi)	100% Modulus (psi)	Elongation at Break (%)	Die B Tear Strength (pli)	Shrinkage (in./in.)
Ecoflex 50	8000	1.07	25.9	18	3	00–50	315	12	980	50	<0.001
Ecoflex 30	3000	1.07	26.0	45	4	00–30	200	10	900	38	<0.001

2.2. Experimental Setup

The system consists of six key components: (1) a motion control platform based on an open-source 3D printer (CoreXY, Guangzhou, China), (2) a syringe-pump extruder with accuracy of 0.05 μm to dispense the material, (3) a double barrel syringe connected to a static mixer using tubing (TIUB05C, SMC Corporation, Tokyo, Japan), (4) stainless nozzles with tip size between 20 and 21 gauges (inner diameter from 0.5 mm to 0.6 mm), (5) heating elements on the printer nozzles and platform with their controllers, as shown in Figure 1.

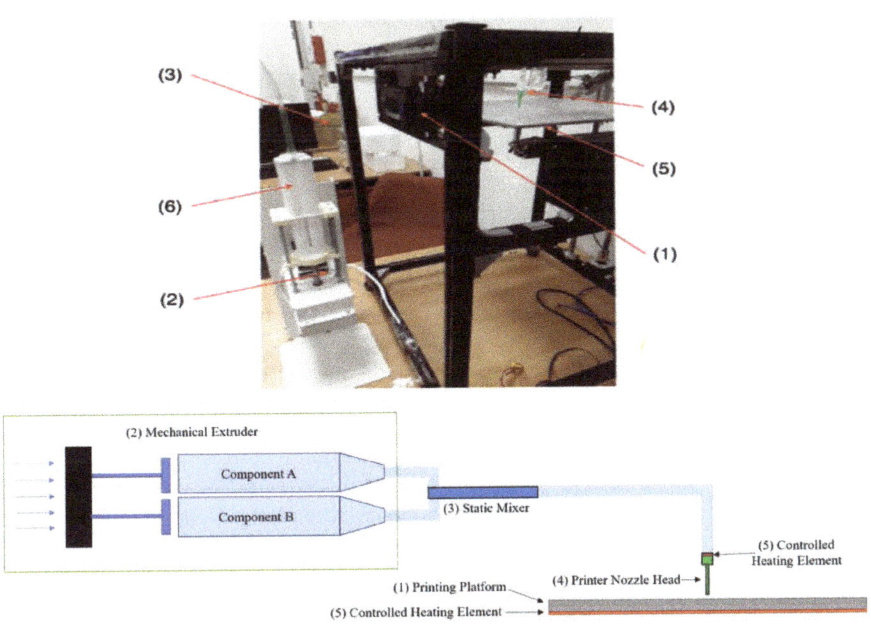

Figure 1. Experimental setup for heat cure extrusion-based additive manufacturing (AM): legends: (1) motion control platform, (2) extruder, (3) static mixer, (4) printhead, (5) heating elements.

Parts A and B of the Ecoflex ® resins were added into the double barrel syringe with mix ratio of 1:1 by volume prior to fabrication. The resins were subsequently passed through the static mixer before extrusion. 3D printing software (Repetier-Host, Hot-World GmbH & Co. KG, Willich, Germany) was used to control the 3D printer while Slic3r was used for tool path generation.

2.3. Process Parameters and Experiment Design

In this study, several key process parameters, such as the solution flow rate (Q), nozzle-to-substrate distance (h), were fixed to simplify the optimization process. Before the 3D printing of meniscus implants, several experiments were conducted to find a suitable range of the process parameters. The analysis of the 3D printed meniscus fabricated under varied temperature of the heating elements was done to find out the best heat cured meniscus.

For the first part, inner nozzle diameter (d) and the print speed (v) were analyzed through fabricating several cylinders and T-bones. Meanwhile, checks were conducted on the layer-layer deformation and whether there is void inside the specimens. Specimens with complicated structure were fabricated to verify whether these ranges of the process parameters are feasible for the meniscus fabrication. In second part, several printing experiments were conducted to find out the optimized combination of the temperatures of the heated platform (T_1) and the heating elements on the print head (T_2). Lastly, comparison between the compressive properties of the 3D printed meniscus and heat molded meniscus was done.

2.4. Rheology Test

Rheology test was conducted to measure the gelation in silicone since it is instantaneous using a TA Texas Instruments DHR2 rheometer (New Castle, DE, USA). The testing geometry used is a 40 mm steel Peltier parallel plate which is fitted after instrument inertia calibration, bearing friction correction and geometry inertia calibration. The rheometer was operated in parallel-plate mode.

For the experiments, the two components Parts A and B of each silicone resin Ecoflex30 and Ecoflex50 (cross-linker with catalyst and polymer) were manually mixed according to the manufacturer's datasheet. Characterization of the rheological behavior of the silicone in triplicates was done in both steady and oscillation modes at different isothermal temperatures (30 °C to 80 °C) according to DIN 53529 which is similar to ASTM D 5289. The selected temperatures were controlled by the Peltier device with a cooling pump at an accuracy of ±0.1 °C. For the dynamic oscillatory rheology investigation, the samples were exposed to increasing strain rate (0.00 to 3.00 s^{-1}) at a constant frequency (1 Hz) to determine the linear viscoelastic range of the samples [28].

The curing times, complex viscosities and complex moduli of the silicone at various temperature are determined using the temperature ramp mode from 30 °C to 80 °C with ramp rate of 5 °C/min. A normal force control of 2 N, with tolerance of 1.75 N is applied on the torsion geometry with gap change limit maintained at 550 µm. The control strain is set at 0.025% and constant angular frequency was set at 3 rad/s. The data are collected and plots of G', G" and tan (delta) against temperature and time are performed [29].

The frequency sweep mode is used to determine the viscoelastic properties of the silicone resin in steady shear and dynamic-mechanical experiments in a frequency range between 0.01 and 100 rad/s. The heat-treated material was first ground and subsequently pressed to dense discs at 70 °C like the untreated material. Before being measured, all samples were dried in a desiccator at room temperature for 12 h.

2.5. Surface Morphology of 3D Printed Silicone Meniscus Implant

Nikon SMZ 1000 (Tokyo, Japan) was used for 6× magnification surface visualization of the overall layer-by-layer step morphology of the 3D printed implants and Eclipse LV 150 (Nikon Metrology NV, Leuven, Belgium), was used for 50× magnification cross-sectional visualization of the 3D printed implants.

2.6. Statistical Analysis

Descriptive statistics (mean ± standard deviation) for all quantitative variables are obtained. Normality of continuous variables were tested by Kolmogorov–Smirnov and Shapiro–Wilk tests. An exploratory analysis of the correlation between meniscus length and width with nozzle and print bed temperatures using Spearman rank correlation due to small sample sizes is done. The univariate linear regression is used to evaluate the effect of nozzle and print bed temperatures on the meniscus length and width. Statistical analyses were carried out using Stata version 14.2 (StataCorp, College Station, TX, USA).

2.7. Computational Fluid Dynamics (CFD) Simulation Studies for Heated Printer Nozzle

The nozzle design and schematic is shown in Supporting Information, Figure S1. Using Autodesk computational fluid dynamics (CFD) 2017 software (Autodesk Inc, San Rafael, CA, USA), the input parameters for materials used are provided in Supporting Information, Table S1.

After applying the material properties, it is necessary to provide the boundary conditions. The flow simulation works only in a closed region. Thus, there is a need to create a computational domain and a fluid sub domain [30]. Boundary conditions are set as follows: Surface Temperature—60 °C, Pressure—0 Pa Gage, Volume Flow Rate—50 mm^3/min, Temperature—25 °C.

2.8. Solidworks Simulation Studies for 3D Printed Silicone Meniscus Implant

Solidworks 2018 (Dassault Systems Solidworks Corporation, Waltham, MA, USA) was used to study the temperature gradient distribution in the 3D printed silicone meniscus implants. Analysis using the stimulation module can reduce the number of product development cycles and time to market (TTM), optimize the meniscus design and reduce cost by testing the model on the computer rather than physical tests. From the simulation task pane, thermal study and the material properties are first entered: Elastic modulus (0.290075×10^6), Poisson's ratio (0.47), shear modulus (0.0029×10^6 Psi), mass density (143.58 lb/ft^3), tensile strength (0.797708 ksi), compressive strength (4.35113 ksi), yield strength (0.758547 ksi), thermal expansion coefficient (540×10^{-6}), thermal conductivity (4.77 BTU ft/h), specific heat (1.00602 BTU/lb.P). Subsequently, this is followed by automatic generation of mesh of finite elements for FEA by the program. To set the boundary conditions, the perimeter of each of the eleven layers sliced by the Slic3r program was used to estimate the amount of time taken to print each layer and the estimated perimeter in millimeters and printing time taken in seconds are expressed below: From the first layer to the final (eleventh) layer, the perimeters are 273, 274, 165.7, 151.8, 138.9, 125.8, 112.3, 97.9, 82.0, 64.0 and 43.0 mm requiring estimated printing times of 330, 330, 183, 167, 153, 138, 124, 108, 90, 70 and 47 seconds, respectively. The perimeters of each layer was used to estimate the total amount of time taken for each layer to be printed as shown in Supporting Information, Table S2. Experimental temperature values at every layer laid down are measured with heat probe model HT with K-type thermistor.

3. Results

3.1. Rheology

3.1.1. Steady Shear Flow Study

The uninterrupted extrusion of silicone is crucial for the 3D printing of the meniscus implants. The extrusion is dependent on their viscous behavior under different shear rates. The rheological fingerprints of both Ecoflex30 and Ecoflex50 silicone samples at the temperature 50 °C are shown in Figure 2. Both Ecoflex30 and Ecoflex50 showed non-Newtonian behavior. The shear stress of Ecoflex30 is proportional to the shear rate at lower shear rates. Its viscosity was around 3 Pa·s, the so-called zero-shear rate viscosity η_0. At shear rates of about 0.5 s^{-1}, the Ecoflex30 viscosity starts to decrease significantly, until it starts to level off at higher shar rates of 3 s^{-1}.

Similar behavior for PDMS solutions is reported [31]. Figure 3 shows a sharp decrease in the viscosity of both Ecoflex30 and Ecoflex50, from about 200 to 20 Pa·s and from 500 to 5 Pa·s, respectively, over the shear rate range 0.01 to 5 s^{-1} At very high shear rates the viscosity may again become independent of shear rates, approaching the infinite-shear rate viscosity η_∞. Polymer degradation becomes a serious problem before sufficiently high shear rates can be obtained which made η_∞ not usually measurable. The behavior of Ecoflex30 under steady shear flow in the range above 3.5 s^{-1} is shear thinning or pseudoplastic behavior. The decreasing viscosity with increasing shear rates is utilized for the current extrusion nozzle design.

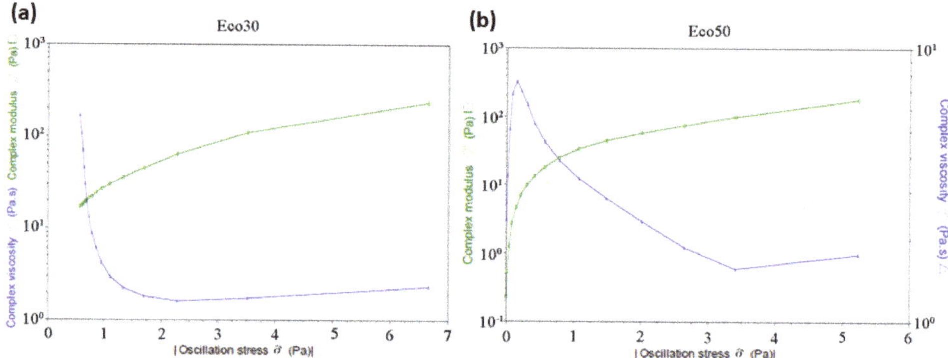

Figure 2. (**a**) Ecoflex30 and (**b**) Ecoflex50 complex viscosity and complex modulus versus shear stress.

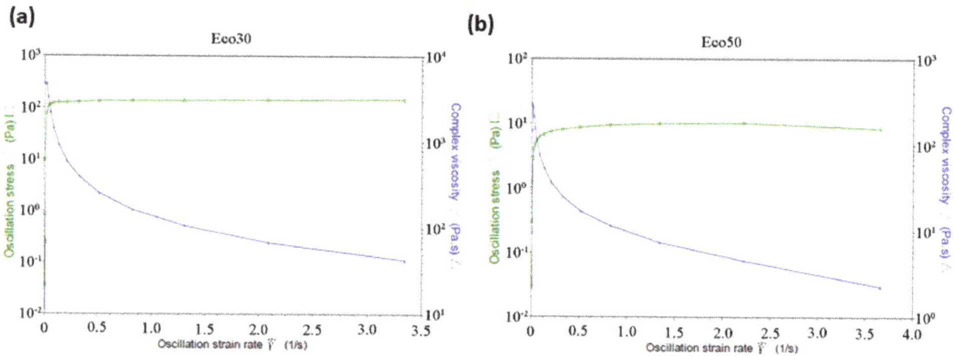

Figure 3. (**a**) Ecoflex30 and (**b**) Ecoflex50 complex viscosity and shear stress versus strain rate.

3.1.2. Transient Shear Stress Response

For rheologically complex materials, understanding the transient shear behavior is important. To examine the transient behavior of Ecoflex30 and Ecoflex50, their shear stress response against time were measured. Both Ecoflex30 and Ecoflex50 show Bingham pseudoplastic shear thinning behavior on the first 250 seconds, as shown in Figure 2.

Over the tested non-Newtonian range of shear rates $0.5 < \dot{\gamma} < 10\ \mathrm{s}^{-1}$, shear thinning behavior was recorded for all temperatures. The shear rate has significant effect on the flow behavior of Ecoflex30. For example at 55 °C, the viscosity of Ecoflex30 decreases from 3000 Pa·s at 0.05 s^{-1} to almost 2 Pa·s at 2.5 s^{-1} (Figure 3a) and the viscosity of Ecoflex50 decreases from 300 Pa·s at 0.05 s^{-1} to almost 2 Pa·s at 2.5 s^{-1} (Figure 3b). This effect represents a depression of 3 orders and 2 orders of magnitude, for Ecoflex30 and Ecoflex50, respectively, over a shear rate range of less than 3 s^{-1}.

3.1.3. Dynamic Test

The storage modulus G' (elastic response) and the loss modulus G'' (viscous response) are measured using a dynamic test where oscillating stresses or strains are applied to the test samples. The total resistance versus the applied strain gives the complex modulus G^*.

At the start of a dynamic test, the linear viscoelastic range is defined by increasing the stress to cover a wide range. The range where complex moduli G^* is constant with stress is the linear viscoelastic range which indicates that the internal bonds of the sample are still intact. The linear viscoelastic range was found to be around 35 to 50 Pa range for Ecoflex30 and 100 to 150 Pa for Ecoflex50.

A frequency sweep test was carried out at the stress value 1.5 Pa to study the viscoelastic behavior Ecoflex30 and Ecoflex50. Both samples demonstrated elastic and viscous responses. Figure 4 show the elastic G' and viscous modulus G'' for Ecoflex30 and Ecoflex50 over the frequency range 1–100 rad/s. Ecoflex30 has both higher elastic and higher viscous modulus (Figure 4a), when compared to those of Ecoflex50 (Figure 4b). Eco30 samples also demonstrated a greater elastic response than viscous response over the entire range of frequencies.

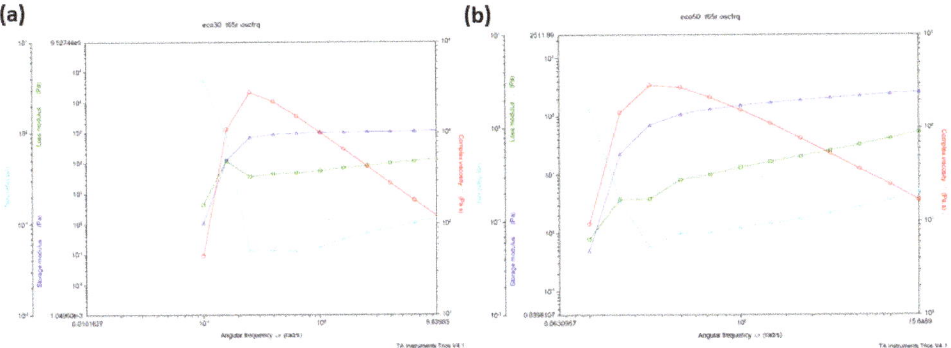

Figure 4. (**a**) Eco30 and (**b**) Eco50 storage modulus, loss modulus, complex viscosity vs angular frequency.

3.1.4. Gelation Times

Ecoflex 30 has a gelation time of 805 s with a crossing modulus of 130 MPa when heated to 40 °C. This is shown in Figure 5. It has a gelation time of 185 s with a crossing modulus of 365 MPa at 50 °C. With the this increase in temperature, the gelation time was shortened by 10 min, the complex modulus was also lowered by from 28 mPa to 8 mPa. However, there was an increase in complex viscosity from 5 Pa·s to over 40 Pa·s.

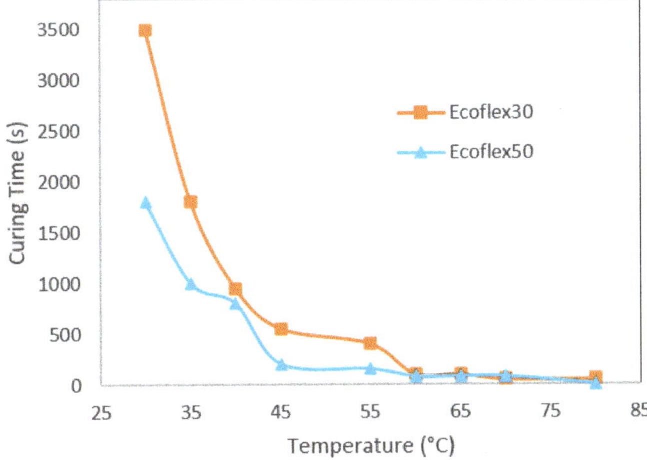

Figure 5. Plot of curing times (s) of Ecoflex30 (orange line) and Ecoflex50 (blue line) versus temperatures (°C).

At below 60 °C, Ecoflex50 demonstrated longer curing times when compared to Ecoflex30. Above 60 °C, both Ecoflex30 and Ecoflex50 have similar curing times. As the temperature increases, the complex viscosity decreases according to the temperature dependence of the viscous properties of the silicone resins. However, at higher temperatures, the crosslinking rates dominate and increases the viscosity. The change in storage and loss modulus over time for Ecoflex30 and Ecoflex50 is shown in Supporting Information, Figure S2.

3.2. Optimization of Print Speed and Nozzle Inner Diameter

An optimized combination of flow rate and print speed is critical in determining printing outcomes. In this study, all flow rates were fixed at 1.2 ml/min. An initial simple three-layer T bone was designed and printed to determine the optimized print speed. The T-bone CAD drawings, shown as Figure 6a,b, were designed with height of 3 mm, overall width of 19 mm, overall length of 55 mm and width of narrow section of 6 mm.

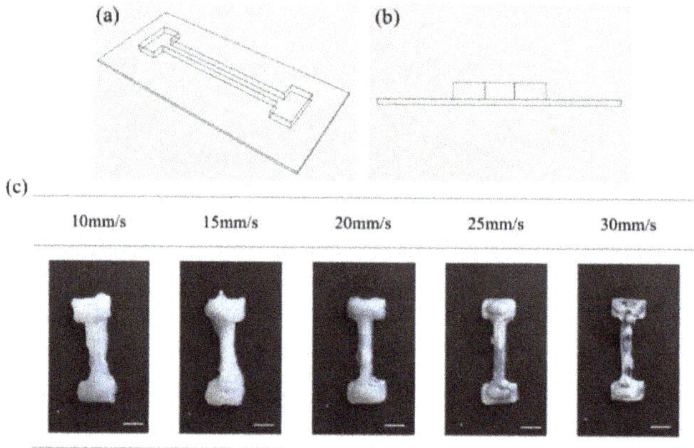

Figure 6. Schematic illustration of T bone: (**a**) overview and (**b**) side view. (**c**) Top view images of the 3D printed T bone under varied print speed. Scale bar: 1 cm.

In this section, print speeds, varying from 10 mm/s to 30 mm/s with 5 mm/s intervals, were used to print the silicone. Other process parameters were fixed ($T_1 = 40\ °C$, $T_2 = 100\ °C$, $Q = 1.2$ ml/min and $d = 0.90$ mm) and 3 specimens were printed with each set of parameters. The top view images of selected 3D printed silicone T-bone fabricated under different print speeds were displayed in Figure 6c. With increasing print speeds, the silicone line diameter gradually decreases until it becomes spotty or discontinuous. At $v = 30$ mm/s, silicone droplets were deposited as the print speed was too high. On the other hand, a low print speed at $v = 10$ mm/s resulted in silicone overflow. The best printout was obtained with a print speed of 20 mm/s.

Subsequently, cylinders were printed to determine the optimal nozzle diameter for more complex printings. The CAD drawing of the cylinders, Figure 7a,b, has a diameter of 30 mm and height of 10 mm. In this section, seven different nozzle diameters, d, (0.41, 0.51, 0.60, 0.70, 0.90, 1.21 mm) were used to print the cylinders. Other process parameters were fixed at $T_1 = 40\ °C$, $T_2 = 100\ °C$, $Q = 1.2$ mL/min and $v = 20$ mm/s.

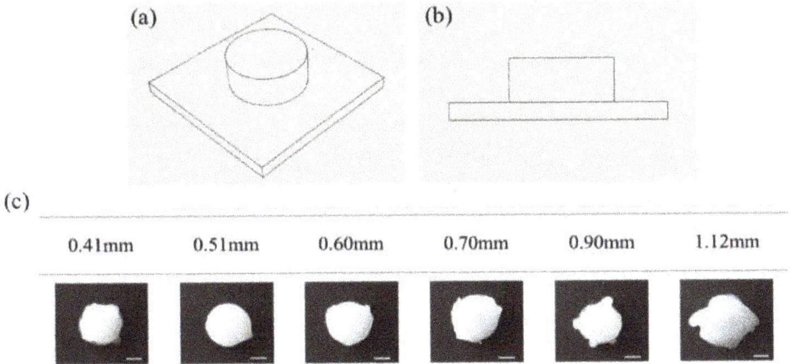

Figure 7. Schematic illustration of 3D printed cylinder (**a**) overview and (**b**) side view. (**c**) Cross-sectional images of the cylinder. Scale bar 1 cm.

Results of printed silicone cylinders are shown in Figure 7c. Using a nozzle diameter of $d = 0.41$ mm resulted in discontinuous print lines, droplets and under extrusion. Best print results, with print heights of 10 mm, were obtained with nozzle diameters of 0.51 mm and 0.6 mm, while keeping other process parameters constant. Silicone overflow due to over-extrusion were observed from $d = 0.70$ mm to $d = 1.12$ mm. Except for $d = 0.51$ mm and 0.60 mm, using all other nozzle diameters may result in under- or over-extrusion. Based on these experiments, a combination of a print speed of 20 mm/s and a nozzle inner diameter of either 0.51 mm or 0.60 mm were selected for subsequent printings. These specific process parameters will need to be modified accordingly for the printing of different specimens.

A more complicated polyhedron with four slopes was printed. The CAD dimensions of the polyhedron are shown in Figure 8a,b, which is approximately 50 mm × 40 mm × 20 mm and 10 mm deep. After slicing, this polyhedron has 20 layers and the slope starts from the 11th layer. Figure 8c shows the top view images of the polyhedron, which fabricated based on the previous process parameters ($T_1 = 40$ °C, $T_2 = 100$ °C, $Q = 1.2$ ml/min, $v = 20$ mm/s, $d = 0.51$ mm and 0.61 mm). Overall, the layers can be clearly observed on both polyhedrons and the dimensions of these two were close to the designed values. However, with d = 0.60 mm, slight overflow was observed due to over-extrusion and inadequate heating. This observation demonstrates that void-less silicone polyhedron with slopes can be reliably achieved.

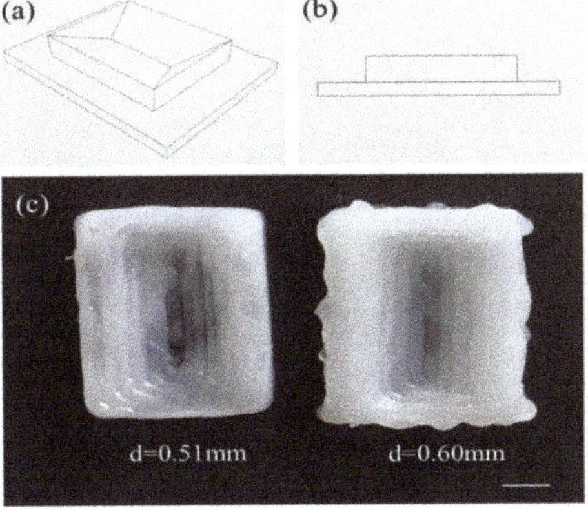

Figure 8. Schematic illustration of a polyhedron in (**a**) overview and (**b**) side view. (**c**) Top view images of the polyhedron. Scale bar: 1 cm.

3.3. Fabrication of Silicone Meniscus

To determine the effects of temperature variation on the silicone meniscus printing, a series of experiments were conducted. The temperature of the heated platform was varied from 80 °C to 110 °C with intervals of 10 °C and the nozzle temperature was varied from 40 °C to 80 °C with intervals of 10 °C. As shown in Figure 9, several combinations of the temperatures were selected to fabricate the silicone meniscus.

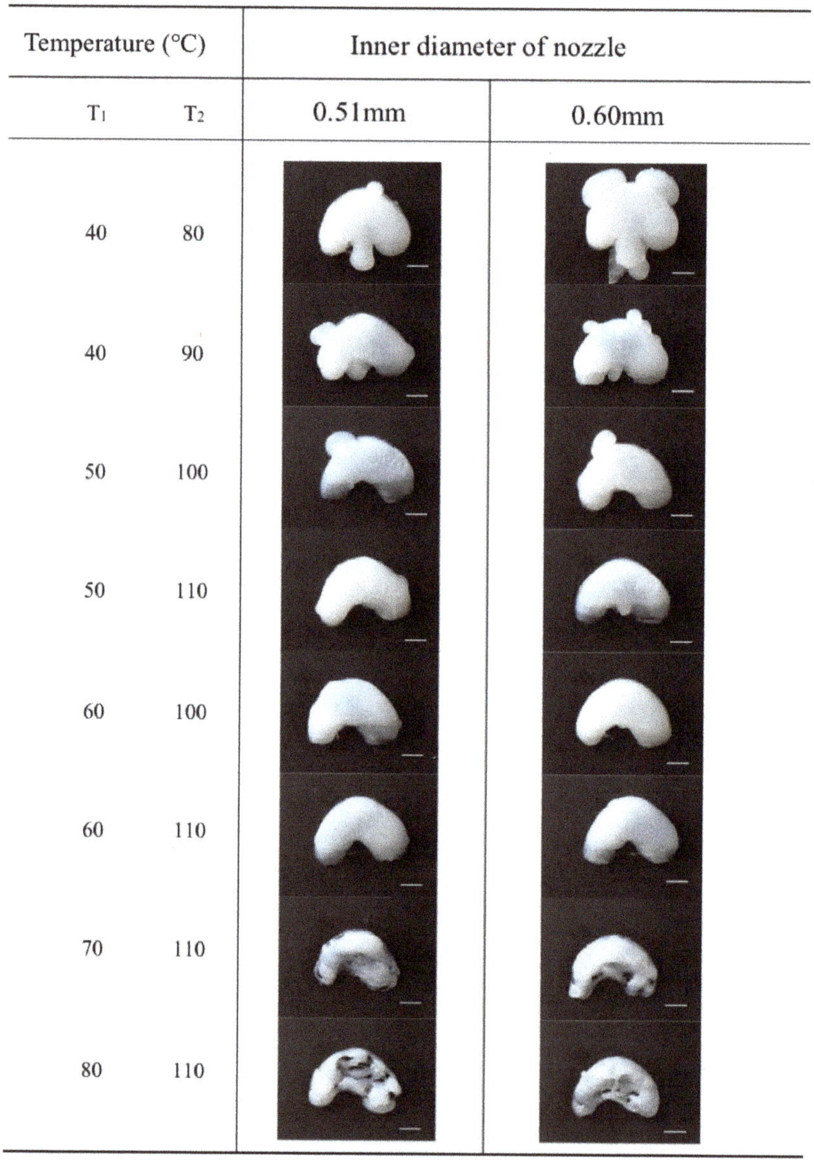

Temperature (°C)		Inner diameter of nozzle	
T_1	T_2	0.51mm	0.60mm
40	80		
40	90		
50	100		
50	110		
60	100		
60	110		
70	110		
80	110		

Figure 9. Top view images of the 3D printed silicone meniscus under varied heating temperature (T_1 = nozzle temperature and T_2 = platform temperature) and inner nozzle diameter. Scale bar: 1 cm.

In general, a larger volume of silicone is extruded with a larger nozzle diameter. This extruded volume decreases with an increase in temperature and is caused by clogging of the nozzle or tubing. Clogging is particularly significant at nozzle temperatures of more than 70 °C and above. Nozzle temperatures of 50 °C and below resulted in inadequate heating, inadequate curing and overflow. With increasing height of meniscus being printed, heat conduction via the thickness of silicone meniscus is impaired and this may similarly result in silicone overflow and collapse of the printout. Therefore, a higher platform temperature of T_2 = 100 °C to 110 °C is required. From Figure 9, a combined nozzle temperature T_1 = 60 °C platform temperature T_2 = 110 °C and nozzle diameter d = 0.51 mm gave the best printout of meniscal implant.

3.4. Association between Meniscus Length and Width with Nozzle and Print Bed Temperatures

Mean meniscus length and width were calculated to be 4.08 ± 0.14 cm (range: 3.90–4.30 cm) and 2.08 ± 0.19 cm (range: 1.90–2.54 cm), respectively. Similarly, mean nozzle and print bed temperatures were calculated as 57 ± 12.52 °C (range: 40–80 °C) and 102 ± 10.33 °C (range: 80–110 °C), respectively. A strong negative correlation has been observed between meniscus length vs. nozzle temperature (rho = −0.93; $p < 0.01$); meniscus length vs. print bed temperature (rho = −0.82; $p < 0.01$); meniscus width vs. nozzle temperature (rho = −0.84; $p < 0.01$); and meniscus width vs. print bed temperature (rho = −0.83; $p < 0.01$). The linear relationship between the meniscus length and width to the nozzle and print bed temperatures is presented in Table 2 and Figures 10 and 11. The univariate regression analysis demonstrated that high amount of variability in the meniscus length and width can be explained by the nozzle and print bed temperatures independently.

Table 2. Univariate linear regression analysis to predict meniscus length and width using nozzle and print bed temperatures.

Independent Variables	Dependent Variables							
	Length (mm)				Width (mm)			
	Adjust R^2	β	S.E.	*p*-Value	Adjust R^2	β	S.E.	*p*-Value
Nozzle temperature (°C)	0.84	−0.01	0.002	$p < 0.001$	0.68	−0.01	0.003	0.002
Print bed temperature (°C)	0.62	−0.01	0.003	$p < 0.001$	0.64	−0.02	0.004	0.003

Adj: adjusted; S.E.: standard error.

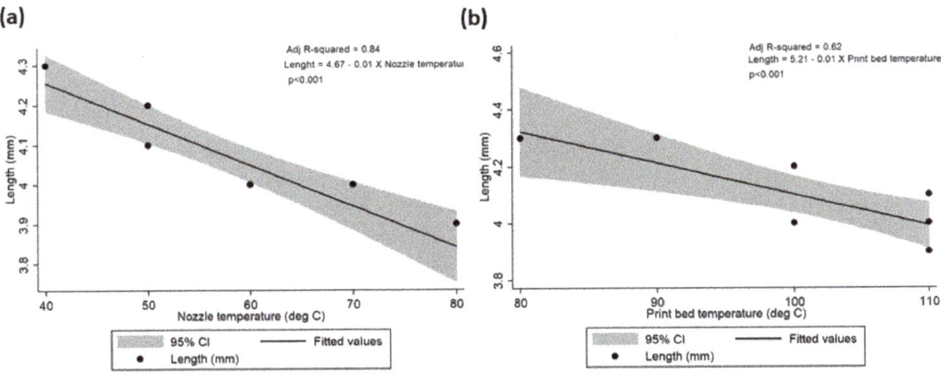

Figure 10. Correlation between meniscus length and (**a**) nozzle temperature and (**b**) print bed temperature.

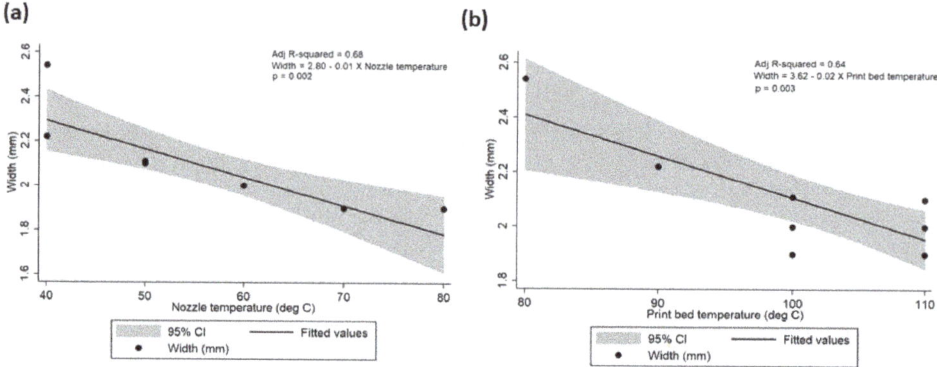

Figure 11. Correlation between meniscus width and (**a**) nozzle temperature and (**b**) print bed temperature.

3.5. Surface and Cross-Sectional Morphology

The surface morphology observed in Figure 12a,d shows relatively smooth surface with pits measuring less than 20 micrometers due to impurities and bubbling. The orderly layer-by-layer stepwise deposition is seen in Figure 12b,c, in both the posterior and anterior horns of the meniscus, respectively. Supporting Information, Figure S3 illustrate the interlayer silicone bonding with clear lamination lines.

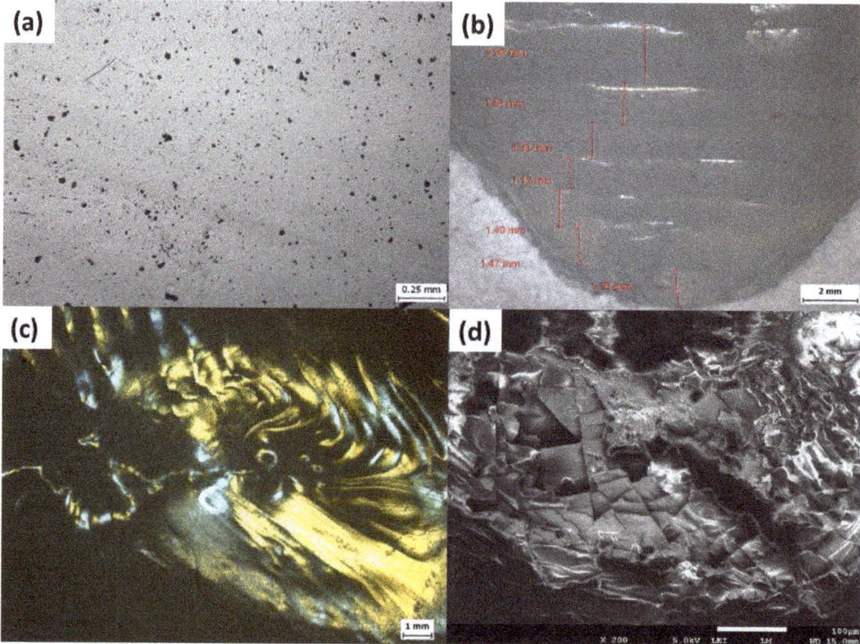

Figure 12. Bright field images of (**a**) surface, (**b**) posterior horn and (**c**) anterior horn of 3D printed silicone meniscus implant. (**d**) Scanning electron microscopy (SEM) image of meniscus surface at ×200 magnification.

3.6. Heated Nozzle Computational Fluid Dynamics (CFD) Simulation Studies

The following simulation tests results illustrate the distribution of temperature, velocity and viscosity of the silicone resin along the nozzle, as shown in Figure 13a–c below, respectively. Figure 13a shows a temperature of 60 °C at the nozzle inlet, shown in red, where it is in direct contact with the heating block. The subsequent reduction in temperature along the nozzle tip to around 40 °C is shown in dark green. These simulation temperatures correlated quite accurately with the actual temperatures of thermocouple readings of aluminum heating block thermocouples and heated platforms. Figure 13b shows a relatively slow traveling speed of silicone resin at 0.5 mm/s at the nozzle inlet, shown in blue. Within the barrel of the nozzle, the velocity of the silicone increases centripetally to about 6 mm/s, shown in yellow and 8 mm/s in the central axis, as shown in orange. Figure 13c shows a relative constant viscosity throughout the nozzle from the inlet to the outlet at 3 Pa·s, shown in red.

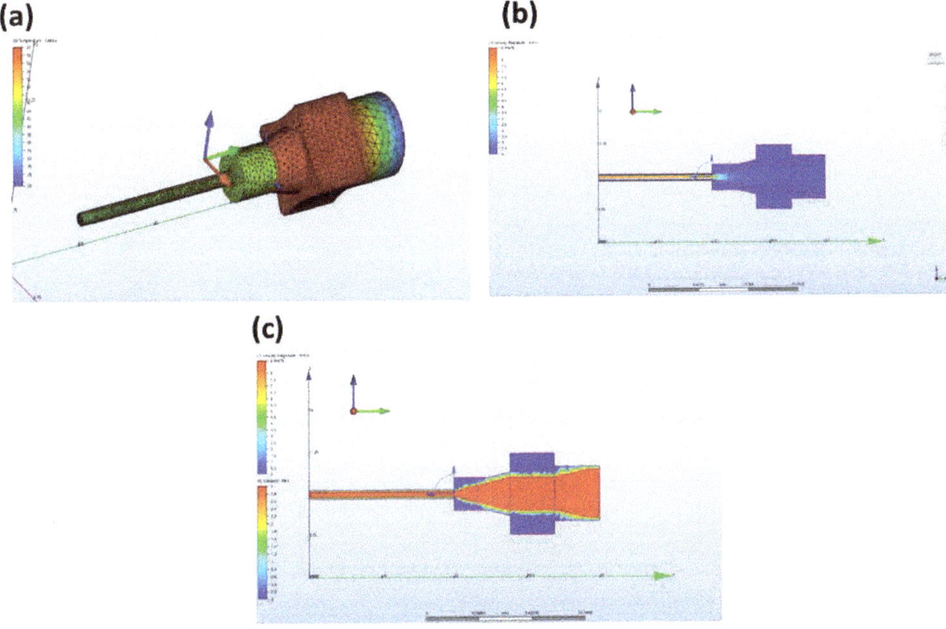

Figure 13. (**a**) Temperature distribution along nozzle, (**b**) velocity and (**c**) viscosity distribution of silicone resin along nozzle.

3.7. Solidworks Meniscus Implant Heat Gradient Simulation Studies

One-convection simulation and on-off layer simulation methods were performed and the results of heat gradients of printed meniscus are shown in Figures 14 and 15, respectively. The simulation temperatures from one-convection model correlates more closely with the experimental results, inferring the adoption of this model for future 3D printing of implants. The comparison between thermal results using one-convection and on-off simulations is shown in Supporting Information, Figure S4.

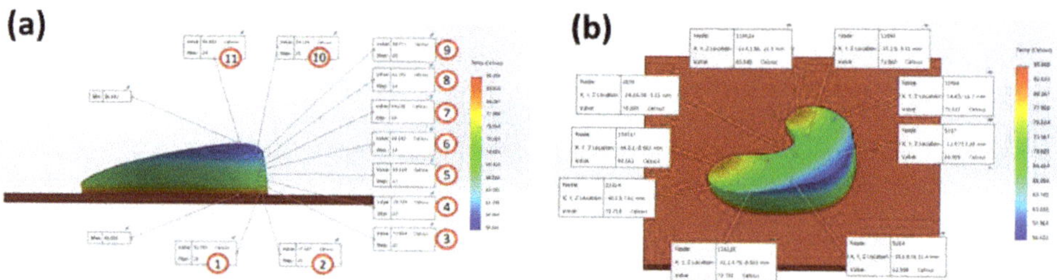

Figure 14. (**a**) Vertical and (**b**) horizontal temperature distribution of the 3D printed silicone meniscus implant using one convection block simulation.

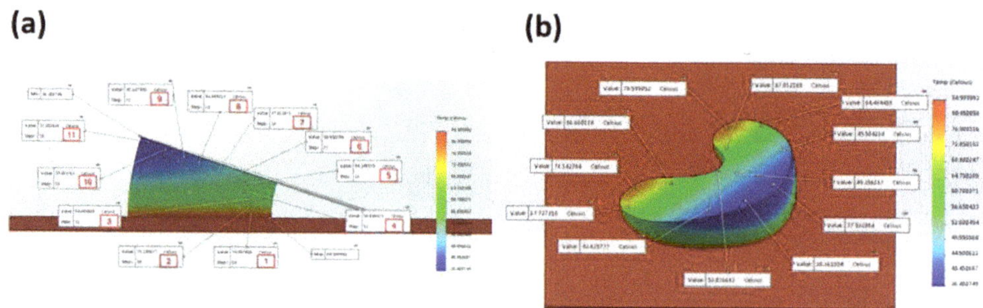

Figure 15. (**a**) Vertical and (**b**) horizontal temperature distribution of on-off layering simulation.

3.8. Comparison of Compression Modulus of 3D Printed and Heat Molded Silicone Meniscus

The student's independent *t*-test was used to compare the compressive modulus of the 3D printed silicone meniscus (0.838 +/− 0.070) MPa vs that of the heat-molded silicone meniscus (0.131 +/− 0.024) MPa. The 95% confidence interval of this difference range from −96,564.16 to 1665.17 MPa. There is no statistical difference between the two groups. The two-tail P is more than 0.05 ($p = 0.058$). The results demonstrate that the 3D printed silicone meniscus has similar compressive mechanical properties as that of the heat-molded silicone meniscus. Figure 16 shows the stress vs strain plot comparison of the 3D printed meniscus versus heat-molded silicone meniscus.

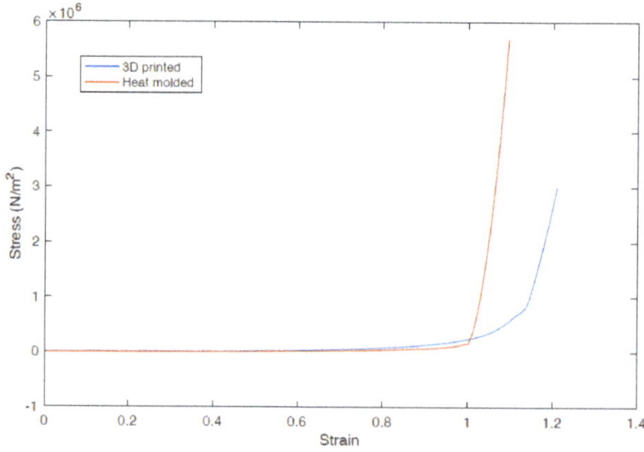

Figure 16. Stress vs strain plot of 3D Printed (blue line) vs heat molded silicone meniscus (red line).

4. Discussions

This study describes the first successful direct 3D printing of heat-cure silicone meniscal implant, using biocompatible and bio-implantable silicone resins. Previous successful works with silicone extrusion, using non-heat curing technology include catalyst extrusion onto a silicone bath (Fripps, Sheffield, UK], multi-materials silicone 3D printing using UV-cured silicone technology (ACEO, Wacker Cheime, Germany] and moisture-cured silicone extrusion actuator [18].

Previously, 3D printing has focused mainly on the printing of bio-models for medical education and preoperative planning and trainings, with special cases of 3D printed titanium calcaneal, spinal and dental implants. To print medical models, implants or devices, one can first use the FDA-approved MIMICS software (Materialise, Belgium] to convert the DICOM files of the patients' CT-MRI to the STL 3D printable format or an open-source Slice3r software for practice.

This new technology certainly opens up the gateway to rapidly 3D print various customizable medical silicone implants and devices for patients and fills the current gap in the AM industry, since the current AM technology have not involved direct 3D printing of medical silicone implants.

The current study has shown that by the precise control of flow rate, nozzle diameter, nozzle temperature and platform temperature, it is possible to accurately print a customized meniscal implant. The starting heat temperatures required for the resins are obtained from the calibration curves shown in the result section. These temperatures ensure adequate degree of gelation prior to extrusion and optimal degree of inter-laminar bonding with previously extruded layers. With the progression of printing, these temperatures require fine modulation with time to cater for different geometries and thicknesses of the end-products.

In addition, the univariate linear regression analysis showed that there is a close correlation between the accuracy and variability of both the lengths and widths of the 3D printed meniscus and both the nozzle and print bed temperatures. A higher optimal temperature not only reduces variability of both the printouts of lengths and widths of the meniscus but also improves the precision and accuracy of printouts.

The simulation results also reflect quite accurately the precise setting of experimental temperature of the heating block at 60 °C to produce a nozzle output of silicone resin at around 40 °C, the extrusion of which provides optimal consistency and lamination with the previous layers. In contrast to the body of the printed meniscus implant, the anterior and posterior horns of the 3D printed meniscus implant retained the highest temperatures of about 80 °C upon completion of printing, making these sites are not suitable for further manipulation or incorporation of micro-channels, drugs or cellular components.

To ascertain the functionality of 3D printed silicone meniscus, the compressive mechanical properties are compared to heat-molded silicone meniscus. The results confirmed that the 3D printed silicone meniscus has similar compressive mechanical properties as that of the heat-molded silicone meniscus. This shows that the new process developed has potential to replace the current heat molding process.

Although several biodegradable and biocompatible scaffolds are available on the market to reconstruct the segmental meniscus defects of previous parts, these scaffolds still need to consider the bulk material properties, structure design and functional requirements and the fabrication process also should be reproducible and reliable. 3D silicone direct printing based on extrusion technique are compared with common fabrication technologies used in meniscal tissue engineering and are tabulated in Table 3.

Table 3. Comparison of direct silicone print with 3D scaffold for meniscus.

	3D Direct Printing Silicone Meniscus	**3D Scaffold for Meniscus**
Material	Silicone	Tissue-derived materials (e.g., periosteal tissue [19], small intestine submucosa (SIS) [20]), synthetic polymers (e.g., polycaprolactone (PCL), Hydrogels, etc.
Fabrication method	Direct silicone extrusion with heat cured Direct silicone extrusion with moisture cured UV cure (ACEO, Germany) Direct catalyst extrusion (Fripps, UK)	For sponge scaffold: leaching, gas foaming Fibrous scaffold: electrospinning, electrohydrodynamic jetting [32]
Pros & cons	(+) Biocompatible material approved by FDA (+) Minimal void inside (+) High strength and elongation (+) Relatively low price (−) Low resolution (−) limited capability to fabricate complex structures	(+) For sponge scaffold: high porosity benefit for cell proliferation, etc. (+) For fibrous scaffold: ability to fabricate nanofibrous architecture and complex structures; wild range of fiber diameter; (−) Using solvent can be toxic (−) Limited fabrication capabilities (−) Limited biochemical composition and biomechanical properties for implant

5. Conclusions

A novel silicone 3D printer was successfully built for the direct 3D Printing of silicone meniscus implants which demonstrate mechanical properties as the conventionally heat-molded meniscus. The nozzle diameter, nozzle and bed temperatures were shown to be critical factors in determining the precision and accuracy of the lengths and widths of the meniscus. In this study, heated printer nozzle and meniscus implant designs were evaluated to determine the thermal distribution along the nozzle and across the meniscus implant by employing CFD and heat simulations using Solidworks.

Incorporating complex interior lattice or micro-channels and composite multi-material 3D silicone printing will be the future challenges in direct silicone 3D printing. Scaffold-based tissue regeneration based on AM technology is another promising approach to meniscus surgical treatment. In theory, the meniscal scaffold should provide appropriate biomechanical functions after implantation to shield cells from damaging compressive or tensile forces, maintain their shape integrity (without shrinkage, etc.), mechanical stability and strength on the defect area until enough host tissue was regenerated, produce mechanical stimuli to promote tissue regeneration.

Unique challenges present in silicone 3d printing are: (1) difficulty in the handling of silicone resins, (2) difficulty in printing multi-materials or different silicone resins, (3) finding suitable post-processing methods and 4) coming up with suitable standards in medical silicone 3D Printing.

Handling of silicone resins require meticulous even mixing to avoid trapping of air bubbles. Two-part resins are prone to disproportionate mixing and uneven curing. One-part resins are susceptible to moisture contamination and premature curing.

Different grades of silicone resins or different materials require different printing parameters and printing processes for optimal output. Consequently, modifications and additions to the 3D printers are necessary to achieve multi-grade silicones or multi-material printing.

Unlike post-processing methods in the 3D printing of other solid, liquid or powder substrates, silicone rubber products are highly susceptible to cuts, fissures, abrasions and lacerations while undergoing post-processing. Conventional processing methods for the above substrates cannot be applied to silicone printing. Therefore, this is a unique case where post-processing should be avoided as much as possible.

Currently no ASTM or similar standards was prescribed for medical silicone 3D printing. It is therefore imperative that one of the major areas of future works focus on the setting of standards in medical silicone 3D printing.

Supplementary Materials: The following are available online at http://www.mdpi.com/2073-4360/12/5/1031/s1, Figure S1. (a) Model dimensions and (b) schematic of CFD design of heated nozzle. Figure S2. Changes in storage and loss modulus over time for (a and b) Eco30 and (c and d) Eco50 under heat curing of 40, 45, 55 and 65 °C.

Figure S3. Bright field images of (a) topmost layer and (b) vertical cross-section of body of 3D printed silicone meniscus implant. Figure S4.. Comparison of thermal results using one-convection (above) and on-off (below) simulation. Table S1. Material Properties of Aluminum, Stainless Steel 304, Ecoflex Silicone. Table S2. Boundary conditions for printed silicone meniscus implant.) and on-off (below) simulation.

Author Contributions: Conceptualization, E.L. and W.Y.Y.; data curation, H.M.P.; formal analysis, R.B.; funding acquisition, W.Y.Y.; investigation, E.L.; methodology, R.B.; resources, S.L.S., J.S. and W.Y.Y.; supervision, S.L.S., J.S. and W.Y.Y.; writing—original draft, E.L.; writing—review & editing, H.M.P., S.L.S. All authors have read and agreed to the published version of the manuscript.

Funding: This research is supported by the National Research Foundation, Prime Minister's Office, Singapore under its Medium-Sized Center Funding scheme and the NTU Start-Up Grant.

Acknowledgments: This research was supported by the National Research Foundation, Prime Minister's Office, Singapore under its Medium-Sized Center Funding Scheme and the NTU Start-Up Grant. The authors thank all administrative and research staff in Singapore Center for 3D Printing, Nanyang Technological University that have contributed to the successful publication of this work.

Conflicts of Interest: The authors declare no conflict of interest.

References

1. Chambers, M.C.; El-Amin, S.F., III. Tissue engineering of the meniscus: Scaffolds for meniscus repair and replacement. *Musculoskelet. Regen.* **2015**, *2*, e998. [CrossRef]

2. Guo, W.; Liu, S.; Zhu, Y.; Yu, C.; Lu, S.; Yuan, M.; Gao, Y.; Huang, J.; Yuan, Z.; Peng, J.; et al. Review article: Advances and Prospects in TE meniscal scaffolds for meniscal regeneration. *Stem Cell Int.* **2015**, *2015*, 517520. [CrossRef]

3. Vaquero, J.; Forriol, F. Meniscus tear surgery and meniscus replacement. *Muscles Ligaments Tendons J.* **2016**, *6*, 71–89. [CrossRef]

4. Lahey, F.H. Comments made following the speech "Results from using Vitallium tubes in biliary surgery," read by Pearse, HE before the American Surgical Association, Hot Springs, VA. *Ann. Surg.* **1946**, *124*, 1027.

5. Clark, G. *Cochlear Implants: Fundamentals and Applications*; Springer: Berlin, Germany, 2006.

6. Braga-Silva, J. The use of silicone tubing in the late repair of the median and ulnar nerves in the forearm. *J. Hand Surg.* **1999**, *24*, 703–706. [CrossRef]

7. FDA Update on the Safety of Silicone Gel-Filled Breast Implants. U.S. Food and Drug Administration, United States. Available online: https://www.fda.gov/medical-devices/breast-implants/update-safety-silicone-gel-filled-breast-implants-2011-executive-summary (accessed on 3 January 2018).

8. Rohrich, R.J.; Wan, D. Working toward a Solution: The Unanswered Questions about Silicone Gel Breast Implants. *Ann. Intern. Med.* **2016**, *164*, 201–202. [CrossRef]

9. Balk, E.M.; Earley, A.; Avendano, E.A.; Raman, G. Long-Term Health Outcomes in Women with Silicone Gel Breast Implants: A Systematic Review. *Ann. Intern. Med.* **2016**, *164*, 164–175. [CrossRef]

10. Singh, N.; Picha, G.J.; Hardas, B.; Schumacher, A.; Murphy, D.K. Five-Year Safety Data for More than 55,000 Subjects following Breast Implantation: Comparison of Rare Adverse Event Rates with Silicone Implants versus National Norms and Saline Implants. *Plast. Reconstr. Surg.* **2017**, *140*, 666–679. [CrossRef]

11. Chugay, N.V.; Chugay, P.N.; Shiffman, M.A. *Body Sculpting with Silicone Implants*; Springer Science & Business Media: Berlin, Germany, 2014.

12. Swanson, A.B. Silicone rubber implants for replacement of arthritic or destroyed joints in the hand. *Surg. Clin. N. Am.* **1968**, *48*, 1113–1127. [CrossRef]

13. Foliart, D.E. Swanson silicone finger joint implants: A review of the literature regarding long-term complications. *J. Hand Surg.* **1995**, *20*, 445–449. [CrossRef]

14. Mazas, F.B. GUEPAR total knee prosthesis. *Clin. Orthop. Relat. Res.* **1973**, *94*, 211. [CrossRef] [PubMed]

15. Harmand, M.F.; Briquet, F. In vitro comparative evaluation under static conditions of the hemocompatibility of four types of tubing for cardiopulmonary bypass. *Biomaterials* **1999**, *20*, 1561–1571. [CrossRef]

16. Robb, W.L. Thin silicone membranes-their permeation properties and some applications. *Ann. N. Y. Acad. Sci.* **1968**, *146*, 119–137. [CrossRef] [PubMed]

17. Available online: https://www.meddeviceonline.com/doc/silicon-a-material-with-huge-potential-for-lab-on-chips-0001 (accessed on 30 March 2020).

18. Zinoviev, K.; Carrascosa, L.G.; Sánchez del Río, J.; Sepulveda, B.; Domínguez, C.; Lechuga, L.M. Silicon Photonic Biosensors for Lab-on-a-Chip Applications. *Adv. Opt. Technol.* **2008**, *2008*, 383927. [CrossRef]

19. Dhanekar, S.; Jain, S. Porous silicon biosensor: Current status. *Biosens. Bioelectron.* **2013**, *41*, 54–64, ISSN 1873-4235. [CrossRef]

20. Takahashi, T.; Kaschta, J.; Münstedt, H. Melt Rheology and Structure of Silicone Resins. *Rheol. Acta* **2001**, *40*, 490–498. [CrossRef]

21. Fripp, L.; Frewer, N.; Green, L. Method and Apparatus for Additive Manufacturing. EP3060380B1, 4 March 2020.

22. Plott, J.; Shih, A. The extrusion-based additive manufacturing of moisture-cured silicone elastomer with minimal void for pneumatic actuators. *Addit. Manuf.* **2017**, *17*, 1–14. [CrossRef]

23. Liravi, F.; Toyserkani, E. A hybrid Additive Manufacturing Method for the Fabrication of Silicone Bio-structures: 3D Printing Optimization and Surface Characterization. *Mater. Des.* **2018**, *138*, 46–61. [CrossRef]

24. Kim, D.S.; Tai, B.L. Hydrostatic support-free fabrication of three-dimensional soft structures. *J. Manuf. Process.* **2016**, *24*, 391–396. [CrossRef]

25. Hinton, T.J.; Hudson, A.; Pusch, K.; Lee, A.; Feinberg, A.W. 3D Printing PDMS Elastomer in a Hydrophilic Support Bath via Freeform Reversible Embedding. *ACS Biomater. Sci. Eng.* **2016**, *2*, 1781–1786. [CrossRef]

26. Luis, E.; Pan, H.M.; Sing, S.L.; Bastola, A.K.; Goh, G.D.; Goh, G.L.; Tan, H.K.J.; Ram, B.; Song, J.; Yeong, W.Y. Silicone 3D Printing: Process Optimization, Product Biocompatibility, and Reliability of Silicone Meniscus Implants. *3D Print. Addit. Manuf.* **2019**, *6*, 319–332. [CrossRef]

27. Bullions, T.A.; McGrath, J.E.; Loos, A.C. Development of a Two-Stage, Dual-Arrhenius Rheology Model for a High-Performance. PhenylEthynyl-Terminated Poly(Etherimide). *Polym. Eng. Sci.* **2002**, *42*, 2182–2192. [CrossRef]

28. Romano, M.R.; Cuomo, F.; Massarotti, N.; Mauro, A.; Salahudeen, M.; Costagliola, C.; Ambrosone, L. Temperature Effect on Rheological Behavior of Silicone Oils. A Model for the Viscous Heating. *J. Phys. Chem. B* **2017**, *121*, 7048–7054. [CrossRef] [PubMed]

29. Stieghorst, J.; Doll, T. Doll Rheological behavior of PDMS silicone rubber for 3D printing of medical implants. *Addit. Manuf.* **2018**, *24*, 217–223.

30. Nithyavathy, N.; Parameshwaran, R.; Suganeswaran, K.; Naveen, C.; Raja, G. Simulation based Sensitivity Study of Tread Pattern and Materials on Cooling Efficiency of M1 Vehicle Tyres. *Int. J. Veh. Struct. Syst.* **2016**, *8*, 179–182.

31. El Kissi, N.; Piau, J.M.; Attané, P.; Turrel, G. Shear rheometry of polydimethylsiloxanes. Master curves and testing of Gleissle and Yamamoto relations. *Rheol. Acta* **1993**, *32*, 293–310. [CrossRef]

32. Gao, D.; Zhou, J. Designs and applications of electrohydrodynamic 3D printing. *Int. J. Bioprinting* **2019**, *5*, 172. [CrossRef]

Article

Mode I Fracture Toughness of Polyamide and Alumide Samples obtained by Selective Laser Sintering Additive Process

Dan Ioan Stoia, Liviu Marsavina and Emanoil Linul *

Department of Mechanics and Strength of Materials, Politehnica University of Timisoara, 1 Mihai Viteazu Avenue, 300 222 Timisoara, Romania; dan.stoia@upt.ro (D.I.S); liviu.marsavina@upt.ro (L.M.)
* Correspondence: emanoil.linul@upt.ro; Tel.: +40-256-40-3741

Received: 30 January 2020; Accepted: 9 March 2020; Published: 11 March 2020

Abstract: Selective Laser Sintering is a flexible additive manufacturing technology that can be used for the fabrication of high-resolution parts. Alongside the shape and dimension of the parts, the mechanical properties are essential for the majority of applications. Therefore, this paper investigates dimensional accuracy and mode I fracture toughness (K_{IC}) of Single Edge Notch Bending samples under a Three Point Bending fixture, according to the ASTM D5045-14 standard. The work focuses on the influence of two major aspects of additive manufacturing: material type (Polyamide PA2200 and Alumide) and part orientation in the building environment (orientations of 0°, 45° and 90° are considered). The rest of the controllable parameters remains constant for all samples. The results reveal a direct link between the sample densities and the dimensional accuracy with orientation. The dimensional accuracy of the samples is also material dependent. For both materials, the angular orientation leads to significant anisotropic behavior in terms of K_{IC}. Moreover, the type of material fundamentally influences the K_{IC} values and the fracture mode. The obtained results can be used in the development of additive manufactured parts in order to obtain predictable dimensional tolerances and fracture properties.

Keywords: Three Point Bending test; mode I fracture toughness; selective laser sintering; polyamide and Alumide; geometrical errors; microstructure.

---

1. Introduction

The geometrical flexibility of the parts produced by Selective Laser Sintering (SLS) represents the top characteristic of this technology [1–3]. Commonly, SLS technology uses the energy of a laser beam in addition to an electrical heating source in order to generate a high enough temperature for particle bonding. Due to the large number of process variables that have to be controlled during the process, some variations of shape, size and mechanical properties of the finite product may occur not only in SLS but also in all additive manufacturing (AM) technologies [4–6]. The SLS process variables include chamber temperature, laser energy density, layer thickness, beam offset, shrinkage at cooling, part position and orientation, material type and powder degradation [7]. The deflection from the nominal size and mechanical properties of parts represents a challenging aspect of AM [8–10].

Some authors have focused their work on the geometrical accuracy of samples obtained at different orientations and/or energy density by measuring the size and shape and inspecting the surface of the parts by stereomicroscopy [11–13] without taking into account the mechanical aspects that are directly influenced by the process parameters. The mechanical characterization of SLS polymer parts was conducted through tensile tests by some authors [14], while in other studies flexural properties were determined according to the manufacturing parameters [15,16]. However, the process parameters greatly influence the mechanical properties of the samples. Efforts in assessing the failure predictions of

PA12 samples obtained by SLS have been conducted. The work was based on implementing the failure criterion for AM parts in tensile, compression and shear [17].

To the best of the authors' knowledge, limited studies cover the fracture behavior of cracked or notched samples manufactured by SLS [18–20]. This is a challenging topic since the presence and growing of cracks will interact with the manufacturing layers, sinterization directions and powder particle scattering. Linul et al. [18] used symmetric and asymmetric four-point bending tests to determine the mode I and mode II fracture toughness of laser sintered polyamide. The authors observed that the K_{IC} values were higher than K_{IIC} ones, while the density of the 3D printed samples was highly dependent on the process energy. Fracture mechanics of laser sintered cracked polyamide as a new method to induce cracks was studied by Brugo et al. [19] on six different configurations of Mode I Compact Tension samples. Their results showed that the samples with better mechanical performance were those in which all the layers contained a portion of the crack. On the other hand, Crespo et al. [20] performed experiments at different loading rates to measure the failure loads of different laser sintered notched samples. The results have been used by the authors to apply the TCD to the failure of PA12 notched samples prepared by AM techniques.

The investigation reported here focuses on determining the dimensional accuracy and mode I fracture toughness values of Single Edge Notch Bending Polyamide PA2200 and Alumide samples through a Three Point Bending fixture, designed according to ASTM D5045–14 [21]. Finding out the fracture properties is a step forward, beside tensile and bending properties, in the characterization of AM products. This is of high industrial interest since AM is considered a key technology for the fourth industrial revolution [22]. The mechanical properties of polymers are particularly important because they are generally used as net-shaped parts, unlike metallic components. The fracture properties were underlined in close relation to part orientation inside the building envelope.

2. Materials and Methods

2.1. Materials and Specimen Manufacturing

The Polyamide PA2200 powder is a commercially available material produced by Electro Optical Systems (EOS GmbH, Krailling, Germany). This is a multipurpose material that has adequate mechanical (high strength and stiffness) and chemical properties [23]. Long term stability and high detail resolution are also important characteristics of PA2200, making it suitable for a wide range of technical applications [23]. In addition, its biocompatible property makes it usable in custom-made disposable guidance elements, such as those used in the surgical field [24]. Alumide, on the other hand, is a homogeneous mixture of fine polyamide PA12 and fine aluminum particles developed by the same producer. The sintered Alumide possesses a porous structure comparable to polyamide but has a higher stiffness, better thermal conductivity and good density-stiffness ratio [24]. More details regarding the properties and advantages of the PA2200 and Alumide materials are presented in [2].

The additive manufacturing process was conducted on an EOS Formiga P100 (EOS GmbH Electro Optical Systems), on separate stages for each material. The geometry of the specimen was constructed in SolidWorks 2013 according to the specifications of the ASTM D5045–14 standard [21]. The design containing no notches was imported in Materialise Magics 10.0 software [25], where the 3D model was positioned and angularly oriented. The positioning was done using 10 mm clearance between each edge of the parts, and between each part and the machine's limit. The orientations were only in the XY plane, at values of 0°, 45° and 90° in respect to transversal axis of the machine (X). Ten samples of each material and orientation were manufactured and post-processed by air blasting. In Figure 1a the sintered parts according to their orientations can be observed while Figure 1b depicts the final aspect of both the Alumide and polyamide samples. There is also a visible rib network that was designed to connect the standard part in order to prevent the curling and twisting phenomena as much as possible.

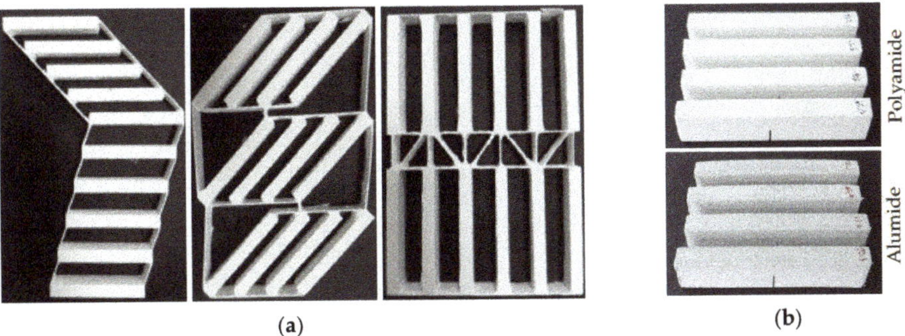

(a)   (b)

Figure 1. Samples after post-processing oriented at 0°, 45° and 90°, respectively (**a**) and notched polyamide and Alumide samples (**b**).

The additive manufacturing of all 60 samples was done using the same process parameters (except the orientation angle) in order to obtain consistent and comparable results. The building chamber and removal chambers were set to 170 °C and 159 °C, respectively. The energy density value was 0.067 J/mm^2, obtained by setting the power to 25 W, the laser beam velocity to 1500 mm/s and the scan spacing to 0.25 mm. These parameters were chosen based on previous studies and the experience of the authors [2,18,26,27]. A scaling factor of 2.3% was applied for all directions of the sample in order to compensate the shrinkage effect at cooling.

2.2. Methods

Prior to crack machining and mechanical testing all samples were measured and weighed. The linear measurements (length L, thickness B and width W) were done using a digital caliper of 0.01 mm precision while the samples masses were measured using a 500 g Kern balance (0.01 g accuracy). In addition, a linear measurement defined as Z deflection was conducted. It characterized the distance from the flat plane supporting the sample to the top of the sample, recorded along the middle line of the sample. The designed samples had a height/width (W) of 20 mm, thickness B = W/2= 10 mm and span length S = 4W = 80 mm (Figure 2). The linear measurements were an indication of curling in the frontal-lateral building planes while the mass and volume of each sample were further used for computing the density.

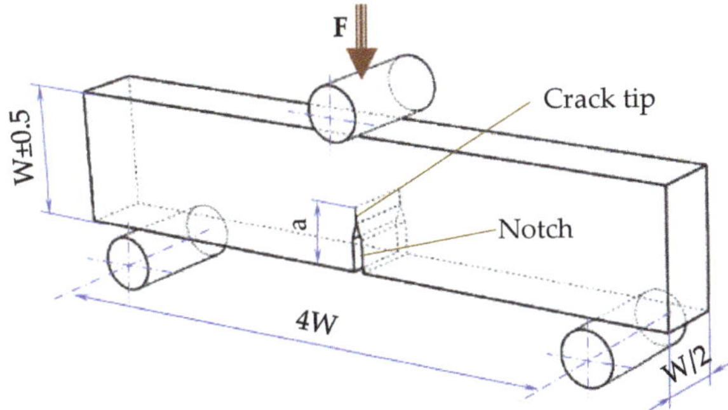

Figure 2. Geometrical parameters of the Single Edge Notch Bend (SENB) sample according to [21].

In order to assess the fracture behavior and to determine the fracture toughness of polyamide and Alumide materials, Single Edge Notch Bend (SENB) samples were adopted in this investigation [21]. The sample notch (8 ± 0.05 mm) was produced using a milling cutting disc with a thickness of 0.6 mm, while the crack (2 ± 0.1 mm) was initiated by using a very sharp and thin razor blade. Therefore, the total length of the crack *a* was about 10 ± 0.15 mm (Figure 2). At least five samples were used for every orientation and each material in the testing phase.

The SENB samples were tested under a Three Point Bending (TPB) fixture using a Zwick Roell Z005 quasi-static universal testing machine having a maximum load-cell capacity of 5 kN (of 0.1% accuracy). The experimental TPB tests were carried out at room temperature according to the ASTM D5045–14 standard [21]. A constant crosshead speed of 5 mm/min was used, until a fracture occurred. Figure 3 shows the fixing of the SENB samples in the TPB device and the form of the samples before and after the test for the two types of investigated materials, respectively.

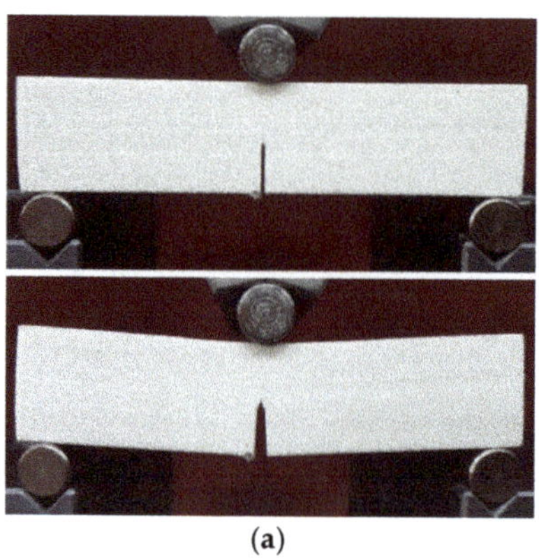
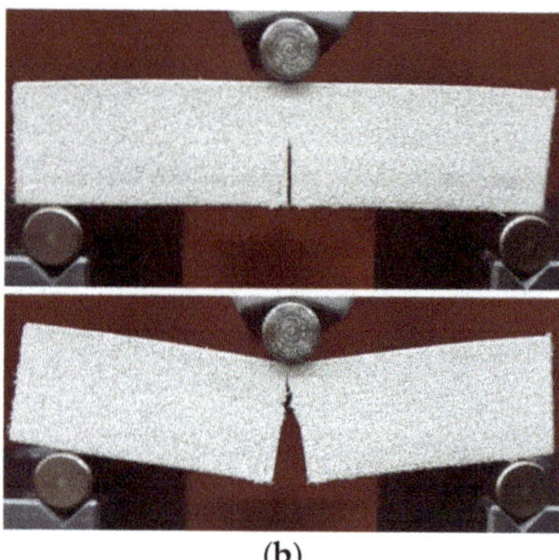

(a) (b)

Figure 3. Polyamide (**a**) and Alumide (**b**) SENB samples before and after the quasi-static TPB test.

The upper loading pin and bottom supports of TPB device were large enough in diameter (10 mm) in order to avoid sample indentation. The load-displacement data were automatically recorded using a sampling frequency of 600 Hz.

Correlations between the orientation input parameter and the output variables (density, fracture toughness and dimensional error) were performed using the conventional computation of Pearson's correlation coefficient (PCC). The PCC assumes a linear relationship between the input and output parameters and is suitable for small data sets. The PCC's value will result in the interval [−1, +1], where negative values indicate an inverse relationship between the parameters while a positive value indicates a direct relationship. A value of 1 represents the best correlation while 0 value will indicate no correlation at all [28,29].

The statistical significance of the data obtained for both materials was put in evidence by a probability test (*P*-value). The null hypothesis H0 was defined as no significant difference between the parameter (ρ, K_{Ic}, Err.L, Err.B, Err.W,) values of both materials. The alternative hypothesis Ha was defined as the parameter values of the two materials being significantly different. The *P*-values greater than 0.05 (95% confidence) would reject H0 and accept the alternative hypothesis Ha [30,31].

3. Results and Discussion

3.1. Accuracy Assessment

The length, thickness and width of the specimens were investigated from a form dimensional accuracy perspective. The nominal dimensions of the sample were customized by the scaling factors used for the manufacturing process of 2.3% on each direction. Design dimensions and actual measurements of the sample size were used to calculate the relative error of length, width, thickness and deflection in the Z axis. Each measurement was conducted three times and the mean value for dimension/material at every orientation was used for graphical representation.

In Figure 4 the relative dimensional errors of the Alumide and polyamide samples are presented as percentages of the design sizes in relation to the angular orientations. Comparable errors were determined in the interval 0.2% to 1.2% for the X-Y manufacturing direction (length and thickness) of both materials. These errors translated in dimensions of 1 to 3 tenths of millimeters, which for an additive manufactured part are acceptable [32,33]. The near error interval was generated by the process symmetry in both sinterization directions and temperature distribution.

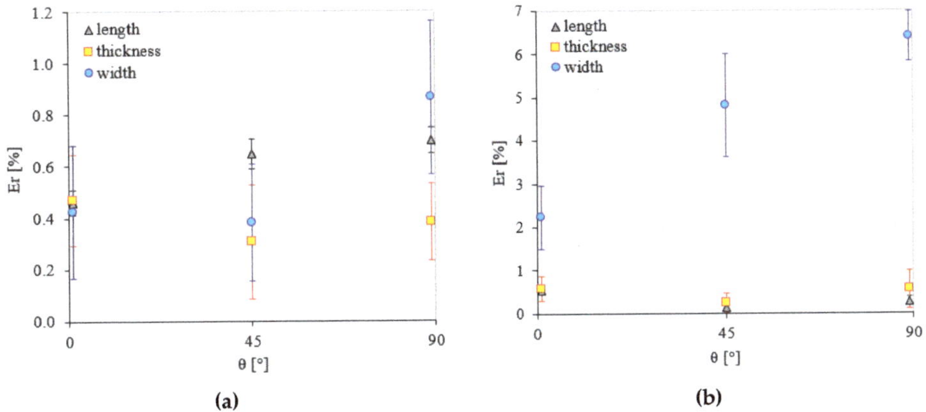

Figure 4. Relative dimensional errors of polyamide (**a**) and Alumide (**b**) samples.

By inspecting the dimensional changes of both individual materials on X and Y directions (length and thickness) no trend could be identified in relation to orientation. However, both materials experienced acceptable size errors (below 1%) for the additive manufacturing process [32,33]. The exception was the width error of Alumide that experienced values as large as 6.5% (Figure 4a,b). These high errors for all three orientations revealed an unstable geometry in the z direction. During the sinterization phase, the large expansion of aluminum particles replaced a significant local volume of polyamide powder such that the particle bonding occurred differently than in pure polyamide powder. The aluminum did not participate in the bonding but influenced the bonding formation of polymer particles, conducting finally to changes in structure, dimensions and properties. Furthermore, the uneven distribution of the aluminum particles in the polymer powder generated different local bonding conditions that led to different samples geometries. The phenomenon occurred particularly on the vertical direction because of the free surface of the top layer.

In Figure 5 the density of both sintered materials at every orientation angle and the sample deflection in Z direction (width dimension) are presented. The density of the Alumide was higher than the PA due to the aluminum particles that possessed a density 2.2 to 2.3 times higher than PA.

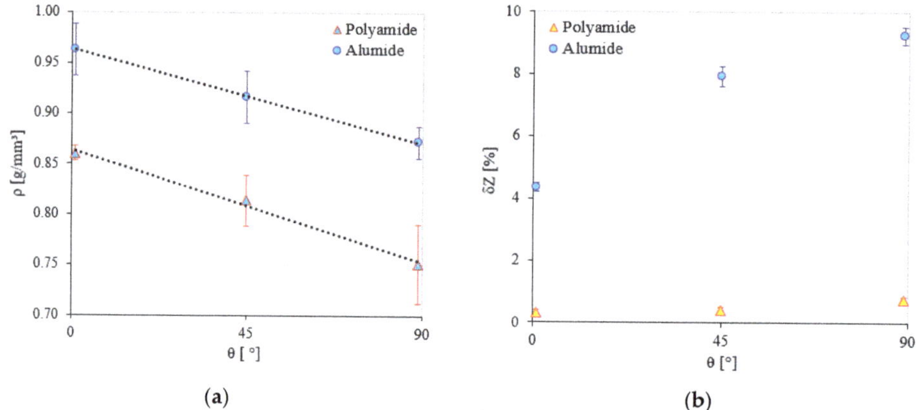

Figure 5. Density (**a**) and Z deflection (**b**) of polyamide and Alumide samples.

Regarding the orientation effect, both materials exhibited the same tendency of decreasing the density of the sintered part with the increasing of the orientation angle. This may have been caused by the powder spreading direction in relation to the sample's directions. A 0° orientation meant that the powder was longitudinally deposited along the length of the samples, while a 90° orientation meant that the powder dispenser blade of the machine was depositing along the thickness of the samples. It appeared that when the dispenser blade spent more time over the object, the powder was better distributed and settled down on the previous layer. The proper powder spreading and slight compaction produced by the dispenser on one hand and the sinterization energy on the other hand led to a higher density of the structure. The deflection error in the Z direction is presented in Figure 5b, showing large values for Alumide (9.1%). The deflection phenomena occurred predominantly during the cooling phase of the process. The larger thermal expansion of the aluminum particles compared with the PA generated larger sample sizes during the sinterization. During the cooling time, the samples shrunk especially in the longitudinal direction. Because the temperature gradient on the top of the samples was different from that on the bottom, the samples bent in Z direction. The differences in the cooling rates on the top and bottom of the samples were generated by the different powder volume (non-sintered) that surrounded the objects and the machine's architecture.

3.2. Fracture Toughness Assessment and Parameter Correlations

Following the quasi-static TPB tests, the load (F)-displacement (Δ) curves were obtained and processed. In Figure 6, the F-Δ curves of notched samples are presented for both investigated materials (polyamide and Alumide) and sample orientation angle (0°, 45° and 90°). The F-Δ curves highlighted a linear-elastic behavior that ended with a maximum load, followed by the final fracture of the sample [21]. Beyond the maximum load, the samples either broke sharply or showed a progressive decrease in load carrying capacity, followed by a final fracture. It was seen that the polyamide samples had higher maximum loads than in the case of the Alumide ones. Furthermore, higher loads were required to fracture the samples obtained at 0° orientation, independent of material type. In addition, when we observed the displacement to break, the Alumide had a much more brittle behavior than the polyamide samples. The higher maximum loads and displacements of polyamide compared with Alumide came from the internal structure of these. The pure polyamide material established better bonds between particles, which in the case of Alumide were restricted by aluminum particles. On the other hand, the higher loads recorded for 0° orientation in both materials had to do with the higher densities recorded for this orientation. An explanation for why the density behaved in this way is presented next to the dimensional results.

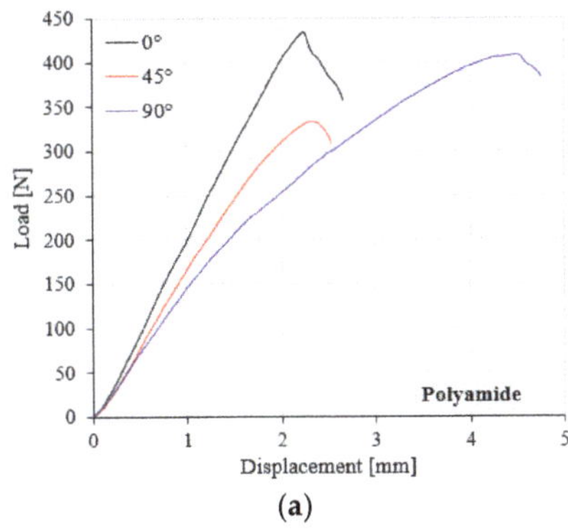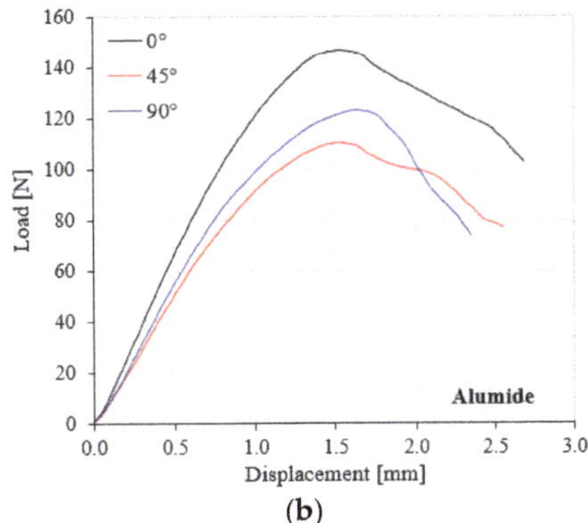

Figure 6. Load-displacement curves of polyamide (**a**) and Alumide (**b**) notched samples.

By using the dimensions of the samples, the mode I fracture toughness (K_{IC}) values were determined according to the ASTM D5045–99 standard [21], applying Equation (1).

$$K_{IC} = \frac{F_{crt}}{BW^{0.5}} f\left(\frac{a}{W}\right) \qquad \left[MPa \cdot m^{0.5}\right] \tag{1}$$

where F_{crt} is the critical load in [N], determined in accordance with [21], B and W are geometrical parameters of the samples (thickness and width) in [mm], a is the initial crack length of the sample in [mm], while $f(a/W)$ is a geometrical function expressed by Equation (2) in terms of a/W ratio [21]:

$$f\left(\frac{a}{W}\right) = 6\sqrt{\frac{a}{W}} \frac{1.99 - \left(\frac{a}{W}\right)\left(1 - \frac{a}{W}\right)\left[2.15 - 3.93\left(\frac{a}{W}\right) + 2.7\left(\frac{a}{W}\right)^2\right]}{\left(1 + 2\frac{a}{W}\right)\left(1 - \frac{a}{W}\right)^{1.5}}. \tag{2}$$

Table 1 presents the computed mode I fracture toughness values for the two materials (PA2200 and Alumide) for every orientation angle (θ).

Table 1. Mode I fracture toughness values of polyamide and Alumide samples according to θ.

Material	Orientation Angle θ [°]	Sample Density ρ [g/cm^3]	Crack Length a [mm]	Maximum Load F_{crt} [MPa]	Displacement at F_{crt} δ [mm]	Fracture Toughness K_{IC} [MPa/m$^{0.5}$]
Polyamide	0	0.855	9.27	335	1.6	2.182
		0.864	9.36	434	2.0	2.902
		0.853	9.63	397	2.3	2.745
		0.861	9.67	407	2.5	2.864
		0.852	9.25	419	2.4	2.753
	45	0.797	8.79	336	2.2	2.063
		0.837	9.20	321	2.0	2.104
		0.766	8.89	287	2.1	1.797
		0.852	9.50	333	2.0	2.276
		0.782	9.93	266	1.6	1.924
	90	0.717	9.36	315	1.6	1.907
		0.737	8.90	350	3.3	2.208
		0.764	8.63	409	3.9	2.464
		0.759	9.76	335	1.7	2.405
		0.781	9.03	427	4.3	2.743

Table 1. *Cont.*

Material	Orientation Angle θ [°]	Sample Density ρ [g/cm³]	Crack Length a [mm]	Maximum Load F_{crt} [MPa]	Displacement at F_{crt} δ [mm]	Fracture Toughness K_{IC} [MPa/m$^{0.5}$]
Alumide	0	0.957	9.52	136	1.5	0.897
		0.979	9.62	147	1.5	0.982
		0.987	9.78	158	1.5	1.079
		0.970	9.48	140	1.6	0.929
		0.982	9.78	148	1.5	1.020
	45	0.854	10.67	99	1.3	0.713
		0.878	10.56	96	1.3	0.713
		0.881	10.24	110	1.6	0.783
		0.886	10.37	118	1.6	0.848
		0.879	10.36	110	1.5	0.755
	90	0.901	9.88	121	1.3	0.834
		0.878	10.30	99	1.2	0.708
		0.939	9.74	127	2.0	0.885
		0.899	10.23	104	1.4	0.739
		0.915	10.02	123	1.6	0.865

It was obvious that the investigated material had a significant influence on mode I fracture toughness values. Values up to about 63.5% higher in the case of polyamide than in the case of Alumide samples for 0° orientation were obtained, while for the 90° orientation this difference increased up to 65.6%. The smallest difference, of about 62.5%, was observed in the case of a sample orientation angle of 45° (Figure 7). The considerable higher fracture toughness values of pure polyamide versus Alumide were directly connected with the bonding formation between the powder particles in the two materials. Weaker bonding in Alumide was caused by the presence of aluminum particles, which restricted the sinterization bridges formation.

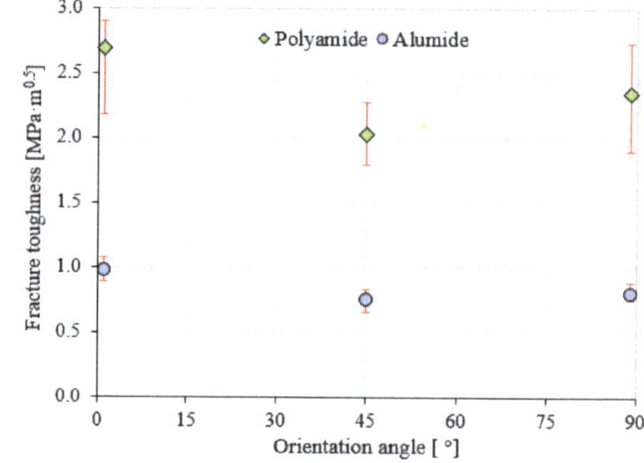

Figure 7. Fracture toughness variation according to the sample orientation angle and material type.

Major differences in K_{IC} values were observed even between the loading directions of the same material. The 0° orientation fracture toughness values were found to be higher than the 90° orientation, with about 12.8% for polyamide and 17.9% for Alumide, respectively. This K_{IC} percentage difference increased by up to 24.4% for polyamide and 22.3% for Alumide between 0° and 45° orientations. Thus, the two materials had approximately the same K_{IC} increases/decreases between different θ. An important aspect to note, which can be seen in Figure 7, is that the polyamide K_{IC} values had much larger error bars compared with the Alumide ones for all sample orientation angles. One of the reasons for these large errors could be related to poor process stability in the case of pure PA compared with Alumide, in which the better thermal transfer conducted a more stable process. Therefore, the investigated materials highlighted a significant anisotropic behavior in terms of mode I fracture toughness, being directly related to the different powder spreading and heat transfer.

Figure 8 presents the crack paths for polyamide and Alumide samples obtained after the TPB fracture test for an orientation angle of 0°. The images were acquired by a Kruss stereo microscope using a magnification of 40X. The crack propagation in both cases followed a relative vertical line, which sustained the mode I fracture.

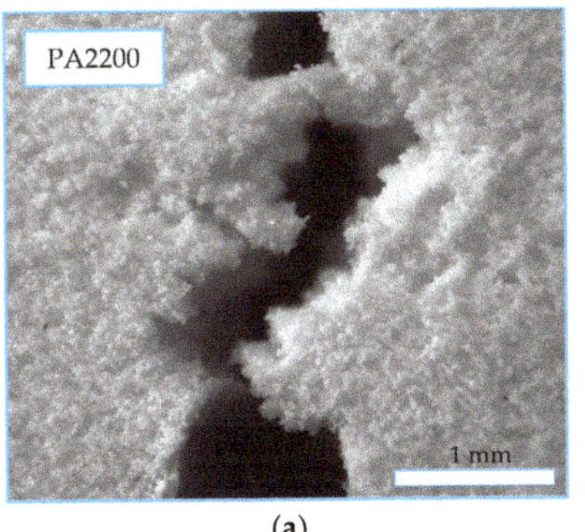

(a)   (b)

Figure 8. Crack propagation in polyamide (**a**) and Alumide (**b**) samples

Structure images obtained by optical microscopy of 200X magnification are presented in Figure 9. These revealed the particle bonding of the PA2200 sample (Figure 9a) in the form of small bridges. The identification of the individual particles in the image revealed the sinterization aspect of the process, where coalescence occurred only partially. The image of the Alumide structure revealed some aluminum particles stuck inside the bonded polyamide particles (Figure 9b). The aluminum was not participating to the bond because of its different melting temperature.

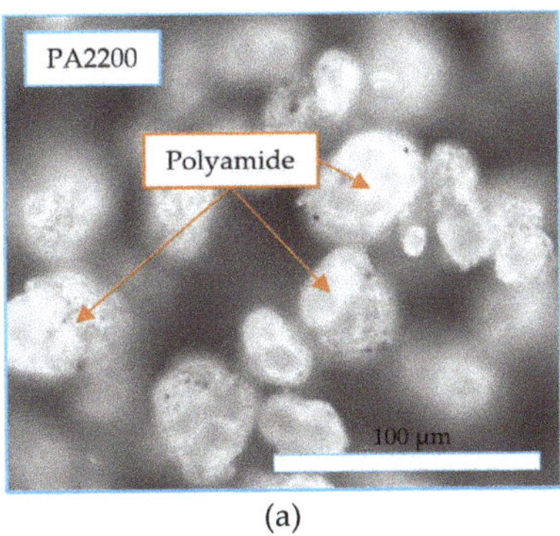

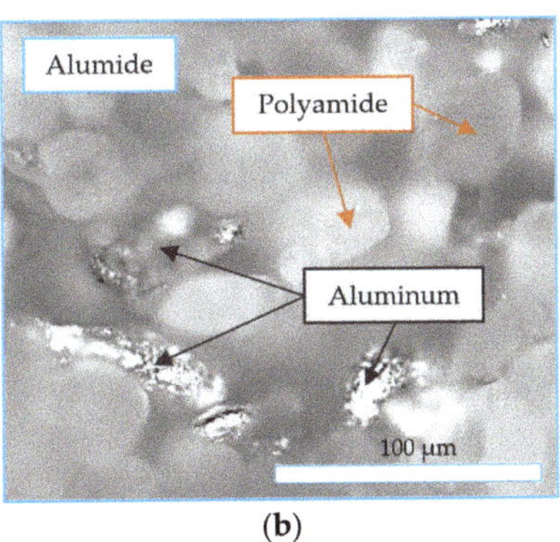

(a)   (b)

Figure 9. Structure of polyamide (**a**) and Alumide (**b**) samples.

In order to establish a relationship between the sample's orientation angle and the outcome parameters (density, length, thickness and width errors and fracture toughness), Pearson's correlation was applied to the data. In Table 2 the correlation values are presented, having the following range where -1 is best negative correlation, 0 represents no correlation and +1 is best positive correlation. A very good negative correlation was observed between the sample orientation and the density of the parts, independent of the material type. This was an important finding because the density is a direct consequence of the structure and subsequently to all mechanical properties. Table 2 presents a direct correlation between the fracture toughness and the density, and also good correlations (positive or negative) between the density and the dimensional error.

Table 2. Pearson's correlation between sample orientation, density, size and fracture toughness.

Material/Parameter	Polyamide		Alumide	
	θ [°]	ρ [g/mm^3]	θ [°]	ρ [g/mm^3]
ρ [g/mm^3]	−0.996	1	−0.999	1
K_{IC} [MPa/m$^{0.5}$]	−0.523	0.448	−0.755	0.764
Err. L [%]	0.946	−0.918	−0.644	0.661
Err. B [%]	−0.530	0.461	−0.098	0.118
Err. W [%]	0.828	−0.869	0.989	−0.992

A low correlation value between ρ and K_{IC} in both materials underlined the fact that fracture toughness is not exclusively dependent on the material structure but on the notch manufacturing and accuracy as well. Poor correlation results of Alumide Err B with rho and theta underlined that the effect of temperature differences on the top and bottom of the samples during the cooling phase and the compaction produced by the dispenser influenced the accuracy of this dimension in a much more significant way than rho and theta.

By statistical evaluation of the data, significant differences (P-value = 0.003 for densities; P-value = 0.007 for fracture toughness) were obtained when comparing the two materials. When the probability test was applied to the dimensional errors, no significant differences were obtained for the investigated materials (P-value = 0.275 for length errors; P-value = 0.411 for thickness error; P-value = 0.07 for width errors). The probability test showed from a statistical perspective that the data we obtained for the two materials were significantly different only for structure and fracture properties, and had no signification when we compared the dimensional error.

4. Conclusions

This paper investigates the effect of material type (Polyamide PA2200 and Alumide) and sample orientation angle (0°, 45° and 90°) on mode I fracture toughness (K_{IC}) and geometrical accuracy in Selective Laser Sintering.

Size measurements and microstructures were conducted on the samples in order to determine the density and dimensional errors. Three Point Bending tests of Single Edge Notch Bending samples were accomplished for computing the fracture toughness based on force displacement data. In order to give objective interpretations of the recorded data, the probability test (P-value) and Pearson's correlation coefficients were determined.

The density of the samples was greatly influenced by the orientation angle of the samples in both materials, and the best density value was recorded for 0° of orientation. The structural differences between PA2200 and Alumide highly affected the fracture toughness, with the Alumide possessing K_{IC} values of about 62%–65% lower (orientation dependent) than PA2200.

The dimensional accuracy was not significantly different between the considered materials. No matter what the manufacturing angle was, the probability values were above the 0.05 limit. Nevertheless, the dimensional accuracy on every direction had a strong dependence on the manufacturing angle.

Author Contributions: Conceptualization, methodology, D.I.S. and L.M.; investigation, writing—original draft preparation, E.L. and D.I.S.; writing—review and editing, D.I.S., E.L. and L.M.; supervision, L.M. All authors have read and agreed to the published version of the manuscript.

Funding: This research was partially funded by a grant of the Romanian Ministry of Research and Innovation, project number 10PFE/16.10.2018, PERFORM-TECH-UPT—The increasing of the institutional performance of the Polytechnic University of Timisoara by strengthening the research, development and technological transfer

Polymers **2020**, *12*, 640

capacity in the field of "Energy, Environment and Climate Change", within Program 1—Development of the national system of Research and Development, Subprogram 1.2-Institutional Performance-Institutional Development Projects—Excellence Funding Projects in RDI, PNCDI III"; by research funding from the European Union's Horizon 2020 research and innovation program under grant agreement No 857124, and from Politehnica University of Timisoara, grant number GNaC2018-ARUT, no.1363/01.02.2019.

Conflicts of Interest: The authors declare no conflict of interest.

References

1. Salmoria, G.V.; Ahrens, C.H.; Klauss, P.; Paggi, R.A.; Oliveira, R.G.; Lago, A. Rapid manufacturing of polyethylene parts with controlled pore size gradients using Selective Laser Sintering. *Mater. Res.* **2007**, *10*, 211–214. [CrossRef]
2. Stoia, D.I.; Linul, E.; Marsavina, L. Influence of manufacturing parameters on mechanical properties of porous materials by Selective Laser Sintering. *Materials* **2019**, *12*, 871. [CrossRef]
3. Hesse, N.; Dechet, M.A.; Bonilla, J.S.G.; Lübbert, C.; Roth, S.; Bück, A.; Schmidt, J.; Peukert, W. Analysis of tribo-charging during powder spreading in Selective Laser Sintering: Assessment of polyamide 12 powder ageing effects on charging behavior. *Polymers* **2019**, *11*, 609. [CrossRef]
4. Feng, L.; Wang, Y.; Wei, Q. PA12 powder recycled from SLS for FDM. *Polymers* **2019**, *11*, 727. [CrossRef]
5. Pereira, A.B.; Fernandes, F.A.O.; de Morais, A.B.; Quintão, J. Mechanical strength of thermoplastic polyamide welded by Nd:YAG Laser. *Polymers* **2019**, *11*, 1381. [CrossRef]
6. García Plaza, E.; López, P.J.N.; Torija, M.Á.C.; Muñoz, J.M.C. Analysis of PLA geometric properties processed by FFF additive manufacturing: Effects of process parameters and plate-extruder precision motion. *Polymers* **2019**, *11*, 1581. [CrossRef]
7. Zeng, Z.; Deng, X.; Cui, J.; Jiang, H.; Yan, S.; Peng, B. improvement on selective laser sintering and post-processing of polystyrene. *Polymers* **2019**, *11*, 956. [CrossRef]
8. Mousa, A.A. Experimental investigations of curling phenomenon in selective laser sintering process. *Rapid Prototyp. J* **2016**, *22*, 405–415. [CrossRef]
9. Hofland, E.C.; Baran, I.; Wismeijer, D.A. Correlation of process parameters with mechanical properties of laser sintered PA12 parts. *Hindawi Adv. Mater. Sci. Eng.* **2017**, *2017*, 4953173. [CrossRef]
10. Robinson, J.; Ashton, I.; Fox, P.; Jones, E.; Sutcliffe, C. Determination of the effect of scan strategy on residual stress in laser powder bed fusion additive manufacturing. *Addit. Manuf.* **2018**, *23*, 13–24. [CrossRef]
11. Mengqi, Y.; Bourell, D. Orientation effects for laser sintered polyamide optically translucent parts. *Rapid Prototyp. J.* **2016**, *22*, 97–103.
12. Sharanjit, S.; Anish, S.; Vishal, S.S. Investigation of Dimensional Accuracy/Mechanical Properties of Part Produced by Selective Laser Sintering. *Int. J. Appl. Sci. Eng.* **2012**, *10*, 59–68.
13. Calignano, F. Investigation of the accuracy and roughness in the laser powder bed fusion process. *Virtual Phys. Prototyp.* **2018**, *13*, 97–104. [CrossRef]
14. Pilipović, A.; Brajlih, T.; Drstvenšek, I. Influence of processing parameters on tensile properties of SLS polymer product. *Polymers* **2018**, *10*, 1208. [CrossRef]
15. Pilipović, A.; Valentan, B.; Šercer, M. Influence of SLS processing parameters according to the new mathematical model on flexural properties. *Rapid Prototyp. J.* **2016**, *22*, 258–268. [CrossRef]
16. Borzan, C.Ş.; Dudescu, M.C.; Berce, P. Bending and compression tests for PA 2200 parts obtained using Selective Laser Sintering method. *Matec Web Conf.* **2017**, *94*, 03010. [CrossRef]
17. Obst, P.; Launhardt, M.; Drummer, D.; Osswald, P.V.; Osswald, T.A. Failure criterion for PA12 SLS additive manufactured parts. *Addit. Manuf.* **2018**, *21*, 619–627. [CrossRef]
18. Linul, E.; Marsavina, L.; Stoia, D.I. Mode I and II fracture toughness investigation of Laser-Sintered Polyamide. *Appl. Fract. Mech.* **2020**, *106*, 102497. [CrossRef]
19. Brugo, T.; Palazzetti, R.; Ciric-Kostic, S.; Yan, X.T.; Minak, G.; Zucchelli, A. Fracture mechanics of laser sintered cracked polyamide for a new method to induce cracks by additive manufacturing. *Polym. Test.* **2016**, *50*, 301–308. [CrossRef]
20. Crespo, M.; Gómez-del Río, M.T.; Rodríguez, J. Failure of SLS polyamide 12 notched samples at high loading rates. *Appl. Fract. Mech.* **2017**, *92*, 233–239. [CrossRef]

21. ASTM D5045-14. *Standard Test Methods for Plane-Strain Fracture Toughness and Strain Energy Release Rate of Plastic Materials*; ASTM International: West Conshohocken, PA, USA, 2014; Available online: www.astm.org (accessed on 21 February 2020).

22. Minetola, P.; Calignano, F.; Galati, M. Comparing geometric tolerance capabilities of additive manufacturing systems for polymers. *Addit. Manuf.* **2020**. [CrossRef]

23. EOS GmbH Product Information. Available online: https://eos.materialdatacenter.com/eo/en (accessed on 24 February 2020).

24. EOS GmbH Product Information. Available online: www.eos.info/material-p (accessed on 24 February 2020).

25. Materialise Magics Product Information. Available online: https://www.materialise.com/en/software/magics/product-information (accessed on 24 February 2020).

26. Stoia, D.I.; Marsavina, L.; Linul, E. Correlations between process parameters and outcome properties of Laser-Sintered Polyamide. *Polymers* **2019**, *11*, 1850. [CrossRef]

27. Stoia, D.I.; Marsavina, L. Effect of aluminum particles on the fracture toughness of polyamide-based parts obtained by Selective Laser Sintering (SLS). *Procedia Struct. Integr.* **2019**, *18*, 163–169. [CrossRef]

28. Lee, S.; Peng, J.; Shin, D.; Choi, Y.S. Data analytics approach for melt-pool geometries in metal additive manufacturing. *Sci. Technol. Adv. Mater.* **2019**, *20*, 972–978. [CrossRef]

29. Baturynska, I. Statistical analysis of dimensional accuracy in additive manufacturing considering STL model properties. *Int. J. Adv. Manuf. Technol.* **2018**, *97*, 2835–2849. [CrossRef]

30. Campanelli, S.L.; Cardano, G.; R.Giannoccaro, R.; Ludovico, A.D.; Bohez, E.L.J. Statistical analysis of the stereolithographic process to improve the accuracy. *Comput.-Aided Des.* **2007**, *39*, 80–86. [CrossRef]

31. Baran, I.; Tutum, C.C.; Hattel, J.H. Probabilistic analysis of thermosetting pultrusion process. *Sci. Eng. Compos. Mater.* **2016**, *23*, 67–76. [CrossRef]

32. Brajlih, T.; Valentan, B.; Balic, J.; Drstvensek, I. Speed and accuracy evaluation of additive manufacturing machines. *Rapid Prototyp. J.* **2011**, *17*, 64–75. [CrossRef]

33. Lee, P.H.; Chung, H.; Lee, S.W.; Yoo, J.; Ko, J. Review: Dimensional Accuracy in Additive Manufacturing Processes. In Proceedings of the ASME 2014 International Manufacturing Science and Engineering Conference MSEC2014, Detroit, MI, USA, 9–13 June 2014; ISBN 978-0-7918-4580-6.

Article

Additively Manufactured Parts Made of a Polymer Material Used for the Experimental Verification of a Component of a High-Speed Machine with an Optimised Geometry—Preliminary Research

Artur Andrearczyk [1,*,†], Bartlomiej Konieczny [2,†] and Jerzy Sokołowski [3]

1   Institute of Fluid Flow Machinery, Polish Academy of Sciences, 80-231 Gdansk, Poland
2   University Laboratory of Material Research, Medical University of Lodz, 92-213 Lodz, Poland; bartlomiej.konieczny@umed.lodz.pl
3   Department of General Dentistry, Medical University of Lodz, 92-213 Lodz, Poland; jerzy.sokolowski@umed.lodz.pl
*   Correspondence: aandrearczyk@imp.gda.pl
†   These authors contributed equally to this work.

Abstract: This paper describes a novel method for the experimental validation of numerically optimised turbomachinery components. In the field of additive manufacturing, numerical models still need to be improved, especially with the experimental data. The paper presents the operational characteristics of a compressor wheel, measured during experimental research. The validation process included conducting a computational flow analysis and experimental tests of two compressor wheels: The aluminium wheel and the 3D printed wheel (made of a polymer material). The chosen manufacturing technology and the results obtained made it possible to determine the speed range in which the operation of the tested machine is stable. In addition, dynamic destructive tests were performed on the polymer disc and their results were compared with the results of the strength analysis. The tests were carried out at high rotational speeds (up to 120,000 rpm). The results of the research described above have proven the utility of this technology in the research and development of high-speed turbomachines operating at speeds up to 90,000 rpm. The research results obtained show that the technology used is suitable for multi-variant optimization of the tested machine part. This work has also contributed to the further development of numerical models.

Keywords: additive manufacturing; polymer resin; material jetting; turbomachinery; optimization

Citation: Andrearczyk, A.; Konieczny, B.; Sokołowski, J. Additively Manufactured Parts Made of a Polymer Material Used for the Experimental Verification of a Component of a High-Speed Machine with an Optimised Geometry—Preliminary Research. *Polymers* **2021**, *13*, 137. https://doi.org/10.3390/polym13010137

Received: 10 November 2020
Accepted: 25 December 2020
Published: 31 December 2020

Publisher's Note: MDPI stays neutral with regard to jurisdictional claims in published maps and institutional affiliations.

1. Introduction

Additive manufacturing (AM) technologies are increasingly used in scientific research. They also offer a practically unlimited number of application possibilities for the industry. Several key emerging trends can be identified, such as the combination of manufacturing techniques and processes, the increasing modularity of machines, mass production and the adaptation of AM techniques for use in selected industrial branches. Currently, AM is most often used in the medical industry [1–3], and its rapid growth is taking place during the COVID-19 [4] pandemic to produce all possible elements used to combat the pandemic and support the production of life-saving components [5]. A wide range of applications can be observed in industries such as stomatology [6,7], power engineering [8], aerospace industry [9] and even the automotive industry [10,11]. AM methods used to manufacture real 3D objects using the layer-by-layer technique use a few types of building materials, namely metals, polymers, ceramics and glass [12,13]. Added missing references Although metal AM technologies offer relatively high manufacturing accuracy, they are rarely used to produce precision parts because surface treatment is required to reduce roughness. AM technologies that use polymers make it possible to print parts with high accuracy without the need for further processing. Unlike AM technologies capable of making parts from metals powders, among which Laser Powder Bed Fusion (LPBF) [14] is currently the most widely used and offer a similar level of manufacturing accuracy, polymer-based AM technologies use a wide range of printing methods such as Material Extrusion(ME), Binder Jetting (BJ), Material Jetting (MJ), Vat Photopolymerization (VP) and hybrid technologies, with different levels of manufacturing accuracy [15]. Some polymer-based technologies, despite their weaker mechanical properties than metal-based 3D printing technologies, successfully find application in the prototyping of machine parts [16,17].

There are three types of polymers used in AM machines, namely thermoplastic [18], thermohardening and light-curing polymers [19,20]. The evaluation of their properties makes it possible to choose the most suitable building material, depending on the required characteristics of the element to be manufactured. The mechanical and chemical properties of the materials make them suitable for industrial use. Thermoplastics are best suited for the production of finished products and testing prototypes. They have good mechanical properties, high impact strength as well as high abrasion and chemical resistance. They can also contain coal, glass or other additives, which improve their physical properties [21]. The most commonly used thermoplastics in the industry are nylons—poliamids (PA) and elastomers—thermoplastic polyurethane (TPU) [22], which have similar mechanical and physical properties (especially ultimate tensile strength which is very important in the design process of impellers) to materials used for 3D printing using ME technologies such as polylactic acid (PLA), acrylonitrile butadiene styrene (ABS), polyethylene terephthalate glycol (PETG), polyetherimide (PEI)—an amorphous material better known by its trade name ULTEM, acrylonitrile styrene acrylate (ASA) [23,24], which are also used due to economic reasons and their shorter printing process. One of their disadvantages is low printing accuracy. Thermohardening plastics (resins) are better suited for applications where aesthetics and high manufacturing accuracy are important, as they can be used to produce parts with smooth surfaces, like during injection moulding [25]. They generally have high stiffness but are more brittle than thermoplastics, so not all of them are suitable for functional applications, even those mimicking the properties of ABS and polypropylene (PP) [26], which are used, for example, in the manufacture of dental prostheses and implants [27]. The last group are light-curing polymers, which currently offer one of the highest printing accuracies (without post-treatment) thanks to the minimum height of a single layer of 16 μm [28]. Depending on the manufacturer and building material used in this technology, it is possible to distinguish aesthetic and functional resins with strictly defined mechanical properties. The strongest of these resins are similar in strength to thermoplastics. In this study, this type of manufacturing technology has been used because of its advantages.

In order to increase the performance and efficiency of both prototypical and commercially available high-speed power generation machines [29], it is essential to optimise their main components in this respect [30]. In turbomachines, the fluid-flow system and the geometry of the blades are usually optimised [31]. Sometimes strength optimisation is also performed on components that must withstand harsh operating conditions to increase performance and achieve higher operating parameters [32]. During the process of optimising the geometrical parameters of the fluid-flow system, a number of optimisation methods are used, which can be divided into the following three groups: deterministic methods (for example, gradient methods, simplex methods and simple search methods) [33], stochastic methods (genetic algorithms and annealing) [34] and hybrid deterministic-stochastic methods [35]. The main approach in the optimisation of blade systems is based on the concept of target function optimisation [36], which represents a certain general property of the fluid-flow system, for example, enthalpy loss in the turbine stage(s). So far, the optimisation of components such as the rotor discs of high-speed turbomachines has been experimentally verified only to the extent that enables the creation of such a geometry using the available manufacturing technologies. The rest of the optimised surfaces were either very difficult or very expensive to produce, for example, blades with less than 1 mm clearance or blades with a very complex profile [37]. Such obstacles prevented any experimental verification, leaving the optimisation outcome in the form of theoretical considerations. The emergence and rapid development of 3D Printing technologies and devices have made it possible to take a closer look at some AM methods, in terms of their applications in this field, and to enhance the process of verifying and optimising the geometries in high-speed machines.

The aim of the paper is to present a tool for verifying the optimised shape of the blades of rotor discs (i.e., turbine disc and compressor disc), the manufacture of which is either very difficult or impossible using the conventional production methods (that is, machining, CNC milling, etc.). The results and considerations presented in the paper constitute a continuation of research on the use of polymer-based AM technologies for the manufacture of prototype parts of high-speed machines [11,38].

2. Materials and Methods

2.1. Manufacturing Technology

After a thorough examination of the available AM technologies, MJ technology was selected and used in our research. AM technology uses additive manufacturing to produce models. The building material, which is deposited on the build platform by a printhead, is automatically cured by a UV lamp. The material is precisely deposited layer-by-layer using printing jets. The device builds wax supports, which must be melted after the printing process. The MJ AM process is fully automated and the role of the machine operator is limited to placing models on a virtual platform and uploading them to an AM machine. A ProJet 3500 HD Max printer (3D Systems, Rock Hill, SC, USA) was used to manufacture rotors in this work.

MJ technology makes it possible to select several types of light-curing resins with different mechanical properties. Due to the specified operating conditions [11], the most resistant material with the catalogue name of VisiJet M3-X (3D Systems, Rock Hill, SC, USA) with a tensile strength of 54 MPa and a softening temperature of 88 °C [39] was used in the research presented in this paper. The material used in the research was chosen from two of the most durable materials offered by the manufacturer of the selected printing technology. The materials were experimentally tested to determine their tensile strength and maximum operating temperature and the results are described in papers [11,39]. The aluminium compressor disc comes originally from an industrial turbocharger used in this research and no design data is available. The original compressor disc was scanned with a 3D laser scanner and subjected to a numerical strength analysis [40] to first determine its operating range. The disc model was then converted to the required .stl format using the Autodesk Inventor environment (Autodesk, California, USA) and the disc was created using the method described above. The disc was printed with the highest possible accuracy (XHD mode) using the mentioned printer, and the height of a single layer was 16 μm at a resolution of 750 DPI. The post-treatment involved melting the wax in an oven at a temperature of 60 °C. The scan result and the 3D printed disc are shown in Figure 1.

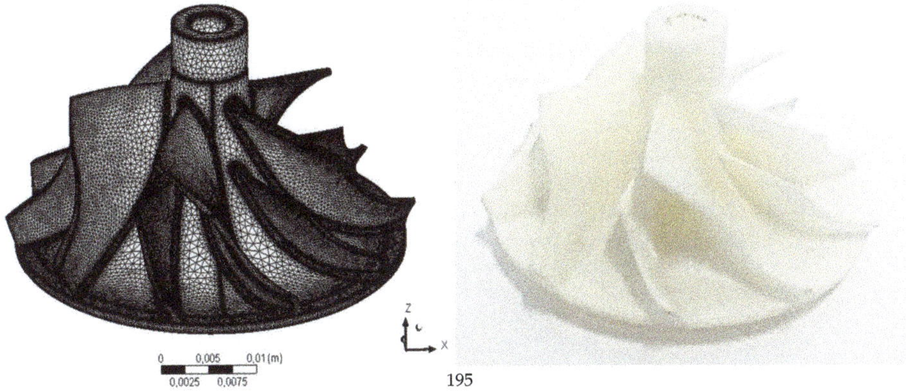

Figure 1. Model of compressor disc obtained by 3D scanning (**left**) and disc created using the MJ method (**right**).

After printing the rotor disc, the main dimensions of both discs (aluminium and polymer disc) were measured using precise micrometres and diameter measuring devices. The external dimensions did not differ by more than 0.01 mm. A surface quality test was also carried out using a surface roughness meter—MarSurf PS1 (MAHR/Unipre, Werl, Germany). Slight differences in the values of the surface roughness coefficient (R_a) were observed, where for the aluminium disc its value was around 0.6 μm and for the polymer disc 0.5 μm. Such a small difference should not affect the operational performance of the discs.

2.2. Numerical Analysis

A numerical model was created using reverse engineering. After performing a 3D laser scan, a 3D point cloud was obtained. This cloud was used to create a 3D model of the machine (using the appropriate software). The model was imported into ANSYS. In the numerical model, the diameter of the compressor disc was the same as in reality (42.5 mm). In order to simulate the performance of the recreated geometry, a RANS (Reynolds-Averaged Navier-Stokes) analysis was performed using the commercial software ANSYS CFX (ANSYS, Academic research 2014, Canonsburg, PA, USA) [41]. A second-order space discretization was applied. The working fluid (air) was modelled as an ideal gas. This was justified because the maximum pressure ratio was approximately 3. Steady-state simulations were conducted with direct interpolation between the rotor nodes and the volute domains. The Shear Stress Transport turbulence model with an automatic wall function was applied. The computational mesh for a single rotor channel was created in ANSYS Turbogrid [41], which generates hexahedral grids. The entire rotor domain was prepared by means of a circular pattern of the obtained mesh that was characterised by matching periodic nodes. The quality of the mesh was sufficient, as the minimum orthogonality angles were greater than 22° and the maximum aspect ratios were lower than 1000 at the boundary layer. The mesh quality was calculated as the minimum ratio of the height to the base length of each side (normalised to 1) of an element. The mesh of the compressor volute consisted of tetrahedral elements of minimum quality equal to 0.05. The numerical model used for the flow analysis is described in detail in [11].

Due to the influence of the inlet air temperature of the compressor on the measurement results, a decision was made to carry out a flow analysis for two temperatures values, taking into account the minimum and maximum ambient temperatures prevailing in the laboratory during the tests. Temperature values of 15 °C and 27 °C were taken into account during the setting of the boundary conditions. The rotational speed range was selected on the basis of previous experiments [11,38]. In the speed range of 60,000–90,000 rpm, the machine with a polymer disc functioned properly compared to the turbocharger with the original aluminium disc.

2.3. Experimental Setup

In order to determine the flow characteristics of the turbocharger compressor with a polymer disc, a test stand was used to supply the turbine with compressed air heated to a temperature of 50–150 °C. The turbocharger used in this research was an industrial automotive turbocharger from Renault Clio II 1.5 dCi 82HP (Renault, Boulogne-Billancourt, France). It was not possible to use combustion gases to power the machine due to the mechanical properties of the polymer resin of which the rotor disc used in the research is made. Therefore, an in-house-designed air preheater (2) was used at the turbine inlet to avoid this situation. The air preheater consists of two heaters connected in parallel (each with a power of 3.3 kW) and is controlled by a voltage regulator. An oil-free compressor with a maximum operating pressure of 10 bar was used to supply the turbine. During the tests, the supply pressure was set to 8 bar. The inlet pressure of the turbine was controlled by an analogue throttle valve (1), which allows for precise control of the rotational speed of the turbine. Since the research presented in this paper was not intended to determine the power of the turbine, it was not necessary to measure its mass flow rate. The mass flow rate measurements were carried out only on the compression side. The test stand used for the research is shown in Figure 2.

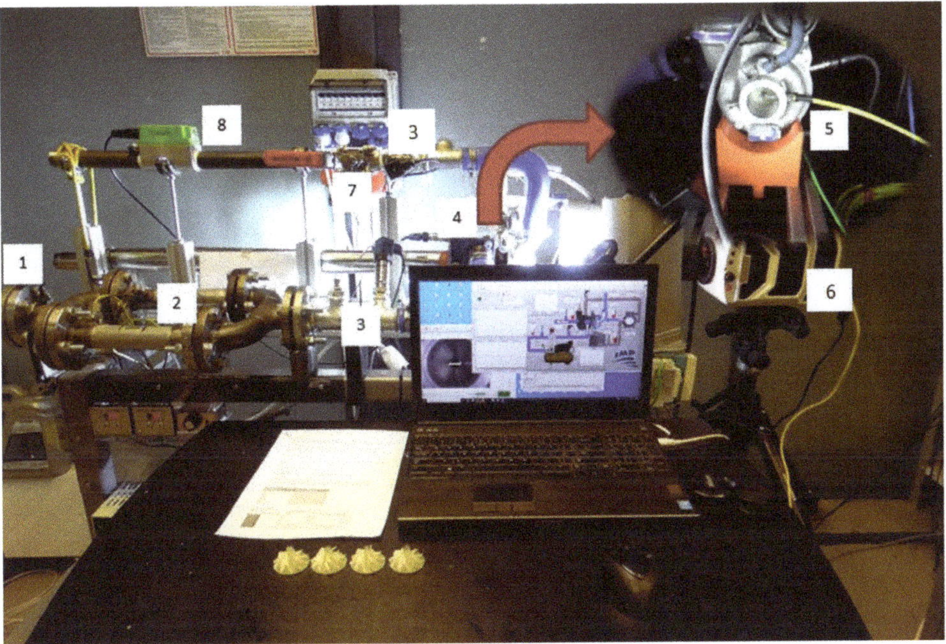

Figure 2. Test stand with a measurement system for registering and determining the flow characteristics of the turbocharger compressor. 1—air supply side with a throttle valve; 2—air pre-heating system; 3— temperature and pressure sensors; 4—turbocharger; 5—speed sensor; 6—high-speed camera; 7—compressor throttle valve; 8—flowmeter.

An extremely important element of the turbocharger is the lubrication system, which, if not properly controlled, may lead to energy losses resulting from friction of the lubricating film and throttling the operation of the machine. The lack of or poor lubrication pressure of bearings used in the machine may lead to its damage. This was achieved by using an oil pump controlled by an inverter, which kept the oil pressure in the range of 2.5–4 bar (depending on the rotational speed). An electric heater was installed in the oil tank in order to maintain an appropriate temperature level of the lubricant through hysteresis (60–80 °C).

In order to make it possible to measure several characteristics points, a throttle valve (7) was used at the compressor outlet and it was possible to change the opening level by 10% at a time. The experimental research assumed to analyse six characteristics points for each rotational speed (0—without throttling, 1–5—with throttling). Due to the lack of reaction of the system to throttling at a 10% valve closure level, throttling began at a 20% closure level and ended at a 60% closure level. The upper limit of the closure level of the valve was adopted for safety reasons. During previous experimental tests, above this level (70%), the turbocharger operated in the surge mode [42], which might have lead to damage to the machine if the operation had been longer.

During testing of this disc, the same temperature was set for the air supplied to the turbine as when testing the aluminium disc. In this way, similar operating temperatures of the machine were obtained in both tests. The measurements were carried out using a program created in the LabVIEW programming environment, coupled with the National Instruments measurement system. The following measuring transducers were used in the measurements: K-type thermocouples (3) with programmable transducers (TMD20) (CZAKI, Rybie, Poland), with an accuracy cold junction compensation ±1 °C; laser rotational speed sensor (5) (Optel Thevon,Montreui, France) that enables measurement in the range up to 1 million rpm; EE741 (Introl, Katowice, Poland) thermal flow meter (8) that allows mass airflow measurement with an accuracy of 1%. Compression level was measured using NPX (Peltron, Wiazownia Koscielna, Poland) pressure transducers (3) with an accuracy of 0.5%.

The test stand, in accordance with the operating conditions, was also adapted to perform destructive tests. For this purpose, a Phantom v2511 (Vision Research, New Yersey, USA) high-speed camera (6) was used to capture fast-changing phenomena (up to 1 million frames per second) and additionally the casing of the turbocharger (4) was equipped with vibration acceleration sensors (PCB Piezotronics, New York, USA). Based on previous analyses and experimental tests [40], it had been determined that at a speed of 120,000 rpm, the disc is torn into pieces. Since experimental tests were carried out on a polymer disc to determine the operating characteristics of the compressor, a decision was made to re-verify the destructive tests (previous tests were performed on unused ten polymer discs).

3. Results

3.1. Numerical Results

The results were determined for four rotational speeds that lie within the range of the correct operation of the polymer compressor disc. The results of the numerical flow analysis are shown in Figure 3.

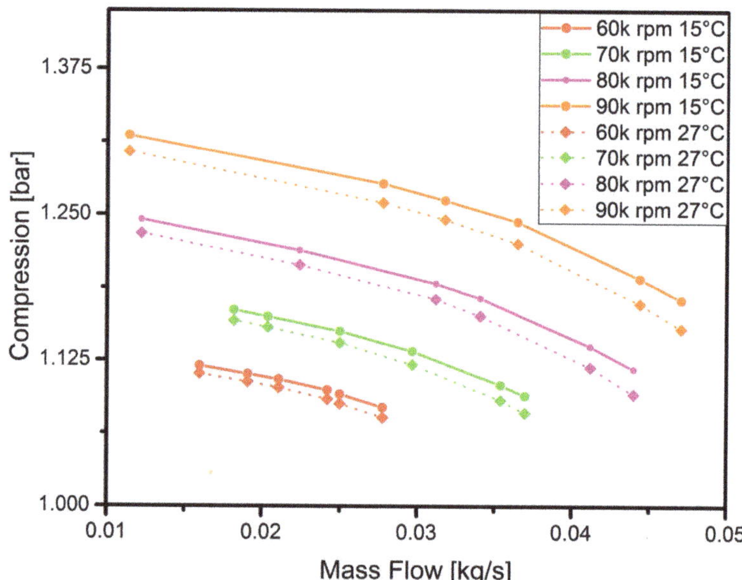

Figure 3. Numerically obtained flow characteristics of the compressor for two air temperatures measured at the inlet: 15 °C (solid line) and 27 °C (dotted line).

It is clear from the analysis that the higher the temperature at the compressor inlet, the greater the compression. At a speed of 60,000 rpm, the effect of temperature is very small or even negligible. It can be noticed that the temperature increases linearly with the rotational speed. This is due to the greater temperature difference resulting from the increase in temperature at the compressor outlet (due to the compression of air). Given the imperfection of the numerical model of the disc in relation to the real disc, such results make it possible to verify whether at least some of the characteristic points fall within the range of the numerical analysis. The model [11] does not take into account details such as blade clearance and needs to be extended and improved, and this is being planned for study in future papers.

3.2. Flow Characteristics

The experimental research was focused on analysing the performance characteristics of a turbocharger with an aluminium or polymer disc. Turbochargers are used in combustion engines to improve their operational performance by adding to them the pressure built up by compression (the so-called 'charge pressure'). A compressor functions properly when the pressure at its outlet is increased while the pressure at the compressor outlet is throttled (i.e., mass flow is reduced). This measurement was carried out to enable the proper functioning of the polymer compressor disc to be verified by comparing its performance with that of a properly functioning aluminium disc. Each performance characteristics consisted of six measurement points. The first tests focused on the aluminium disc and its operation in the selected range of rotational speeds. Then, tests were carried out under the same operating conditions using the polymer disc. Each of the measurement points was recorded after the operating parameters became stable at a predefined rotational speed and throttling level. The stabilisation period varied but was always within a 60-second time interval. The results obtained vary considerably depending on the rotational speed and the throttling level at the compressor outlet. The results of the above-mentioned experiments are shown in Figure 4.

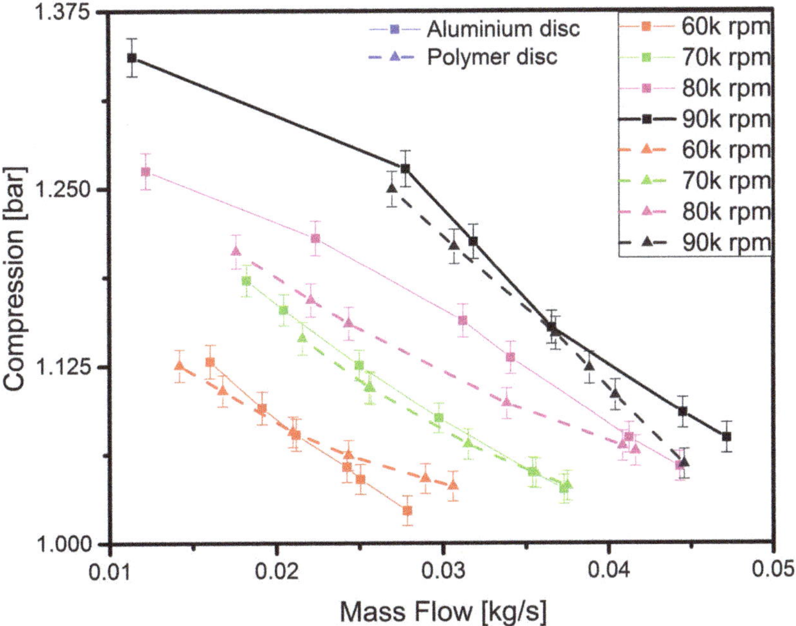

Figure 4. Compressor characteristics obtained experimentally with an aluminium (solid line) and polymer (dotted line) disc.

In the figure, the curves obtained from the the experiment carried out with a polymer and aluminium disc are similar for each rotational speed and the results are the most similar between the second and fourth points (throttling level). Based on previous research regarding the dynamic analysis of the turbocharger operating with the polymer disc, it was at these points that the machine's operation was most stable in terms of vibration. However, in the 80,000–90,000 rpm speed range, a decrease in the performance of the compressor blades can be observed, which is reflected in lower compression levels recorded at the last two compressor outlet throttling levels. In previous research conducted on a polymer disc, similar results were obtained, which indicated the malfunction of the rotor disc blades resulting from the appearance of deformations at their tips.

To discuss the results in more detail, each curve representing the operating characteristics of the disc for a given rotational speed has been analysed separately. In order to verify the model and the experiment, simulation curves determined earlier were added to the results obtained. The results are shown in Figures 5–8.

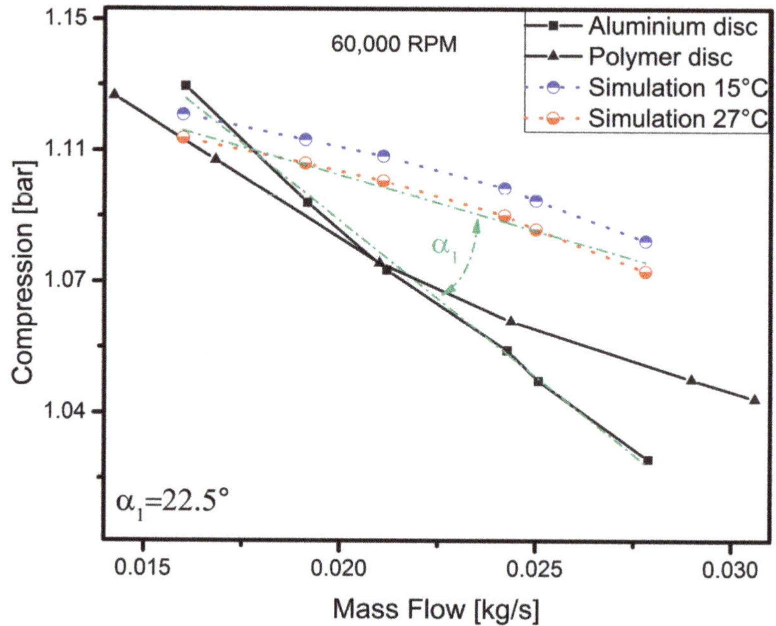

Figure 5. Simulation vs. experimental results for a speed of 60,000 rpm.

The results presented in Figure 5 show the operating characteristics of the compressor with an aluminium and polymer disc recorded at a speed of 60,000 rpm. The nature of the experimental curves slightly differs from each other (the maximum difference between the measured values does not exceed 3%), which makes it possible to affirm that the machine with the polymer disc behaves in a similar way to the one with the aluminium disc under these operating conditions. At this speed, both discs achieve very similar charge pressure values. However, there is a clear difference between the results of the experiment and the simulation. At the last point (the last throttling level), the experimental curve intersects the simulation curves. Since numerical calculations were performed for the aluminium disc (due to the lack of complete data regarding the polymer disc model in the literature), the inclination angles of the curves obtained from the experimental results regarding the aluminium disc were compared to the determined range of curves obtained from numerical calculations. To do this, the linearity of the nature of the operation was assumed and the crossed lines (green dash-dotted lines) were determined based on a least-squares regression fit. In the case of numerically determined characteristics, the regression fit were assumed between the curves for temperatures of 15 °C and 27 °C. The angle of inclination between the curves (α) is 22.5°. It is probably due to the fact that the numerical model does not take blade clearance into account. Due to the dynamically stable operation of the machine, a decision was made to analyse the variations in the angle of inclination between the curves obtained for the particular rotational speeds.

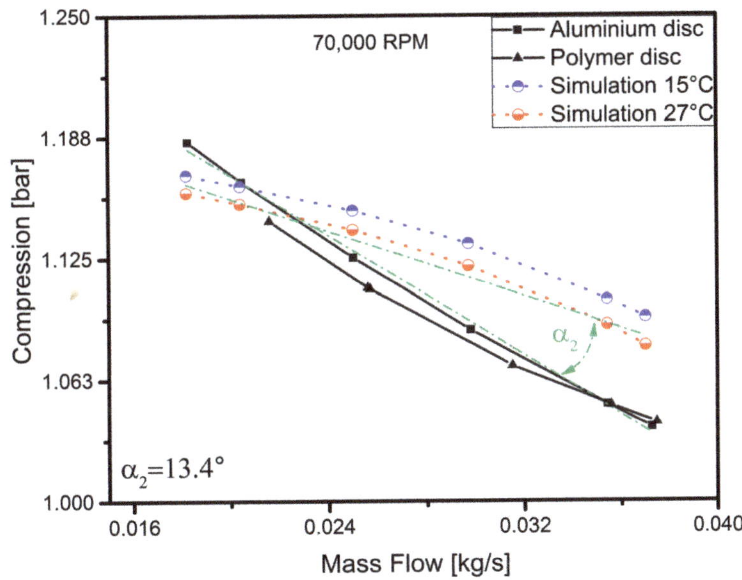

Figure 6. Simulation vs. experimental results for a speed of 70,000 rpm.

At a speed of 70,000 rpm (Figure 6), it can be observed that the nature of the operation of both discs is very similar. Under these operating conditions, the results obtained for the polymer disc are to a large extent similar to those obtained for the aluminium disc. For these operating parameters, the operation of the machine is still stable. In this graph, there are more points where the simulation curves intersect the experimental curves. The angle between the lines decreased to 13.4° as the rotational speed was increased.

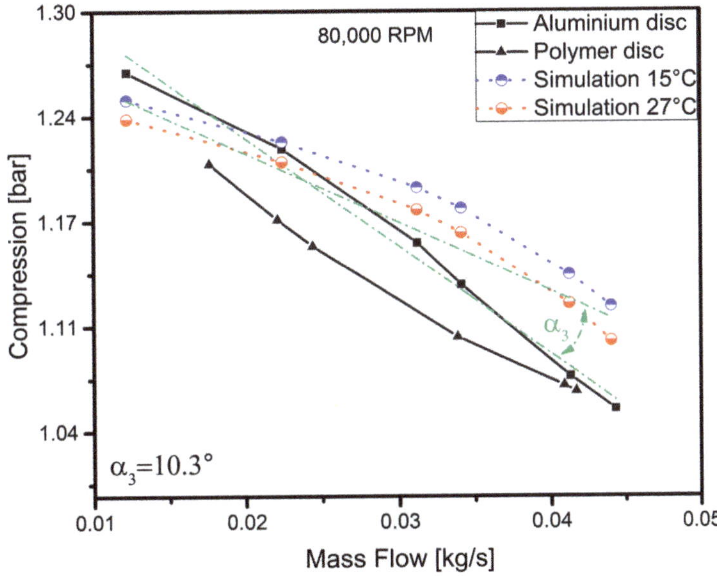

Figure 7. Simulation vs. experimental results for a speed of 80,000 rpm.

The curve representing the operating characteristics of the machine with the polymer disc obtained for a speed of 80,000 rpm (Figure 7) has a concave shape—as in the previous cases, while the curve obtained for the aluminium disc has changed in nature and took a slightly convex shape. This shape demonstrates that the nature of the operation of the turbocharger compressor is typical [43]. Nevertheless, at this speed, the parameters achieved by both discs are similar. It can also be seen in this graph that the curves determined experimentally and by simulation, representing the nature of the operation of the machine, are broadly similar. Once again, there is a decrease to 10.3° in the angle of inclination of the curves compared. This may indicate that this parameter depends on the speed of rotation. In addition, the experimental curve and the simulation curves are more similar than in the previous cases. For these operating parameters, there was also a slight increase in the level of vibration resulting from the bearing system used. From a dynamic point of view, the operation of the turbocharger was still stable and its vibration level was safe.

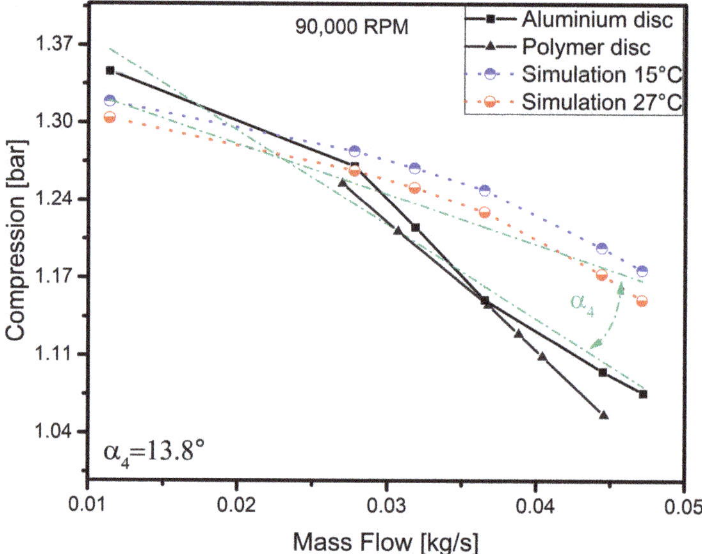

Figure 8. Simulation vs. experimental results for a speed of 90,000 rpm.

The last curve, obtained for a speed of 90,000 rpm, is shown in Figure 8. As can be seen from the graph, the polymer disc only operates efficiently in the first half of the mass flow range and achieves a much lower charge pressure than the aluminium disc. Both experimental curves are slightly convex in shape, indicating that the nature of the operation of the two discs is similar. At this speed, the dynamic tests of the machine showed an increase in vibrations resulting from the operation of the bearings and their overlap with the radial vibrations stemming from the rotational speed (unbalance), which caused that the operation of the machine was unstable. The start of the unstable operation of the turbocharger is marked with a red circle in Figure 9. The figure also shows the dynamic results regarding the entire run-up of the turbocharger up to the rupture of the polymer rotor. But for this to have become possible, the vibration amplitude of the machine had to be presented as a function of the frequency components (order) for each rotational speed. A slight increase (to 13.8°) in the inclination angle of the simulation curves was also observed, which according to the authors, may indicate that the unstable operation of the machine affected this parameter. However, to confirm this, further testing is needed. Despite the increase in the angle of inclination, the nature of the simulation curves and the experimental curves is still broadly similar and results from not taking into account blade clearance in the numerical model.

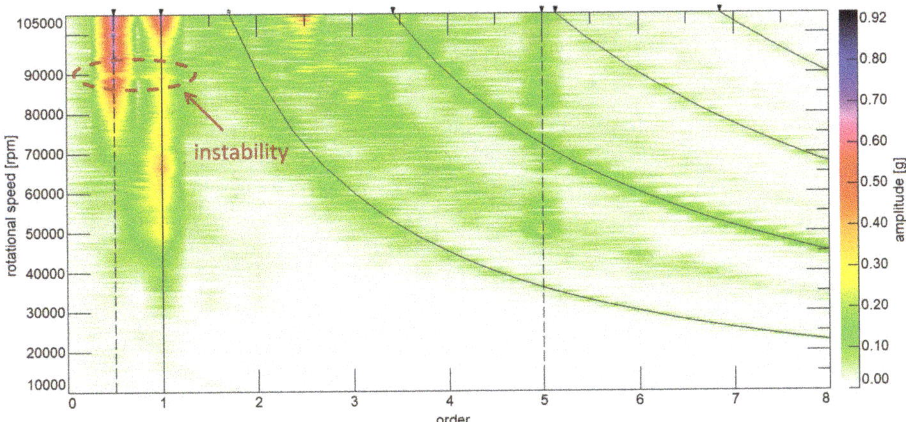

Figure 9. Dynamic characteristic of turbocharger with aluminium compressor disc presented in the form of a colourmap (speed range: from 10,000 rpm to 105,000 rpm).

3.3. Destructive Tests

Paper [40] discusses a strength analysis performed on a polymer disc and dynamic tests in which an unused rotor was ruptured. A decision was made to repeat this research using the rotor on which the above-mentioned tests were carried out. In order to carry out destructive tests, the same numerical model and the same measuring apparatus were used as in the case of the first series of destructive tests described in paper. As dynamic testing was not the purpose of the studies described in this section, a decision was made to only present the rupture result and the data regarding the rotational speed at which the polymer disc was destroyed. In order to conduct a thorough verification, seven rotors were printed using a polymeric material and subjected to tests. As the results were very similar, it was decided to present only one case study. Figure 10 shows the frames recorded by an ultra-high-speed camera (Phantom v2511). The recording speed was 150,000 frames per second.

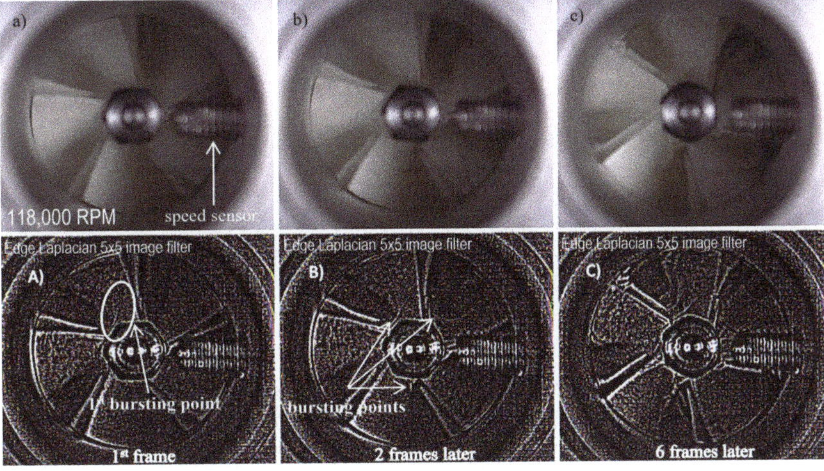

Figure 10. Images obtained at a speed of 118,000 rpm—original images recorded by a camera (top images: (**a**) 1st frame, (**b**) 2 frames later, (**c**) 6 frames later) and images obtained after using the edge Laplacian filter (bottom images: (**A**) 1st frame, (**B**) 2 frames later, (**C**) 6 frames later).

The value of the rotational speed at which the centrifugal force caused the permissible stresses to be exceeded (and the rupture of the disc) was also registered; under experimental conditions, this speed was between 115,000 and 120,000 rpm. The use of the edge Laplacian 5×5 image filter made it possible to highlight the first and other crack spots. As can be seen in Figure 10A, looking at the first bursting point, the disc broke in its central part, indicating that the maximum stresses stemming from the centrifugal force were exceeded at this point. The propagation of the crack is illustrated in Figure 10B,C. Compared to the results obtained previously, there was a slight change in the value of the speed at which the polymer disc broke. Subjecting the disc to other tests before subjecting it to destructive tests affected its strength.

4. Discussion

After analysing the obtained results, the polymer disc was found to function properly, as opposed to the aluminium disc. The polymer disc performed best in the speed range of 60,000 to 80,000 rpm. At a speed of 90,000 rpm, it had a lower performance and its operating characteristics were closest to those of the aluminium disc. It can therefore be concluded that the MJP AM technology with a selected polymer can be used to verify designed geometries in high-speed machines. The mechanical properties and design of the disc require some modifications, as seen in the estimated results. To improve performance, it would be necessary to optimise the disc in terms of its strength, particularly the strength of the compressor blades. In this study, the disc was tested for the possibility of using MJP technology to verify new and optimised geometries in high-speed turbomachines. The results of the experimental studies confirmed the aims of the paper.

The operating properties of elements manufactured using the AM technology make it possible to extend the experimental verification. Destruction tests performed on the discs helped to determine the maximum speed at which the compressor disc, made of light-cured polymer resin, can operate safely (100,000 rpm). This value was determined based on the stress values that occur under these operating conditions and the speed at which the rotor is destroyed (118,000 rpm).

Nowadays, machines and their components are being optimised in a way that makes it possible to create modelled parts using available conventional manufacturing methods such as machining, CNC milling or casting. The authors of this paper, after analysing the measurement data, propose to use MJP technology as the most accurate AM technology for optimising precision parts as an experimental verification. First of all, it is worth pointing out that this manufacturing method can be somewhat universal in this field, having very many advantages, but also limitations. One of the disadvantages of this method is the limitation of the operation of the manufactured part in the above-mentioned range, while in rotating machines of similar dimensions, it is possible to successfully test the printed model in the range of 0–100,000 rpm. Despite this limitation, it is a very wide range. Another disadvantage is the operating temperature of the polymer used in this technology. According to the manufacturer's specifications, the VisiJet M3-X material is not capable of operating at a temperature of 88 °C and stresses of around 0.45 MPa. Based on experimental studies, its maximum operating temperature is estimated to be equal to 70 °C when the stresses are around 40 MPa. Apart from these limitations, when it comes to producing prototypical parts for verification tests, it is a method that is much cheaper and faster in terms of creating very complex geometries compared to conventional methods.

Additive manufacturing techniques are currently used to optimise production processes that are based on conventional methods [44], i.e., casting, etc. The method proposed by the authors could be used to produce prototypical parts that would be ready for testing. A result of such an optimisation can be a simple geometry like in [45,46], where the geometry was optimised to improve the flow performance and thus the efficiency of the machine (a, b). In such cases, optimisations can be carried out simultaneously in several variants, taking into account economic aspects such as a cheaper and faster manufacturing process. An example of a very complex geometry is the rotor disc optimisation described in [37,47]. Geometry optimisation based on conventional methods can be very difficult, time-consuming and expensive (c) or even impossible (d). As for the last type of optimisation, this method makes it possible to verify geometries that have been the subject of theoretical considerations until now. Examples of application of the proposed method, collected based on the current state of knowledge, are shown in Figure 11.

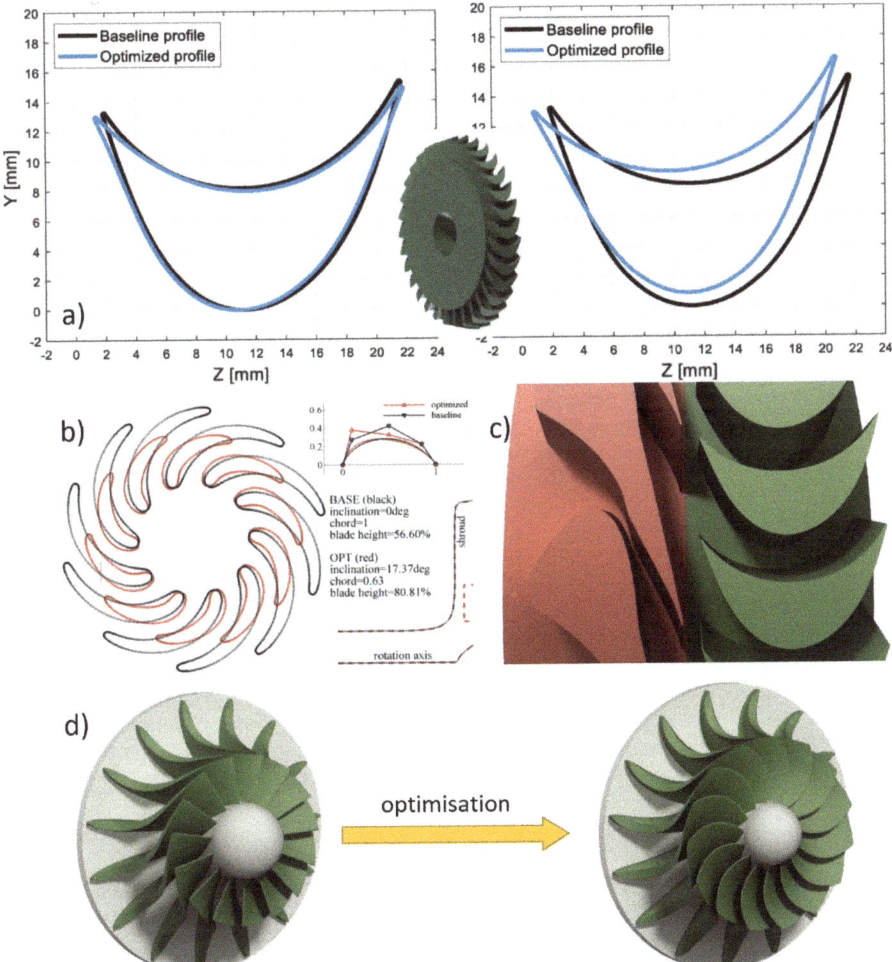

Figure 11. Examples of geometry optimisations carried out on fluid-flow machines: (**a**) optimisation of the geometry of the blades of a gas microturbine [45]; (**b**) optimisation of the rotor of an air-breathing radial outflow turbine [46]; (**c**) optimisation of the guide vanes of an ORC microturbine [47]; (**d**) optimisation of the disc of a radial-axial turbine [37].

The future optimisation of the geometry associated with the current paper will be performed using three variants to carry out an experimental verification. The geometry of the polymer disc of the compressor rotor will be optimised in terms of its strength according to the first variant, and in terms of the flow performance according to the second variant, in order to improve the operating characteristics of the machine. In the last variant, the two previous assumptions will be taken into account for the optimisation, with the emphasis put on improving the operating characteristics obtained and their stability over the entire range of rotational speeds mentioned in the paper.

5. Conclusions

The paper presents the results of some preliminary studies conducted to validate the use of the parts of high-speed machines, manufactured using the selected AM method (MJP). For this purpose, a car turbocharger was used in the research. The original compressor disc (made of aluminium), which is the object of research, was scanned and then, using a 3D printer, a disc made of light-cured polymer resin was printed. The polymer disc was tested to verify if its nature of operation reflects that of the aluminium disc. Before initiating the experimental research, numerical analyses were conducted regarding the strength of the printed element and the flow characteristics of the turbocharger compressor used.

The results of experimental research were used to determine the extent of the usefulness of the polymer disc in experimental work associated with the design of rotor discs for fluid-flow machines. The range of proper operation of the car turbocharger was 60,000–80,000 rpm. At a speed of 90,000 rpm, small losses due to the poor performance of the blades were observed. Due to the nature of operation of the compressor of this machine, it was not possible to obtain the operating characteristics for speeds below 50,000 rpm (only after reaching 50,000 rpm could the variations in flow rate and pressure on the compression side be measured). For machines whose rotors have similar dimensions, for example, gas turbines (in the case of a polymer disc, these machines can be supplied, for example, with air), the operating range is very wide and ranges from 0 to 80,000 rpm.

Some discrepancies were observed between the results of the experiment and simulation. The different nature of the curves resulted from not taking into account the value of the blade clearance in the numerical model. The numerical model needs to be improved in future studies. Nevertheless, the inclination angles of the curves decreased with increasing speed during the dynamically stable operation of the turbocharger (at speeds between 60,000 and 80,000 rpm), and the results of the experiment and simulation were similar to some extent.

The destructive tests carried out on the polymer disc made it possible to determine both its maximum and permissible rotational speed. The maximum and permissible speed is 118,000 rpm and 100,000 rpm respectively, taking into account the permissible stresses. An additional advantage of the polymer disc is the fact that after the disc was torn into pieces, the remaining components of the turbocharger remained intact and it was possible to continue operating the machine.

This paper has made it possible to examine whether MJP technology can be used for the experimental verification of numerically optimised models. The results confirmed that this technology is well suited for research in this field. The proposed method can be used to carry out multi-variant optimisations of the geometry of the blades of existing high-speed machines.

Future papers will focus on the optimisation of the geometry of the polymer disc used in terms of strength and performance (flow optimisation), as well as on the verification of the theoretical hypotheses based on the experimental results. The results obtained in this paper will also be used to prepare a physical model of a disc made of VisiJet M3-X material in order to perform necessary analyses.

Author Contributions: Conceptualization, A.A. and B.K.; methodology, A.A.; software, A.A.; validation, A.A., B.K. and J.S.; formal analysis, A.A. and B.K.; investigation, A.A.; resources, B.K.; data curation, A.A.; writing—original draft preparation, A.A.; writing—review and editing, A.A., B.K. and J.S.; visualization, B.K. All authors have read and agreed to the published version of the manuscript.

Funding: This research was funded by National Centre for Research and Development (NCBR) BIOSTRATEG strategic research and development programme under Grant Agreement Number BIOSTRATEG3/344128/12/NCBR/2017.

Institutional Review Board Statement: Not applicable.

Informed Consent Statement: Not applicable humans.

Data Availability Statement: Data sharing not applicable.

Conflicts of Interest: The authors declare no conflict of interest.

References

1. Gaharwar, A.K.; Cross, L.M.; Peak, C.W.; Gold, K.; Carrow, J.K.; Brokesh, A.; Singh, K.A. 2D nanoclay for biomedical applications: Regenerative medicine, therapeutic delivery, and additive manufacturing. *Adv. Mater.* **2019**, *31*, 1900332. [CrossRef] [PubMed]
2. Guzzi, E.A.; Tibbitt, M.W. Additive manufacturing of precision biomaterials. *Adv. Mater.* **2020**, *32*, 1901994. [CrossRef] [PubMed]
3. Tejo-Otero, A.; Buj-Corral, I.; Fenollosa-Artés, F. 3D printing in medicine for preoperative surgical planning: A review. *Ann. Biomed. Eng.* **2020**, *48*, 536–555. [CrossRef] [PubMed]
4. Choong, Y.Y.C.; Tan, H.W.; Patel, D.C.; Choong, W.T.N.; Chen, C.H.; Low, H.Y.; Tan, M.J.; Patel, C.D.; Chua, C.K. The global rise of 3D printing during the COVID-19 pandemic. *Nat. Rev. Mater.* **2020**, *5*, 637–639. [CrossRef]
5. Liu, D.; Koo, T.; Wong, J.; Wong, Y.; Fung, K.; Chan, Y.; Lim, H. Adapting re-usable elastomeric respirators to utilise anaesthesia circuit filters using a 3D-printed adaptor-a potential alternative to address N95 shortages during the COVID-19 pandemic. *Anaesthesia* **2020**. [CrossRef]
6. Wang, F. Research progress of 3D printing materials in stomatology. In Proceedings of the IOP Conference Series: Earth and Environmental Science, Singapore, 21–23 June 2019; IOP Publishing: Bristol, UK, 2019; Volume 332, p. 032013.
7. Bartkowiak, T.; Walkowiak-Śliziuk, A. 3D printing technology in orthodontics–review of current applications. *J. Stomatol.* **2018**, *71*, 356–364. [CrossRef]
8. Muravyev, N.V.; Monogarov, K.A.; Schaller, U.; Fomenkov, I.V.; Pivkina, A.N. Progress in additive manufacturing of energetic materials: Creating the reactive microstructures with high potential of applications. *Propellants Explos. Pyrotech.* **2019**, *44*, 941–969. [CrossRef]
9. Froes, F.H.; Boyer, R. *Additive Manufacturing for the Aerospace Industry*; Elsevier: Amsterdam, The Netherlands, 2019.
10. Böckin, D.; Tillman, A.M. Environmental assessment of additive manufacturing in the automotive industry. *J. Clean. Prod.* **2019**, *226*, 977–987. [CrossRef]
11. Andrearczyk, A.; Baginski, P.; Klonowicz, P. Numerical and experimental investigations of a turbocharger with a compressor wheel made of additively manufactured plastic. *Int. J. Mech. Sci.* **2020**, *178*, 105613. [CrossRef]
12. Sasan, K.; Lange, A.; Yee, T.D.; Dudukovic, N.; Nguyen, D.T.; Johnson, M.A.; Herrera, O.D.; Yoo, J.H.; Sawvel, A.M.; Ellis, M.E.; et al. Additive Manufacturing of Optical Quality Germania–Silica Glasses. *ACS Appl. Mater. Interfaces* **2020**, *12*, 6736–6741. [CrossRef]
13. Lakhdar, Y.; Tuck, C.; Binner, J.; Terry, A.; Goodridge, R. Additive manufacturing of advanced ceramic materials. *Prog. Mater. Sci.* **2020**, *116*, 100736. [CrossRef]
14. Wani, Z.; Abdullah, A. A Review on metal 3D printing; 3D welding. In Proceedings of the IOP Conference Series: Materials Science and Engineering, Penang, Malaysia, 2–3 December 2019; IOP Publishing: Bristol, UK, 2020; Volume 920, p. 012015.
15. Wickramasinghe, S.; Do, T.; Tran, P. FDM-based 3D printing of polymer and associated composite: A review on mechanical properties, defects and treatments. *Polymers* **2020**, *12*, 1529. [CrossRef] [PubMed]
16. Chuang, K.C.; Grady, J.E.; Arnold, S.M.; Draper, R.D.; Shin, E.; Patterson, C.; Santelle, T.; Lao, C.; Rhein, M.; Mehl, J. A Fully Nonmetallic Gas Turbine Engine Enabled by Additive Manufacturing, Part II: Additive Manufacturing and Characterization of Polymer Composites. Available online: http://snari.arc.nasa.gov/sites/default/files/Grady_TM-2015-218749.pdf (accessed on 10 December 2020).
17. Murr, L.E. Frontiers of 3D printing/additive manufacturing: From human organs to aircraft fabrication. *J. Mater. Sci. Technol.* **2016**, *32*, 987–995. [CrossRef]
18. Luo, C.; Wang, X.; Migler, K.B.; Seppala, J.E. Upper bound of feed rates in thermoplastic material extrusion additive manufacturing. *Addit. Manuf.* **2020**, *32*, 101019. [CrossRef]
19. Shi, J.; Chen, H.; Jia, S.; Wang, W. Rapid and low-cost fabrication of thermoelectric composite using low-pressure cold pressing and thermocuring methods. *Mater. Lett.* **2018**, *212*, 299–302. [CrossRef]
20. Mendes-Felipe, C.; Oliveira, J.; Etxebarria, I.; Vilas-Vilela, J.L.; Lanceros-Mendez, S. State-of-the-art and future challenges of UV curable polymer-based smart materials for printing technologies. *Adv. Mater. Technol.* **2019**, *4*, 1800618. [CrossRef]
21. Heidari-Rarani, M.; Rafiee-Afarani, M.; Zahedi, A. Mechanical characterization of FDM 3D printing of continuous carbon fiber reinforced PLA composites. *Compos. Part B Eng.* **2019**, *175*, 107147. [CrossRef]
22. Mardis, N.J. Emerging technology and applications of 3D printing in the medical field. *Mo. Med.* **2018**, *115*, 368.
23. Salentijn, G.I.; Oomen, P.E.; Grajewski, M.; Verpoorte, E. Fused deposition modeling 3D printing for (bio) analytical device fabrication: Procedures, materials, and applications. *Anal. Chem.* **2017**, *89*, 7053–7061. [CrossRef]
24. Haryńska, A.; Kucinska-Lipka, J.; Sulowska, A.; Gubanska, I.; Kostrzewa, M.; Janik, H. Medical-grade PCL based polyurethane system for FDM 3D printing-Characterization and fabrication. *Materials* **2019**, *12*, 887. [CrossRef]
25. Carneiro, V.; Rawson, S.; Puga, H.; Meireles, J.; Withers, P. Additive manufacturing assisted investment casting: A low-cost method to fabricate periodic metallic cellular lattices. *Addit. Manuf.* **2020**, *33*, 101085. [CrossRef]
26. Wu, H.; Fahy, W.; Kim, S.; Kim, H.; Zhao, N.; Pilato, L.; Kafi, A.; Bateman, S.; Koo, J. Recent developments in polymers/polymer nanocomposites for additive manufacturing. *Prog. Mater. Sci.* **2020**, *111*, 100638. [CrossRef]
27. Tunchel, S.; Blay, A.; Kolerman, R.; Mijiritsky, E.; Shibli, J.A. 3D printing/additive manufacturing single titanium dental implants: A prospective multicenter study with 3 years of follow-up. *Int. J. Dent.* **2016**, *2016*. [CrossRef] [PubMed]
28. Quan, H.; Zhang, T.; Xu, H.; Luo, S.; Nie, J.; Zhu, X. Photo-curing 3D printing technique and its challenges. *Bioact. Mater.* **2020**, *5*, 110–115. [CrossRef]

29. Z Kaczmarczyk, T.; Żywica, G.; Ihnatowicz, E. Vibroacoustic diagnostics of a radial microturbine and a scroll expander operating in the organic Rankine cycle installation. *J. Vibroeng.* **2016**, *18*, 4130–4147. [CrossRef]

30. Baginski, P.; Zych, P.; Zywica, G. Stress analysis of the discs of axial-flow microturbines. *J. Vibroeng.* **2020**, *22*, 1519–1533. [CrossRef]

31. Kaczmarczyk, T.Z.; Żywica, G.; Ihnatowicz, E. Experimental study of a low-temperature micro-scale organic Rankine cycle system with the multi-stage radial-flow turbine for domestic applications. *Energy Convers. Manag.* **2019**, *199*, 111941. [CrossRef]

32. Zych, P.; Żywica, G. Optimisation of stress distribution in a highly loaded radial-axial gas microturbine using FEM. *Open Eng.* **2020**, *10*, 318–335. [CrossRef]

33. Floudas, C.A. *Deterministic Global Optimization: Theory, Methods and Applications*; Springer Science & Business Media: Berlin/Heidelberg, Germany, 2013; Volume 37.

34. Reddi, S.J.; Sra, S.; Poczos, B.; Smola, A.J. Proximal stochastic methods for nonsmooth nonconvex finite-sum optimization. *Adv. Neural Inf. Process. Syst.* **2016**, *29*, 1145–1153.

35. Friedlander, M.P.; Schmidt, M. Hybrid deterministic-stochastic methods for data fitting. *SIAM J. Sci. Comput.* **2012**, *34*, A1380–A1405. [CrossRef]

36. Kondrashov, Y.N.; Ermakov, A. Optimization of planned solutions based on network models for large-scale problems. In Proceedings of the IOP Conference Series: Materials Science and Engineering, Krasnoyarsk, Russia, 16–18 April 2020 ; IOP Publishing: Bristol, UK, 2020; Volume 862, p. 052074.

37. Lampart, P.; Witanowski, Ł.; Klonowicz, P. Efficiency optimisation of blade shape in steam and ORC turbines. *Mech. Mech. Eng.* **2020**, *22*, 553–564. [CrossRef]

38. Andrearczyk, A.; Bagiński, P. Vibration analysis of a turbocharger with an additively manufactured compressor wheel. *Sci. J. Silesian Univ. Technol. Ser. Transp.* **2020**, *107*, 5–17.

39. Mieloszyk, M.; Andrearczyk, A.; Majewska, K.; Jurek, M.; Ostachowicz, W. Polymeric structure with embedded fiber Bragg grating sensor manufactured using multi-jet printing method. *Measurement* **2020**, *166*, 108229. [CrossRef]

40. Andrearczyk, A.; Mieloszyk, M.; Bagiński, P. Destructive Tests of an Additively Manufactured Compressor Wheel Performed at High Rotational Speeds. In *International Conference on Applied Human Factors and Ergonomics*; Springer: Berlin/Heidelberg, Germany, 2020; pp. 117–123.

41. Menter, F.R.; Kuntz, M.; Langtry, R. Ten years of industrial experience with the SST turbulence model. *Turbul. Heat Mass Transf.* **2003**, *4*, 625–632.

42. Sharma, S.; Broatch, A.; García-Tíscar, J.; Nickson, A.K.; Allport, J.M. Acoustic and pressure characteristics of a ported shroud turbocompressor operating at near surge conditions. *Appl. Acoust.* **2019**, *148*, 434–447. [CrossRef]

43. Sharma, S.; Broatch, A.; García-Tíscar, J.; Allport, J.M.; Nickson, A.K. Acoustic characteristics of a ported shroud turbocompressor operating at design conditions. *Int. J. Engine Res.* **2020**, *21*, 1454–1468. [CrossRef]

44. Sivarupan, T.; Upadhyay, M.; Ali, Y.; El Mansori, M.; Dargusch, M.S. Reduced consumption of materials and hazardous chemicals for energy efficient production of metal parts through 3D printing of sand molds. *J. Clean. Prod.* **2019**, *224*, 411–420. [CrossRef]

45. Klonowicz, P.; Lampart, P.; Suchocki, T.; Zaniewski, D.; Klimaszewski, P.; others. Optimization of an axial turbine for a small scale ORC waste heat recovery system. *Energy* **2020**, *205*, 118059.

46. Kavas, I.; Saracoglu, B.H.; Arts, T. Numerical optimization of air breathing radial outflow turbines. In Proceedings of the 13th European Conference on Turbomachinery Fluid Dynamics & Thermodynamics, Lausanne, Switzerland, 8–12 April 2019.

47. Klonowicz, P.; Witanowski, Ł.; Suchocki, T.; Jędrzejewski, Ł.; Lampart, P. Selection of optimum degree of partial admission in a laboratory organic vapour microturbine. *Energy Convers. Manag.* **2019**, *202*, 112189. [CrossRef]

MDPI
St. Alban-Anlage 66
4052 Basel
Switzerland
Tel. +41 61 683 77 34
Fax +41 61 302 89 18
www.mdpi.com

Polymers Editorial Office
E-mail: polymers@mdpi.com
www.mdpi.com/journal/polymers